In Vitro Application in Crop Improvement

In Vitro Application in Crop Improvement

Editors

A. Mujib
New Delhi, India

Myeong-Je Cho
Berkeley, USA

S. Predieri
Bologna, Italy

S. Banerjee
Victoria, Australia

Science Publishers, Inc.
Enfield (NH), USA Plymouth, UK

SCIENCE PUBLISHERS, INC.
Post Office Box 699
Enfield, New Hampshire 03748
United States of America

Internet site: *http://www.scipub.net*

sales@scipub.net (marketing department)
editor@scipub.net (editorial department)
info@scipub.net (for all other enquiries)

Library of Congress Cataloging-in-Publication Data

In vitro application in crop management / editors, Abdul Mujib ... [et al.].
p. cm.
Includes bibliographical references and index.
ISBN 1-57808-300-1
1. Plant micropropagation. 2. Plant tissue culture. 3. Crop improvement. I. Mujib, Abdul.

SB123.6.I43 2004
631.5'233--dc22

2004042884

ISBN 1-57808-300-1

Published by Science Publishers, Inc., Enfield, NH, USA
Printed in India.

Foreword

Plant Biotechnology as a rapidly developing field needs rapid dissemination of results in order to exploit the full potential: the problems yet to solve in securing food supply in a sustainable way are ever demanding: We must enable the best use of the best technology by as many researchers as possible. Although scientific journals and the internet are suitable tools for communicating the latest results, from time to time it is neccesary to compile recent advances in a more reflective way, i.e. in a book. The reason is simple: we have experienced in the past too many claims with respect to "successful" regeneration or "efficient" transformation of crop or ornamental plants, which were published but seldom found a serious follow-up. In addition, the progress in the meanwhile "classical" techniques like haploid production, in vitro selection or approaches to develop biotechnological tools for overcoming constraints in the so-called orphan crops need to be documented in a sincere way. Such a thematic textbook should then serve for some time as a standard source for researchers in the field of plant biotechnology, in related fields like botany, agri- and horticulture as well as advanced students in these disciplines. These developments should also reach the teachers in universities, who do not have laboratories suited for plant biotechnology: these teachers must be in a position, to train undergraduate students with appropriate and actual research papers. To my understanding this book has the potential of being a ressource book for the years to come, providing solid up-to-date information to the scientific community.

Hans-Jörg Jacobsen
Professor & Head
Department of Molecular Genetics
University of Hannover
Hannover, Germany
and
President
German Biologists Association

Preface

In recent years a large number of *in vitro* and laboratory techniques have been developed in plant tissue culture, molecular biology and genetic engineering, collectively referred to as Plant Biotechnology. This branch of science has several basic and practical applications. It has a direct role in crop genetic improvement in which plant cell and an organ's ability to transform into a complete plant has been successfully utilized.

Although it is always very difficult to keep researches abreast of the latest developments, in the present volume an effort has been made to assess the current status and trend in these important areas of biological sciences, particularly in the field of crop improvement. In this volume eighteen chapters are presented which deal with various facets of the present problems and possible future approaches to the problems in some field crops and other species.

Subjects covered include: somatic embryogenesis and plant regeneration; anther and microspore culture; protoplast culture and fusion; somaclonal variation; induced mutation in relation to crop improvement; and the role of plant tissue culture in plant improvement in general.

A few chapters are particularly devoted to genetic transformation methods for transgenic plants. Christian Walter discusses the gene stability in transgenic conifers. Genes for disease resistance/tolerance for improved yield have also been discussed.

We hope that this book will be useful to a vast international audience, including researchers, graduate students, teachers and other professionals working in the areas of plant biotechnology, botany, agriculture, horticulture, forestry, etc.

A. Mujib
New Delhi, India
Myeong-Je Cho
Berkeley, USA
S. Predieri
Bologna, Italy
S. Banerjee
Victoria, Australia

January 2004

List of Contributors

Abdin, M.Z.
Centre for Biotechnology, Faculty of Science, Hamdard University, New Delhi 110 062, India.

Ahmad, Islam S.
Molecular Cell and Development Biology, University of Texas, Austin, TX 78713, USA

Ali, Shakir
Department of Biochemistry, Faculty of Science, Hamdard University, New Delhi 110 062, India

Andersen, S.B.
The Royal Veterinary and Agricultural University, Plant Breeding, Thorvaldsensvej 40 DK-1871 Frederiksberg C, Denmark.
E-mail: sba@kvl.dk

Bansal, K.C.
National Research Centre for Plant Biotechnology, Indian Agricultural Research Institute, New Delhi 110 012, India.

Bapat, V.A.
Plant Cell Culture Technology Section, Nuclear Agriculture and Biotechnology Division, Bhabha Atomic Research Centre, Trombay, Mumbai 400 085, India
Email: vabapat@magnum.barc.ernet.in

Bekesiova, I.
The Royal Veterinary and Agricultural University, Plant Breeding, Thorvaldsensvej 40 DK-1871 Frederiksberg C, Denmark.

Castillo, A.M.
Departamento de Genética y Producción Vegetal. Estación Experimental de Aula Dei. Consejo Superior de Investigaciones Científicas. Apartado 202. 50059 Zaragoza, Spain.

Cho, Myeong-Je
Department of Plant and Microbial Biology, University of California, Berkeley, CA 94720, USA
E-mail: mjcho@nature.berkeley.edu

Choi, Hae-Woon
Department of Plant and Microbial Biology, University of California, Berkeley, CA 94720, USA

Cistué, L.
Departamento de Genética y Producción Vegetal. Estación Experimental de Aula Dei. Consejo Superior de Investigaciones Científicas. Apartado 202. 50059 Zaragoza, Spain.

Clark, Gregory
Molecular Cell and Development Biology, University of Texas, Austin, TX 78713, USA

Deng, Xiuxin
National Key Laboratory of Crop Genetic Improvement, Huazhong Agricultural University, Wuhan 430070, People's Republic of China

Devi, Jyotsna
Department of Plant Breeding & Genetics, Assam Agricultural University, Jorhat, India

Do, Geum-Sook
Department of Biology, Kyungpook National University, Taegu 702-701, Republic of Korea

Eapen, Susan
Plant Biotechnology and Secondary Products Section, Nuclear Agriculture and Biotechnology Division, Bhabha Atomic Research Centre, Mumbai 400 085, India.

Echávarri, B.
Departamento de Genética y Producción Vegetal. Estación Experimental de Aula Dei. Consejo Superior de Investigaciones Científicas. Apartado 202. 50059 Zaragoza, Spain.

Edoardo, Gatti
Istituto di Biometeorologia, IBIMET-BO - CNR Via P. Gobetti, 101 I-40129 Bologna, Italy

Ganapathi, T.R.
Plant Cell Culture Technology Section, Nuclear Agriculture and Biotechnology Division, Bhabha Atomic Research Centre, Trombay, Mumbai 400 085, India

Ghosh, S.B.
Plant Cell Culture Technology Section, Nuclear Agriculture and Biotechnology Division, Bhabha Atomic Research Centre, Trombay, Mumbai 400 085, India

Hansen, A.L.
The Royal Veterinary and Agricultural University, Plant Breeding, Thorvaldsensvej 40 DK-1871 Frederiksberg C, Denmark.

Hasan, M.S.
Cereal Research Non-profit Company, Wheat Genetics and Breeding Department, 6701 Szeged, POB 391, Hungary.
and
St. István University, Department of Genetics and Plant Breeding, 2103 Gödöllö, Pater K. u. l., Hungary.

Hassan, Fathi
Directorate of Agricultural Scientific Research, DASR-Douma-P.O. Box 113, Syria

Holme, I.B.
The Royal Veterinary and Agricultural University, Plant Breeding, Thorvaldsensvej 40 DK-1871 Frederiksberg C, Denmark.

Hoque, M. Imdadul
Department of Botany, University of Dhaka, Dhaka - 1000, Bangladesh

Israr, M.
Centre for Biotechnology, Faculty of Science, Hamdard University, New Delhi 110 062, India.

Jain, S. Mohan
FAO/IAEA Joint Division, Plant Breeding and Genetics Section, International Atomic Energy Agency, Box 100, A-1400, Wagramerstrasse 5, Vienna, Austria
E-mail: s.m.jain@iaea.org

Kertész, Z.
Cereal Research Non-profit Company, Wheat Genetics and Breeding Department, 6701 Szeged, POB 391, Hungary.

Kiesecker, Heiko
Department of Molecular Genetics, University of Hannover, Herrenhauser Str. 2, 30419 Hannover, Germany

Jacobsen, Hans Joerg
Department of Molecular Genetics, University of Hannover, Herrenhauser Str. 2, 30419 Hannover, Germany

Kim, Yong-Bum
Department of Plant and Microbial Biology, University of California, Berkeley, CA 94720, USA

Lantos, Cs.
Cereal Research Non-profit Company, Wheat Genetics and Breeding Department, 6701 Szeged, POB 391, Hungary.
and
St. István University, Department of Genetics and Plant Breeding, 2103 Gödöllö, Pater K. u. l., Hungary.

Le, Khanh Van
Department of Plant and Microbial Biology, University of California, Berkeley, CA 94720, USA

Lemaux, Peggy G.
Department of Plant and Microbial Biology, University of California, Berkeley, CA 94720, USA

Liu, Jihong
National Key Laboratory of Crop Genetic Improvement, Huazhong Agricultural University, Wuhan 430070, People's Republic of China
E-mail: Jhliu26@public.wh.hb.cn or liujihong@mail.hzan.edu.cn

Maiti, Aparna
Biotechnology Laboratory, Central Agricultural Research Institute, Port Blair, 744 101, India

Maluszynski, M.
FAO/IAEA Joint Division, Plant Breeding and Genetics Section, International Atomic Energy Agency, Box 100, A-1400, Wagramerstrasse 5, Vienna, Austria

Mandal, A.B.
Biotechnology Laboratory, Central Agricultural Research Institute, Port Blair, 744 101, India

Matuz, J.
Cereal Research Non-profit Company, Wheat Genetics and Breeding Department, 6701 Szeged, POB 391, Hungary.

Mesterházy, Á.
Cereal Research Non-profit Company, Wheat Genetics and Breeding Department, 6701 Szeged, POB 391, Hungary.

Mihály, R.
Cereal Research Non-profit Company, Wheat Genetics and Breeding Department, 6701 Szeged, POB 391, Hungary.

Mujib, A.
Department of Botany, Hamdard University, New Delhi 110 062, India
E-mail: mujibabdul@hotmail.com

Okamoto, Dorothy
Department of Plant and Microbial Biology, University of California, Berkeley, CA 94720, USA

Pauk, J.
Cereal Research Non-profit Company, Wheat Genetics and Breeding Department, 6701 Szeged, POB 391, Hungary.

Puolimatka, M.
Plant Production Inspection Centre, Seed Testing Department, 32201 Loimaa, POB 111, Finland.

Rehman, R.U.
Centre for Biotechnology, Faculty of Science, Hamdard University, New Delhi 110 062, India.

Sanz, M.
Departamento de Genética y Producción Vegetal. Estación Experimental de Aula Dei. Consejo Superior de Investigaciones Científicas. Apartado 202. 50059 Zaragoza, Spain.

Sarker, R.H.
Department of Botany, University of Dhaka, Dhaka-1000, Bangladesh
E-mail: rhsarker@bd.drik.net

Seo, Bong-Bo
Department of Biology, Kyungpook National University, Taegu 702-701, Republic of Korea

Seraj, Zeba I.
Department of Biochemistry and Molecular Biology, University of Dhaka, Dhaka-1000, Bangladesh

Siddiqua, Murshida K.
Department of Botany, University of Dhaka, Dhaka-1000, Bangladesh

Srivastava, P.S.
Centre for Biotechnology, Faculty of Science, Hamdard University, New Delhi 110 062, India.

Stefano, Predieri
Istituto di Biometeorologia, IBIMET-BO - CNR Via P. Gobetti, 101 I-40129 Bologna (Italy)
E-mail: s.predieri@ibimet.cnr.it

Torp, A.M.
The Royal Veterinary and Agricultural University, Plant Breeding, Thorvaldsensvej 40 DK-1871 Frederiksberg C, Denmark.

Vallés, M.P.
Departamento de Genética y Producción Vegetal. Estación Experimental de Aula Dei. Consejo Superior de Investigaciones Científicas. Apartado 202. 50059 Zaragoza, Spain.

Walter, Christian
New Zealand Forest Research Institute Ltd., Sala Street, Rotorua, New Zealand
E-mail: christian.walter@forestresearch.co.nz

Contents

1

Production of Barley Doubled Haploids by Anther and Microspore Culture

L. Cistué, M.P. Vallés, B. Echávarri, M. Sanz and A.M. Castillo

Departamento de Genética y Producción Vegetal. Estación Experimental de Aula Dei. Consejo Superior de Investigaciones Científicas. Apartado 202. 50059 Zaragoza, Spain.

ABSTRACT

Barley anther and microspore culture have been used largely to produce doubled haploids. However, genotypes still exist with a low embryogenic capacity and/or a high albinism rate. Although a certain number of doubled haploid lines can be obtained from these genotypes inoculating a high number of anthers, further studies are needed to improve these limiting factors.

The introduction of 40 mM Ca^{2+} in the pre-treatment medium significantly increased the numbers of divisions, embryos, and total plants in genotypes with different androgenic responses. The use of 400 g/l Ficoll along the whole culture process not only significantly raised the numbers of divisions, embryos, green and total plants in cvs Volga and Nevada—both of them with a low androgenic response—but it also allowed the production of green plants from cultivar Volga.

Transgressive lines were produced in a doubled haploid population between the cultivars Igri (two rows, high anther culture response) and Dobla (six rows, medium anther culture response) for some important agronomic characters such as head-ing date and 100-kernel weight.

INTRODUCTION

Haploidy refers to the condition of any organism, tissue or cell having the chromosome constitution of the normal gametes of the species involved (Chase 1952). Haploid plants can be obtained either spontaneously or induced through androgenesis, gynogenesis or wide hybridization. Doubling the haploid chromosome number will produce a doubled haploid (DH).

Of the different methods used for DH production in barley, interspecific crossing with *Hordeum bulbosum* (Kasha and Kao 1970) and androgenesis (Clapham 1973) are regularly used for practical applications. Androgenesis consists of production of plants from anthers or microspores cultured in vitro.

Although the *Hordeun bulbosum* method is still being used widely, the potential of anther/microspore culture is remarkable when we consider the efficiency of DH production per flower. It needs to be mentioned that a flower of barley contains only one ovule against roughly 8000–9000 microspores. Great strides have been made during the last years in androgenesis; however, the low embryogenesis capacity and the high albinism rate of some genotypes still remain obstacles to be overcome before the total potential of the technique can be exploited.

APPLICATIONS OF DOUBLED HAPLOID PLANTS

Doubled haploids have many applications in plant research. They have been particularly attractive to plant breeders because of their potential as a time-saving tool for producing pure lines at any stage in the breeding programme, and increasing the selection efficiency, therefore reducing the time to release new cultivars. Another advantage is the avoidance of environmental selection during inbreeding (Snape and Simpson 1984; Kasha et al., 1997).

Currently, the utilization of barley DH plants has already proved to be very useful for the release of new cultivars using the *Hordeum bulbosum* and/or androgenesis methods, both by private companies as well as public institutions. It is difficult to know the exact number of new cultivars released worldwide. Devaux et al., (1996) showed 59 cultivars, but presumably this number has increased considerably during the last years, especially those cultivars derived from androgenesis. As an example, in the years 1999 and 2000, cultivars Belén y Lola were registered by the Technical Institute from Albacete (Spain) selected from DH lines produced in our laboratory by anther culture.

Doubled haploids are also very valuable for genetic analysis (Pickering and Devaux 1992). The use of DH populations has been fundamental for the production of linkage maps of barley (Graner et al., 1991; Kleinhofs et al., 1993), and QTL analysis, particularly with multiple environments. QTL analysis for many quantitative and qualitative traits such as different agronomic traits and diseases resistance (Marquez-Cedillo et al., 2001; Toojinda et al., 2000) have been performed on the basis of DH populations.

DH techniques can also be used together with other biotechnological tools. The application of mutagenic agents to single haploid cells offers the possibility of screening recessive mutants in the first generation, avoiding quimerism and rapidly fixing the selected genotype (Maluszynski et al., 1996). In barley, a protocol has been reported for efficient production of mutants from anthers and isolated microspores cultured in vitro (Castillo et al., 2001). When a selective agent is available, the probability of identifying the beneficial mutants from a large microspore population increases.

Microspores were considered to be an attractive target for transformation since their use could lead to the production of plants with the transgene fixed in the first generation. Many laboratories have attempted to transform barley microspores but a low efficiency of transformation even with the model genotype Igri was achieved (Jähne et al., 1994; Yao and Kasha 1997).

ANDROGENESIS

In the production of plants from microspores cultured in vitro, there are a number of factors influencing, to some extent, the efficiency in the process of the three important steps: division, embryogenesis and albinism. These factors can be grouped as genotypic and physiological in nature.

Genotype

Of the different factors affecting androgenesis, genetic effect is one of the most important. The great variation among cultivars for plant production is well documented. Table 1.1 shows the androgenetic response from three cultivars of great agronomic importance in Spain and the model cv. Igri. Upto 233 green plants were produced from 100 anthers of Igri, whereas this number was reduced considerably from Hispanic (87), Blanche (35) and Nevada (4). Igri had also the highest number of dividing microspores (2793). However, the percentage of embryogenesis of Hispanic (27.9) is almost three times than that of Igri. The low percentage of green plant production in cultivar Nevada accounts for the high rate of albinism.

The genetic control of androgenesis is complex and several nuclear genes, with additive and dominance effects, act independently in each of the steps of the procedure (Foroughi-Wehr et al., 1982; Powell 1988a; Hou et al., 1994). Apart from the nuclear genes, the presence of maternal and cytoplasmic effects have also been described (Powell 1988a; Hou et al., 1994).

Physiological Factors

Culture of Donor Plants

The effect of the donor plant environment must be minimized in order to ensure a repeatable culture response. Important environmental factors such as photoperiod, intensity and quality of light, temperature and nutrition must be optimal. Furthermore, the donor plants should be free of pests and diseases (Luckett and Darvey 1992), since application of

Table 1.1 Anther culture response of four of the most cultivated genotypes in Spain

Cultivar	*Anthers*	*No. of dividing microspores*	*No. of embryos*	*Embryos (%)*	*No. of green plants*	*No. of total plants*	*Green plants (%)*
Blanche	100	524	56	10.7	35	40	87.5
Hispanic	100	402	112	**27.9**	87	93	93.5
Igri	100	**2793**	303	10.8	233	242	96.3
Nevada	100	330	42	12.7	4	28	**14.3**

fungicides and insecticides reduces the number of green plant, especially in microspores culture (Kasha et al., 1990). In general, the variation in the regeneration capacity among different batches of plants is very high, even with plants cultivated in growth chambers (Cho, 1991). Ritala et al., (2001) showed that the best material for microspore culture was obtained from March to October.

Stage of the Microspores and Stress Pre-treatment

From dissolution of the tetrad to the first mitosis in the microsporogenesis of barley, three major stages, namely, early, mid and late uninucleate, can be differentiated based on the size of the vacuole and the position of the nucleus in the microspore. Microspores at the mid-late to late uninucleate stage have the highest embryogenic ability in barley (Hoekstra et al., 1992).

In order to switch the gametophytic programme of the microspore development to the sporophytic pathway, stress pre-treatment of the anthers is needed. Cold-shock pre-treatment has showed to be very effective for induction of androgenesis. The spikes are placed in a plate at 4°C for 14–28 days at relative high humidity in the dark (Huang and Sunderland 1982). Another stress pre-treatment which is widely used is the carbohydrate starvation. This system is based in the substitution of a metabolizable sugar (sucrose, maltose) by a non-metabolizable sugar like mannitol. The anthers are cultured for 4 days at 24°C on 0.3 M liquid mannitol (Roberts-Oehlschlager and Dunwell, 1990).

The optimal length of cold and mannitol pre-treatment as well as the concentration of mannitol are genotype dependent (Powell, 1988b; Luckett and Smithard, 1995; Lezin et al., 1996; Cistué et al., 1994). Recalcitrant cultivars rendered higher numbers of green DH on using higher mannitol concentrations and/or longer periods of starvation (Cistué et al., 1994, 1999).

Hoeskstra et al., (1997) showed that the incorporation of Ca^{2+} and abscisic acid (ABA) in the mannitol pre-treatment medium considerably increased the number of green plants in cv. Igri. Our results demonstrated that the introduction of 40 mM Ca^{2+} in the pre-treatment medium increased the number of divisions, the number of embryos and the total number of green plants not only in Igri, but also in those cultivars with a very low androgenic response (Table 1.2).

Induction and Regeneration Media

Several basal media have been assayed successfully for the production of barley DH plants from anther or microspore culture: Klc (Kao 1981); P-9 (Xu and Sunderland 1981); N-9 (Chu 1978); FC (Foroughi-Wehr et al., 1976); and BAC3 (Szarejko and Kasha 1989). However, the medium most

Table 1.2 Influence of mannitol plus Ca^{++} pre-treatment over three barley genotypes; anther culture response

Cultivar	*Anthers*	*Calcium added**	*No. of dividing microspores*	*No. of embryos*	*Embryos (%)*	*No. of green plants*	*No. of total plants*	*Green plants (%)*
Igri	100	yes	390	126	32.3	88	107	82.2
	100	no	274	82	29.9	62	72	86.1
Albacete	100	yes	198	40	20.2	1	14	7.1
	100	no	80	13	16.2	0	5	0
Steptoe	100	yes	71	16	22.5	1	15	6.7
	100	no	37	4	10.8	0	2	0

*40 mM Ca^{2+}

commonly used is the FHG (Hunter, 1988), a modified MS medium (Murashige and Skoog 1962), where sucrose has been replaced by maltose in order to avoid the fast degradation of sucrose into glucose. This medium also contains a high concentration of a non-toxic nitrogen organic source as glutamine, and a reduced amount of ammonium nitrate (Olsen 1987). The main advantage of this medium is the development of embryoids directly from the microspores.

Several auxins and cytokinins have been used to induce microspore embryogenesis. The auxins most commonly used are 2,4-dichloropheno-xyactict acid (2,4-D), α tnaphthaleneacetic acid (NAA), 2,3,5-triiodobenzoic acid (TIBA) (Cho, 1991) and phenylacetic acid (PAA) (Ziauddin et al., 1992) and the citokinins *6-benzylaminopurine* (BAP), kinetin and zeatin riboside. However, it seems that BAP by itself can produce a high rate of induction. It is important to mention that an interaction between genotypes and growth regulator combination has been described (Luckett and Smithard 1995), confirming the results reported by Cho (1991) that the anthers of each genotype have different concentrations of endogenous auxins and cytokinins.

Another important aspect in androgenesis is the gelling agent used in the culture medium. Different supports have been used for anther culture such as agar, agarose (Lyne et al., 1986), starch from barley or wheat (Sorvari and Schiedler 1987), and Phytagel (Hou et al., 1993). On using a liquid medium, the dividing microspores or the embryos have a tendency to sink, growing in anaerobic conditions. To avoid this, plates should be either shaked or membrane raft technologies (Luckett et al., 1991) or high molecular weight polysaccharide, Ficoll (Kao 1981) can be used. Although the use of Ficoll is expensive (Hou et al., 1993; Devaux et al., 1993), the introduction of Ficoll considerably increases the number of embryos (Devaux 1992; Cistué et al., 1999). Table 1.3 shows that a high concentration of Ficoll along the whole culture process raised the numbers of divisions, embryos and green plants in the two cultivars assayed.

In the microspore culture, since the microspores are not protected by the anther wall, the osmotic pressure of the medium can be a problem. Either a liquid medium or one with Ficoll seems to be the most appropriate for microspore culture. The osmotic pressure should be adjusted along the culture by replenishment of the medium with higher concentrations of Ficoll (Cistué et al., 1995).

A big improvement in the number of green DH plants has been produced during the last years in microspore culture with model genotypes of barley such as Igri (Hunter 1988; Ziauddin et al., 1990; Cistué et al., 1995; Hoekstra et al., 1997; Davies and Morton 1998). However, limited success has been achieved with recalcitrant genotypes.

Table 1.3 Influence of induction medium, with four concentrations of Ficoll, over two barley genotypes; anther culture response

	Experiment	*Anthers*	*No. of dividing microspores*	*No. of embryos*	*Embryos (%)*	*No. of green plants*	*No. of total plants*	*Green plants (%)*
	A*	100	818	61	7.5	0	27	0
VOLGA	**B**	100	896	68	7.6	0	32	0
	C	100	1824	123	6.7	0	48	0
	D	100	5334	601	11.3	7	347	2.0
	A	100	109	8	7.3	4	12	33.3
NEVADA	**B**	100	111	12	10.8	9	16	56.2
	C	100	156	23	14.7	9	18	50.0
	D	100	518	65	12.5	15	40	37.5

A: Fic 200 + Fic 200
B: Fic 200 + Replenishment Fic 200 + Fic 200
C: Fic 200 + Replenishment Fic 400 + Fic 400
D: Fic 400 + Fic 400

Recently, Li and Devaux (2001) reported a great improvement in the embryogenesis rate of recalcitrant genotypes by using ovary co-culture.

Although less attention has been given to the regeneration medium, its composition can influence the vigour of the plants and hence their survival on transfer to soil. Generally, the regeneration medium is the same as the induction medium with some modifications. A reduction of maltose concentration, the elimination of organic nitrogen and the use of the auxins IAA or NAA in this media, favoured the production of plants with a higher number of leaves and roots (Castillo et al., 2000).

Albinism

The high rate of albinism of some genotypes is still a drawback in DH production. As mentioned before, albinism has strong genotypic dependence. However, the culture conditions, the stage of microspores at culture initiation and the physiological state of the donor plants are also known to affect the rate of albinism (Kasha et al., 1990; Pickering and Devaux 1992). Length of time in the culture was highly correlated with the production of a higher proportion of albino plants regenerated (Cistué et al., 1995). Recently, Caredda et al., (2000) suggested that regeneration of only albino plants in the spring cultivar Cork might be due to degradation of microspore plastid DNA during early pollen development, preventing the differentiation of plastids into chloroplasts under culture conditions.

Chromosome Doubling

Barley microspores exhibit the capacity of spontaneous doubling of the chromosome number during the first days of culture; therefore doubled haploid plants are obtained directly from the culture. The rate of spontaneous doubling varied in different reports, from 27% (Finnie et al., 1989) to over 90% (Huang and Sunderland 1982; Olsen 1987). Presently, we obtain between 80% and 90% of spontaneous doubling. These differences in the proportion of doubled haploids seem to depend on the genotype and the culture system used.

Haploid plants should be treated with an agent for chromosome doubling. The most commonly-used agent in barley is colchicine, as described by Jensen (1975).

DOUBLED HAPLOIDS IN PLANT BREEDING

To be used successfully in a breeding programme, any particular doubled haploid system should fulfil the following criteria: (1) production of a large number of DH from all the genotypes supplied by breeders; (2) the population should contain a random sample of the parental gametes; and (3) plants produced must be phenotypically normal, stable and fertile.

Production of DH from all the Genotypes in the Breeding Programme

Although the strong genotype dependence limited the practical application of androgenesis to plant breeding in the past, the improvements achieved during the last few years have led to the production of DH from most genotypes. In fact, some years ago results were expressed as the number of green plants per spike; later as green plants per 100 anthers, and nowadays as green plants per anther. However, it should be remarked that there are some genotypes or crosses that still render a low number of green plants. In such cases, in order to obtain a certain number of DH lines, a higher number of anthers should be inoculated (Cistué et al., 1999).

Parental Gametes Sampled at Random

In order to use the total variability of a cross, the population of DH should contain a random sample of the parental gametes. Some reports have described a significant deviation from the expected segregation of morphological traits and molecular markers in DH populations. When the androgenesis system used favoured calli formation or the parental genotypes had a very different androgenetic response, a high rate of segregation distortion was described (Thompson et al., 1991; Devaux and Zivy 1994). However, when the culture system favours embryogenesis or the parentals have similar response, the distortion is lower (Heun et al., 1991).

For some agronomic characters, it is even possible to obtain lines with higher or lower values than the mid-parental (transgressive lines) in the DH populations. Agronomic performance was studied in a DH population from the cross between cv. Dobla (six rows, early heading date and medium anther culture response) with cultivar Igri (two rows, normal heading date and good anther culture response) (Fig. 1.1). In both the traits, heading date and 1000-kernel weight, several lines were observed with extreme values from the mid-parental. Even a two-row line (2r) with an earlier heading date than the six-row Dobla (D) was obtained. Also, a six-row line (6r) was produced with a 1000-kernel weight close to the value for two row Igri. Jui et al., (1997) working with a cross (six × two-rows), also found a high number of transgressive lines, indicating that some useful genes could be transferred from six-row to two-row varieties.

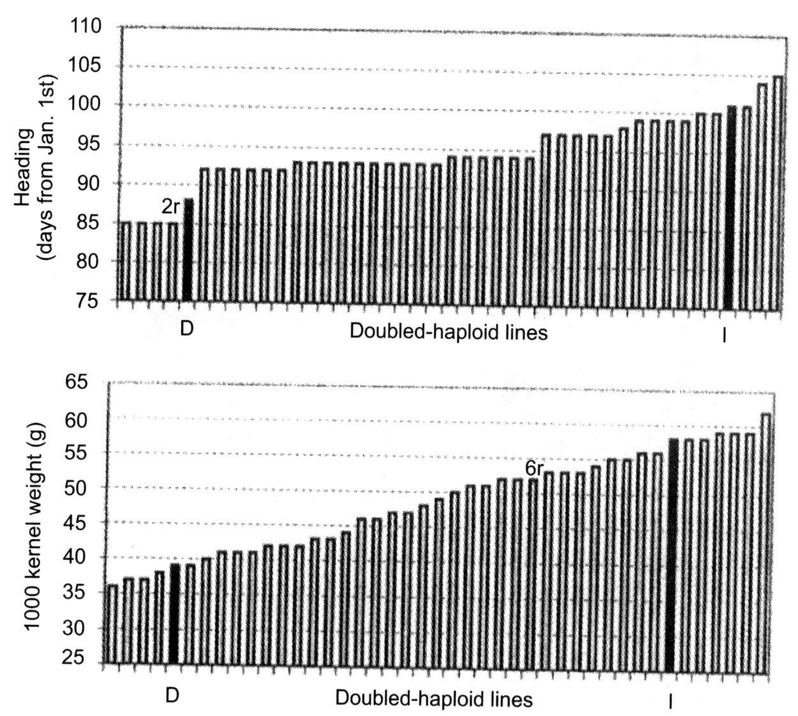

Fig. 1.1 Heading date and 1000-kernel weight of 41 doubled haploids produced from the crossing between the cultivars Dobla (D) and Igri (I)

Characteristics of the DH Lines Produced

The field performance of barley DH derived from anther culture has been compared with lines developed from the same cross using the pedigree and SSD (Single Seed Descent) methods, and in general, there are no significant differences for the agronomic characters among the various methods (Powell et al., 1992; Bjϕrsnstad et al., 1993). No significant gametoclonal variation has been described on using protocols that favour embryogenesis for the production of the DH lines (Szarejko et al., 1997).

IMPORTANT ASPECTS

There are several critical aspects that require emphasis. There is a strong genotypic interaction with different physiological factors such as growing conditions of the donor plants, pre-treatment and composition of induction medium. However, when seeking to obtain doubled haploid plants from different cultivars or F_1, it is not possible to ascertain the

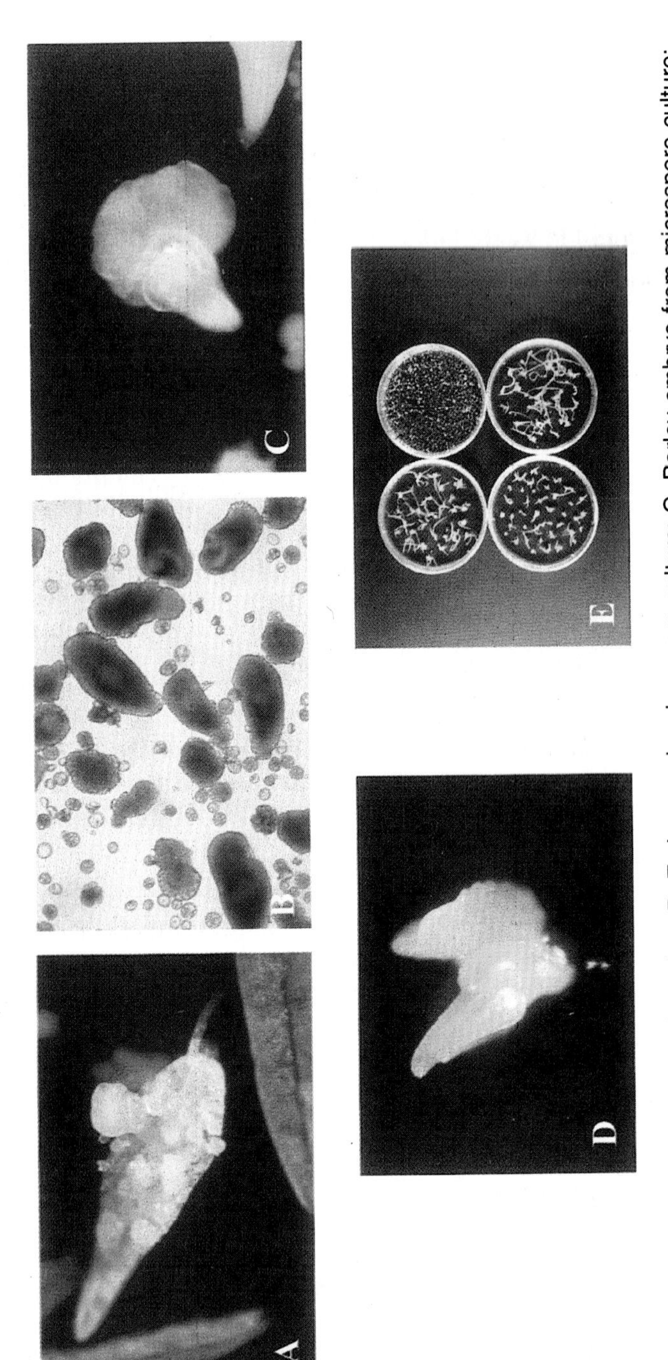

Fig. 1.2 A. Barley embryogenic anther; B. Embryogenic microspore culture; C. Barley embryo from microspore culture; D. Green embryo on regeneration medium; E. Barley microspore culture. 150 green haploid plantlets plus the Petri dish with the embryogenic microspore culture.

optimal physiological factors for every F_1; so a compromise should be established.

Several private companies as well as public institutions are producing a high number of barley DH using different protocols. From our point of view is essential to establish an androgenetic system that favours development of a high number of good quality embryos from anther or isolated microspores cultures (Figs. 1.2a, b, c), avoiding callus formation, and thus increasing the probabilities of plant regeneration (Figs. 1.2d, e).

A pre-treatment with mannitol-calcium normally synchronizes microspores in the late uninucleate stage. These microspores, growing in aerobic conditions with high osmotic pressure, undergo early division and grow actively, to finally develop well-formed embryos. This system produces high amount of green doubled haploids from most of the barley genotypes.

REFERENCES

Bjφrnstad A., Skinnes H., Thoresen K., 1993, Comparisons between doubled haploid lines produced by anther culture, the *Hordeum bulbosum*-method and lines produced by single seed descent in barley crosses. Euphytica **66**: 135–144.

Caredda S., Doncoeur C., Devaux P., Sangwan R.S., Clément C., 2000, Plastid differentiation during androgenesis in albino and non-albino producing cultivars of barley (*Hordeum vulgare* L.) Sex Plant Reprod. **13**: 95–104.

Castillo A.M., Vallés M.P., Cistué L., 2000, Comparison of anther and isolated microspore cultures in barley. Effects of culture density and regeneration medium. Euphytica **113**: 1–8.

Castillo A.M., Cistué L., Vallés M.P., Sanz J.M., Romagosa I, Molina-Cano J.L., 2001, Efficient production of androgenic doubled-haploid mutants in barley by the application of sodium azide to anther and microspore cultures. Plant Cell Rep. **20**: 105–111.

Chase S.S., 1952, Monoploids in maize. In: Gowen J.W. (Ed.) Heterosis. pp. 389–399. Iowa State College Press, Ames, Iowa.

Cho U.H., 1991, Hormonal aspects of androgenic induction in barley (*Hordeum vulgare* L.). Ph.D Thesis, University of Guelph, Canada.

Chu C.C., 1978, The N6 medium and its application to anther culture of cereal crops. In: Proceedings of Symposium on Plant Tissue Culture, pp. 43–50. Science Press, Peking, China.

Cistué L., Ramos A., Castillo A.M., Romagosa I., 1994, Production of large number of doubled haploid plants from barley anthers pre-treated with high concentrations of mannitol. Plant Cell Rep. **13**: 709–712.

Cistué L., Ziauddin A., Simion E., Kasha K.J., 1995, Effects of culture conditions on isolated microspores response of barley cultivar Igri. Plant Cell Tiss. Org. Cult. **42**: 163–169.

Cistué L., Ramos A., Castillo A.M., 1999, Influence of anther pre-treatment and culture medium composition on the production of barley doubled haploids from model and low responding cultivars. Plant Cell Tiss. Org. Cult. **55**: 159–166.

Clapham D., 1973, Haploid *Hordeum* plants from anthers in vitro. Z. Pflanzenzüchtg **69**: 142–155.

Davies P.A., Morton S., 1998, A comparison of barley isolated microspore and anther culture and the influence of cell culture density. Plant Cell Rep. 17: 206–210.

Devaux P., 1992, Haploidy in barley and wheat improvement. In: Dattee Y., Dumas C., and Gallais A. (Eds) Reproductive Biology and Plant Breeding. pp. 139–151, Springer-Verlag, Berlin.

Devaux P., Hou L., Ullrich S.E., Huang Z., Kleinhofs A., 1993, Factors affecting anther culturability of recalcitrant barley genotypes. Plant Cell Rep. **13**: 32–36.

Devaux P., Zivy M., 1994, Protein markers for anther culturability in barley. Theor. Appl. Genet. **88**: 701–706.

Devaux P., Zivy M., Kilian A., Kleinhofs A., 1996, Doubled haploids in barley. In: Slinkard A., Scoles G., Rossnagel B. (Eds) Proceedings of Vth International Oat Conference and VIIth International Barley Genetics Symposium, Vol. 1. pp. 213–222). University Extension Press, University of Saskatchewan, Saskatoon.

Finie S.J., Powell W., Dyer A.F., 1989, The effect of carbohydrate composition and concentration on anther culture response in barley (*Hordeum vulgare* L.). Plant Breeding **103**: 110–118.

Foroughi-Wehr B., Mix G., Gaul H., Wilson H.M., 1976, Plant production from cultured anthers of *Hordeum vulgare* L. Z. Pflanzenzücht **77**: 198–204.

Foroughi-Wehr B., Friedt W., Wenzel G., 1982, On the genetic improvement of androgenetic haploid formation in *Hordeum vulgare* L. Theor. Appl. Genet. **72**: 233–239.

Graner A., Jahoor A., Schondelmaier J., Siedler H., Pillen K., Fischbeck G., Wenzel G., Herrmmann R.G., 1991, Construction of an RFLP map of barley. Theor. Appl. Genet. **83**: 250–256.

Heun M., Kennedy A.E., Anderson J.A., Lapitan N.L.V., Sorrells M.E., Tanksley S.D., 1991, Construction of a restriction fragment length polymorphism map of barley. Genome **34**: 437–447.

Hoekstra S., van Zijderveld M.H., Louwerse J.D., Heidekamp F., van der Mark F., 1992, Anther and microspore culture of *Hordeum vulgare* L. cv. Igri. Plant Sci. **86**: 89–96.

Hoekstra S., van Bergen S., van Brouwershaven I.R., Schilperoort R.A., Heidekamp F., 1997, Androgenesis in *Hordeum vulgare* L.: Effects of mannitol, calcium and abscisic acid on anther pretreatment. Plant Sci. **126**: 211–218.

Hou L., Ullrich S.E., Kleinhofs A., Stiff C.M., 1993, Improvement of anther culture methods for doubled haploid production in barley breeding. Plant Cell Rep. **12**: 334–338.

Hou L., S.E. Ulrich, A. Kleinhofs, 1994, Inheritance of anther culture traits in barley. Crop Sci. **34**: 1243–1247.

Huang B., Sunderland N., 1982, Temperature stress pretreatment in barley anther culture. Ann. Bot. **49**: 77–88.

Hunter C.P., 1988, Plant regeneration from microspores of barley, *Hordeum vulgare*. Ph.D Thesis. Wye College, University of London.

Jähne A., Becker D., Brettschneider R., Lörz H., 1994, Generation of transgenic, microspore-derived, fertile barley. Theor. Appl. Genet. **89**: 525–533.

Jensen C.J., 1975, Barley monoploids and doubled monoploids: techniques and experience. In: Gaul H. (Ed.). Barley Genetics III, Verlag Karl Thieming Munchen. pp. 316–345.

Jui P.Y., Coo T.M., Ho K.M., Konishi T., Martin R.A., 1997, Genetic analysis of a two-row × six-row cross of barley using doubled-haploid lines. Theor. Appl. Genet. **94**: 549–556.

Kao K.N., 1981, Plant formation from barley anther cultures with Ficoll media. Z. Pflanzenphysiol. **103**: 437–443.

Kasha K.J., Kao K.N., 1970, High frequency haploid production in barley (*Hordeum vulgare* L.). Nature (London) **225**: 874–876.

Kasha K.J., Ziauddin A., Cho U.H., 1990, Haploids in cereal improvement. In: Gustafson J.P. (Ed.) Gene Manipulation in Plant Improvement II. Proc. 19th Stadler Genetics Symp, pp. 213–235. Plenum Press, New York.

Kasha K.J., Ziauddin A., Cho U.H., 1997, Haploids in cereal improvement: anther and microspore culture. In: Gustafson J.P. (Ed.), Gene Manipulation in Plant Improvement II, pp. 213–235. Proc. 19th Stadler Genetics Symp. Plenum Press, New York.

Kleinhofs A., Kilian A., Kudrna D., Bollinger J., Hayes P., Chen F.Q., Saghai Maroof M.A., Lapitan N., Fenwick A., Blake T.M., Kanazin V., Ananiev E., Dahleen L., Liu B., Sorrells M., Heun M., Franckowiak J.D., Hoffman D., Skadsen R., Steffenson B.J., Knapp S.J., 1993, A molecular, isozyme and morphological map of barley (*Hordeum vulgare*) genome. Theor. Appl. Genet. **86**: 705–712.

Lezin F., Sarrafi A., Alibert G., 1996, The effects of genotype, ploidy level and cold pre-treatment on barley anther culture responsiveness. Cereal Res. Communication **24**: 7–13.

Li, H., Devaux P., 2001, Enhancement of microspore culture efficiency of recalcitrant barley genotypes. Plant Cell Rep. **20**: 475–481.

Luckett D.J., Venkatanagappa S., Darvey N.L., Smithard R.A., 1991, Anther culture of Australian wheat germplam using modified C17 medium and membrane rafts. Australian Journal of Plant Physiology **18**: 357–367.

Luckett D.J., Darvey N.L., 1992, Utilisation of microspore culture in wheat and barley improvement. Aust. J. Bot. **40**: 807–828.

Luckett D.J., Smithard R.A., 1995, A comparison of several published methods for barley anther culture. Plant Cell Rep. **14**: 763–767.

Lyne R.L., Bennett R.I., Hunter C.P., 1986, Embryoid and plant production from cultured barley anthers In: Withers L.A. and Alderson P.G. (Eds) Plant Tissue Culture and its Agricultural Application. University of Nottingham. pp. 405–411. Butterworth, Pub. Guilford.

Maluszynski M., Szarejko I., Sigurbjörnsson B., 1996, Haploidy and mutation techniques. In: Mohan Jain S., Sopory S.K., Veilleux R.E. (Eds) In Vitro Haploid Production in Higher Plants. Vol. 1: Fundamental Aspects and

Methods, pp. 67–93. Kluwer Academic Publisher, Dordrecht, The Netherlands.

Marquez-Cedillo L.A., Hayes P.M., Kleinhofs A., Legge W.G., Rossnagel B.G., Sato K., Ulrich S.E., Wesemberg D.M., 2001, The North American Barley Genome Mapping Project. QTL analysis of agronomic traits in barley based on the doubled haploid progeny of two elite North American varieties representing different germplasm groups. Theor. Appl. Genet. **103**: 625–637.

Murashige T., Skoog F., 1962, A revised medium for rapid growth and bioassays with tobacco tissue cultures. Physiol. Plant **15**: 473–497.

Olsen F.L., 1987, Induction of microspore embryogenesis in cultured anthers of *Hordeum vulgare*. The effects of ammonium nitrate, glutamine and asparagine as nitrogen sources. Calsberg Res. Commun. **52**: 393–404.

Pickering R.A. and Devaux P., 1992, Haploid production: approaches and use in plant breeding. In: Shewry P.R. (Ed.), Biotechnology in Agriculture No. 5. Barley: Genetics, Biochemistry, Molecular Biology and Biotechnology, pp. 202–227. Alden Press Ltd., Oxford.

Powell W., 1988a, Diallel analysis of barley anther culture response. Genome 30: 152–157.

Powell W., 1988b, The influence of genotype and temperature pre-treatment on anther culture response in barley (*Hordeum vulgare* L.). Plant Cell Tiss. Org. Cult. **12**: 291–297.

Powell W.T., Thomas W.T.B., Thompson D.M., 1992, The agronomic performance of anther culture derived plants of barley produced via pollen embryogenesis. Ann. Appl. Biol. **120**: 137–150.

Ritala A., Mannonen L., Oksman-Caldentey K.M., 2001, Factors affecting the regeneration capacity of isolated barley microspores (*Hordeum vulgare* L.). Plant Cell Rep. **20**: 403–407.

Roberts-Oehlschlager S.L., Dunwell J.M., 1990, Barley anther culture: pretreatment on mannitol stimulates production of microspore-derived embryos. Plant Cell Tiss. Org. Cult. **20**: 235–240.

Snape J.W., Simpson E., 1984, Early generation selection and rapid generation advancement methods in autogamous crops. In: Lange W., Zeven A.C., Hoogendoorn N.G. (Eds), Efficiency in Plant Breeding, pp. 82–86. Pudoc, Wageningen.

Sorvari S., Schieder O., 1987, Influence of sucrose and melibiose on barley anther culture in starch medium. Plant Breeding **99**: 164–171.

Szarejko I., Kasha K.J., 1989, Induction of anther culture derived doubled haploids in barley. In: Maluszynski M., Barabas Z. (Eds) Proc. FAO/IAEA Meeting. Cereal Res Commun **19**: 219–237.

Szarejko I., Falk D.E., Janusz A., Nabialkowska, 1997, Cytological and genetic evaluation of anther culture-derived doubled haploids in barley. J. Appl. Genet. **38**: (4); 437–452.

Thompson D.M., Chalmers K., Waugh R., Foster B.P., Thomas W.T.B., Caligari P.D.S. and Powell W., 1991, The inheritance of genetic markers in microspore-derived plants of barley *Hordeum vulgare* L. Theor. Appl. Genet. **81**: 487–492.

Toojinda T., Broers L.H., Chen X.M., Hayes P.M., Kleinhofs A., Korte J., Kudrna D., Leung H., Line R.F., Powell W., Ramsay L., Vivar H., Waugh R., 2000,

Mapping quantitative and qualitative disease resistance genes in a doubled haploid population of barley (*Hordeum vulgare*). Theor. Appl. Genet. **101**: 508–589.

Xu Z.H., Sunderland N., 1981, Glutamine, inositol and conditioning factor in the production of barley pollen callus in vitro. Pl Sci. Lett. **23**: 161–168.

Yao Q.A., Kasha K.J., 1997, Potential of biolistic transformation of barley microspores based on viability and transient β-glucuronidase activity. Genome **40**: 639–643.

Ziauddin A., Simion E., Kasha K.J., 1990, Improved plant regeneration from shed microspore culture in barley (*Hordeum vulgare* L.) cv. Igri. Plant Cell Rep. **9**: 69–72.

Ziauddin A., Marsolais A., Simion E. and Kasha K.J., 1992, Improved plant regeneration from wheat anther and barley microspore culture using phenylacetic acid (PAA). Plant Cell Rep. **11**: 489–498.

2

In vitro Techniques and Physical Mutagens for the Improvement of Fruit Crops

Predieri Stefano and Gatti Edoardo
Istituto di Biometeorologia
IBIMET - BO - CNR Via P. Gobetti, 101 I-40129 Bologna (Italy)

ABSTRACT

The improvement of vegetatively-propagated plants can be pursued by mutation induction, which can alter only one or a few selected traits, but can save the overall characteristics of an already first-rate cultivar. Induced mutations have well-defined limitations in fruit-breeding applications, but their possibilities may be expanded by the use of in vitro techniques. Tissue culture increases the efficiency of mutagenic treatments for variation induction, handling of large populations, use of ready selection methods, and rapid cloning of selected variants. This manuscript describes the existing methodologies for in vitro mutation induction through the use of physical mutagens, and in particular γ-ray technology on fruit tissue culture. The first step of a breeding program with physical mutagens is based on the assessment of LD50, on which level the choice of the most suited dose is commonly based. Some experiments conducted to determine the LD50 of pear and apple tissue cultures are presented and results showing some differences in radiosensitivity among cultivars of the same species have been discussed in this chapter.

INTRODUCTION

Plant cell and tissue culture techniques have been developed and advanced for most of the commercially-important fruits, which provides a number of additional opportunities to conventional plant breeding, including an effective use of mutation induction protocols. Fruit crops are in general vegetatively propagated, since many of them are either self-incompatible and/or complex polyploids. Mutation breeding is advisable either when the desired traits cannot be found in the existing germplasm or they cannot be transferred to a species or cultivar of interest. Mutagenesis, offering the possibility of altering only one or a few characters, has already been used to introduce useful traits (Briggs and

---

For Correspondence: E-mail: S.Predieri@ibimet.cnr.it

Constanin, 1977) and shows a potential for improving many fruit species. For some crops, induced mutation is the most suitable breeding technique. In dessert banana, conventional breeding failed to produce new cultivars because of no or extremely reduced seed production, while in vitro gamma-ray irradiation proved to be an effective tool (Novak et al., 1990; Tan et al., 1993; Bhagwat and Duncan 1998; IAEA 1998; Mak et al., 1998). Traits induced by mutagenesis include plant size, blooming time and fruit ripening, fruit color, self-compatibility, self-thinning, and resistance to pathogens (Visser et al., 1971; Janick and Moore 1975; Donini 1982; Lapins 1983; Spiegel-Roy 1990; Brunner and Keppl 1991; Janick and Moore 1996; van Harten 1998; Sanada and Amano 1998; Hartman and Vuylsteke, 1999; Predieri 2001). The number of cultivars successfully obtained through mutagenesis increases every year, by the end of 2001, it consisted of more than 2200, including about 50 cultivars of fruits belonging to more than 20 different species (FAO/IAEA Mutant Varieties Database; <http://www-infocris.iaea.org/MVD/> October 2001).

Mutation induction can be empowered by in vitro techniques; many examples related to different vegetatively-propagated species show that the combination of in vitro culture and mutagenesis is relatively inexpensive, simple and efficient (Ahloowalia 1998). Tissue culture allows mutagenic treatment of large populations, a prerequisite for a successful breeding program. Selection and cloning of selected variants can be performed in small spaces, short time periods, and on a year-round basis, maintaining high phytosanitary conditions throughout the entire process. Following mutagen treatments, cultures are chimaeras, composed of non-mutated cells as also cells carrying different mutations. The identification and selection of the desired mutants should be performed on non-chimaeric plants. Tissue culture is more suited—as compared to methodologies involving treatment of in vivo buds—to the generation of homohistonts. In fact, in vitro culture provide a wide choice of plant material composed of a few or even just one cell for mutagenic treatment (Novak 1991; Heinze and Schmidt 1995; Maluszynski et al., 1995; Maluszynski et al., 1996), resulting in a lesser risk of obtaining chimaeric plants and a higher probability for mutated cells to express the mutation in the phenotype (D'Amato 1977). In vitro techniques also provide the possibility of rapidly executing the propagation cycles of subculture aimed to separate the mutated from non-mutated sectors (Ahloowalia 1998; Cassells 1998). Although only genetic or molecular evidence can establish the status of mutant of a selected phenotypic variant, in vitro mutation induction has proven to be effective in inducing genetic variation, since in many cases the in vitro selected variants were found to be mutants (Ahloowalia 1998). Furthermore, the molecular marker technology available today already provides tools to assist mutation-induction protocols by investigating both genetic variation within populations and early detection of mutants with desired traits. Cost is a major limitation to their application.

Among the techniques and sources of genetic variation available for tissue culture mutation induction (Predieri 2001), physical mutagens have already shown a potential for application in fruit breeding. The types of radiation suitable for mutagenesis are ultraviolet radiation (UV) and ionizing radiation (X-rays, γ-rays, α and β particles, protons, and neutrons). Ultraviolet radiation has a moderate capacity to penetrate plant tissues, which limits its use, but they can be used in certain cases such as the treatment of leaf tissues prior to regeneration, when the aim is to reduce the risk of obtaining chimaeras by irradiating only epidermal cells (Pinet-Leblay et al., 1992). Ionizing radiation penetrates deeper into the tissue and can induce a vast variety of chemical changes (Ahnström 1977). X-rays and γ-rays are the most convenient and easiest radiations to use regarding application methods and handling (Sanada and Amano 1998), and have been the most widely-used ionizing radiation types as also the most effective for fruit breeding purposes. Chronic exposure has largely been used, but has not shown definite advantages over acute irradiation (Sigurbjörnsson 1977), which is designed for performing mutagenic treatments on tissue cultures. Physical mutagens have some technical advantages over chemical mutagens. As related to safety and environmental issues, they avoid the manipulation of hazardous substances and the production of toxic residuals. Furthermore, not requiring the rinsing of plant material after treatment, they avoid some side effects of plant stress and allow a more precise determination of the time of exposition.

The routine procedure in assessing the most appropriate dosage for treatment is based on radiosensitivity, which is estimated through the physiological response of the irradiated material. Radiosensitivity varies with the species and the cultivar, with the physiological condition of plant and organs, and with the manipulation of the irradiated material before and after mutagenic treatment (Briggs and Konzak 1977; D'Amato 1992). A preliminary step in a mutation induction plan is the establishment of a dose-response curve, and the determination of the dose that causes a 50% reduction of vegetative growth of the treated material (LD50), when compared to the control in the first vegetative cycle (vM2) after treatment (Gaul 1977). The dose to be applied in a particular breeding project is then chosen on the basis of the breeder's experience with the specific plant material, its genetics, and its physiology, with the aim of having the highest probability of useful mutant rescue. Heinze and Schmidt (1995) suggest using as a starting point for the experimental protocol doses giving LD50 (±10%), or a dose resulting in 20% survival of the treated material. Several studies have been conducted on the radiosensitivity of in vitro cultures of fruits (see Predieri 2001).

Some vegetatively-propagated species are recalcitrant to plant regeneration. Although this can be a limit for the application of gene transfer biotechnology, it is not the same for mutation-induction breeding. Actually, when the aim of the breeding program is to maintain all the traits of a cultivar and improve only one or a few specific traits, the irradiation and propagation of in vitro axillary shoots may be the most adequate and ready method, and efficient micropropagation protocols are available for nearly all species. Furthermore, without the passage through undifferentiated growth, it is possible to avoid the undesired influence of somaclonal variation (van Harten 1998). With the aim of mutating single traits of pear cultivars, Predieri et al., (1997) irradiated in vitro cultures of pear with gamma-rays. Only the shoots produced by axillary buds were propagated, without any passage through undifferentiated growth. The result was a frequency of variants of about 1–2%, depending on the trait and cultivar that can be adapted into a useful selection protocol. Nearly 98% of the populations of self-rooted trees derived from originally-irradiated shoots had normal fruit production in terms of amount, consistency and quality. This allowed for the selection of some variants with reduced tree stature and a standard production of fruits, which was the main goal of the research (Predieri 1998). Preliminary molecular analyses indicated the state of mutants of some selected variants (Schilirò et al., 2001).

Mutagens can be also used on undifferentiated tissues and organs without preformed axillary buds either prior to regeneration, or in different stages of adventitious meristem differentiation; thus adding somaclonal variation source to the mutagen effects and, when regeneration is achieved from single cells, reducing the risk of obtaining chimaeras. Somaclonal variation and in vitro- induced mutagenesis have been studied in the grapevine for their capacity to induce changes in ploidy (Kuksova et al., 1997), showing that somaclonal variation and mutagens can be combined to increase the frequency of induced mutations.

MATERIALS AND METHODS

Microcutting Radiosensitivity Assessment

Buds were taken from certified virus-free mother plants of the pear cultivars ''Abbé Fetel'' and "Conference" and established in vitro. Cultures were grown on a proliferation medium (PM) containing MS salts (Murashige and Skoog 1962), LS vitamins (Linsmaier and Skoog 1965), 4.4 μM benzyladenine (BA), 0.49 μM indole-3-butyric acid (IBA), 2% (w/v) sucrose and 0.65% (w/v) commercial agar (B&V, Italy). The pH was adjusted to 5.75 with 0.1 N KOH before autoclaving at 120°C and 138 kPa for 20 minutes. Cultures were incubated in a growth chamber at 23±2°C with a 16-h photoperiod provided by cool white fluorescent tubes (Philips TLM 40W/33RS) at a photosynthetic photon flux of 48±3 μM m-2s-1.

From thirty-day-old proliferating cultures, microcuttings 1.5–2 cm long were cut and placed horizontally in plastic Petri dishes (10 cm diameter) with a few drops of sterile water added to protect shoots from dehydration during treatment. Acute irradiation (39.60 Gy/min) with γ rays from a cobalt (Co60) source was provided by "Gammacell 220" (Atomic Energy Canada Limited, Ottawa, Ontario, Canada) at the Institute of Photochemistry and High Energy Radiations (FRAE)-CNR Bologna). Total doses of 0, 20, 40, or 60 Gy were given. The shoots derived were labelled respectively: control, 20 Gy, 40 Gy, and 60 Gy. Thirty microcuttings per cultivar per treatment were subjected to irradiation. From shoots irradiated in Petri dishes, jars with 10 shoots each were prepared and placed in the growth chamber following a completely randomized experimental design. Proliferation rate was recorded after 30 days of culture on PM. LD50 was calculated as the doses of γ-radiation that reduces the proliferation rate of irradiated shoots to 50% of unirradiated control shoots, based on linear regression (Wu et al., 1978).

Assessment of radiosensitivity of leaf tissues based on regeneration rate

Tissue cultures of four apple (*Malus pumila* L.) cultivars "Gala", "Empire", "Golden Delicious", and "Stark Delicious" were used as sources of leaves for adventitious bud regeneration. Regeneration protocols were developed elsewhere (see Fasolo and Predieri 1990) and a regeneration medium containing MS salts (Murashige and Skoog 1962), LS vitamins (Linsmaier and Skoog 1965), 22.2 μM benzyladenine (BA), 1.1 μM α-naphthaleneacetic acid (NAA), 2% (w/v) sucrose and 0.7% (w/v) commercial agar (B&V, Italy) was used. The pH was adjusted to 5.7 with 0.1 N KOH before autoclaving at 120°C and 138 kPa for 20 minutes. Leaves were taken from 30-day-old cultures, where only the first three apical unfurled leaves were excised from the shoot. Three transverse cuts were made to the midrib and the petioles were removed. Each leaf blade was then placed with the adaxial face touching the medium. Six leaves were placed in each 10 cm diameter disposable Petri dish containing 20 mL of medium. Cultures were incubated in the dark in a growth chamber at 23±2°C. Acute irradiation (42.7 Gy/min) with gamma rays from a cobalt (Co60) source was provided by "Gammacell 220" (Atomic Energy Canada Limited, Ottawa, Ontario, Canada) at the Institute of Photochemistry and High Energy Radiations (FRAE)-CNR Bologna). Total doses of 0, 10, 20, 30, or 40 Gy were given. Thirty leaves per cultivar per treatment were subjected to irradiation. After treatment, the leaves were transferred to a fresh medium, and the Petri dishes placed in the growth chamber following a completely randomized experimental design. The number of adventitious shoots regenerated after a 45-day culture on regeneration medium was recorded. LD50 was calculated as the dose of γ-radiation that reduces the number of shoots regenerated per irradiated leaf to 50% of unirradiated control leaves, based on linear regression (Wu et al., 1978).

RESULTS

The proliferation rate of control "Abbé Fetel" cultures was 7.06, the irradiation with 20 Gy reduced it to 3.13, causing a total growth decrease of 55%, which was higher than the desired LD50. However, on the basis of linear regression, LD50 was estimated to be 27.47 Gy (Fig. 2.1). The proliferation of control "Conference" shoots was 7.10. The irradiation with 20 Gy reduced it to 6.43 causing a decrease of 10%; the irradiation with 40 and 60 Gy reduced proliferation to 4.36 and 2.86 respectively; with a decrease as compared to the control cultures of 49% for 40 Gy and 60% for 60 Gy. On the basis of linear regression, LD50 was estimated to be achieved with 45.3 Gy (Fig. 2.2). Beside total growth reduction, radiation caused primary damages that were evidenced in the form of apical necrosis, leaf chlorosis, and phylloptosis (Fig. 2.3).

In relation to irradiation of leaves prior to regeneration, "Gala" apple cultures presented the highest regeneration rate (Fig. 2.4), with 8.8 shoot per leaf, LD50 was calculated to be 17.87 Gy (Fig. 2.5a). LD50 of the other cultivars were 17.20 for "Golden Delicious" (Fig. 2.5b); 15.82 for "Empire" (Fig. 2.5c); and 16.56 for "Stark Delicious" (Fig. 2.5d).

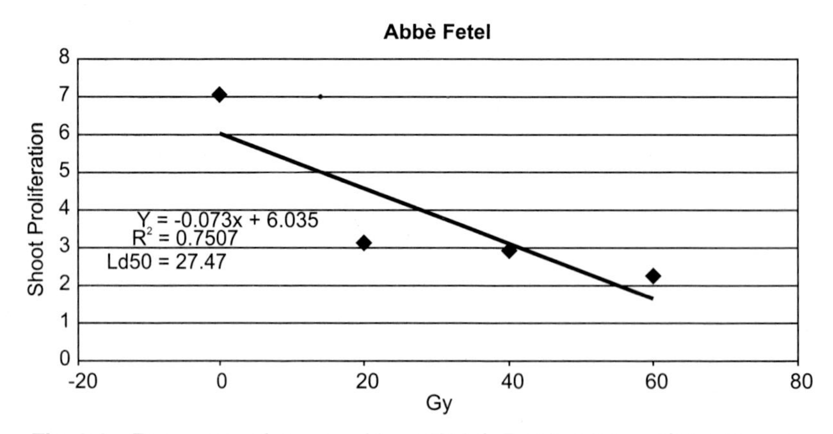

Fig. 2.1 Response of pear cultivar 'Abbé Fetel' microcuttings to γ-ray irradiation with 0, 20, 40 and 60 Gy.

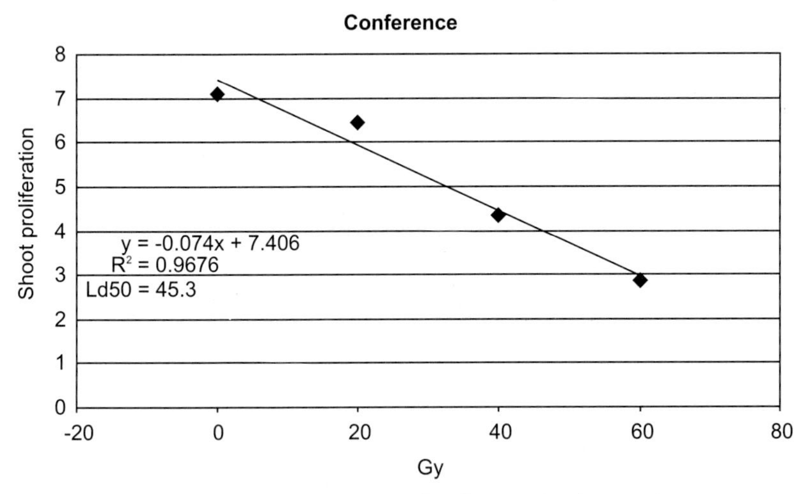

Fig. 2.2 Response of pear cultivar 'Conference' microcuttings to γ-ray irradiation with 0, 20, 40 and 60 Gy.

Fig. 2.3 Symptoms caused by γ-ray irradiation on tissue cultures of pear cultivar 'Conference' microcuttings irradiated with 60 Gy, after 30-day culture.

DISCUSSION

When facilities for physical mutagenic treatments are available, the use of acute irradiation allows rapid treatment of the plant material; less than 1 minute for reaching the LD50 in the cases described in this report. This factual data suggests that besides the primary effects of radiations,

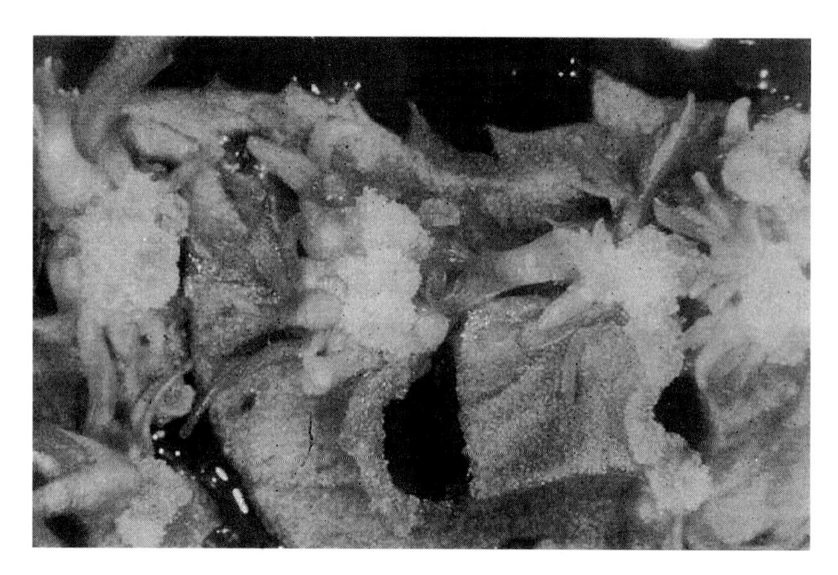

Fig. 2.4 Adventitious bud regeneration from leaves of the apple cultivar 'Gala' without γ-ray irradiation.

accurate handling of plant material can provide the desirable limitation of physiological effects that favors the emergence of phenotypic effects determined by genetic changes. Cultures irradiated either during proliferation or during regeneration should be transferred rapidly to a fresh medium so as to avoid the formation of toxic compounds (Ahloowalia 1998; Predieri and Gatti 2000). The irradiation of a high number of meristems in easy-to-manipulate and transport vessels such as Petri dishes allows space saving and easy execution of the required post-irradiation handling. To allow that all treated meristems could develop shoots and that all of them undergo at least three subcultures, the protocol proposed by Predieri (2001) may be adopted and adapted to the requirements of the different species of interest (Fig. 2.6). Material handling after irradiation should sustain plant growth at all stages, therefore requiring the definition of convenient protocols especially since undesired physiological effects can be observed even at the rooting stage, thus negatively affecting the microcutting's rooting capacity (Predieri et al., 1986; Rosati et al., 1990; Lima da Silva and Daozan 1995; Jain 1997; Predieri and Gatti 2000). The use of certified virus-free mother plants helps in all the steps of the protocol, from useful mutant identification to its possible release as a cultivar.

The assessment of radiosensitivity of two pear cultivars showed that "Abbé Fetel" has a higher sensitivity to irradiation, on the basis of the lower LD50 recorded, as compared to "Conference". The difference in radiosensitivity may be explained not only by inherent genetic differences

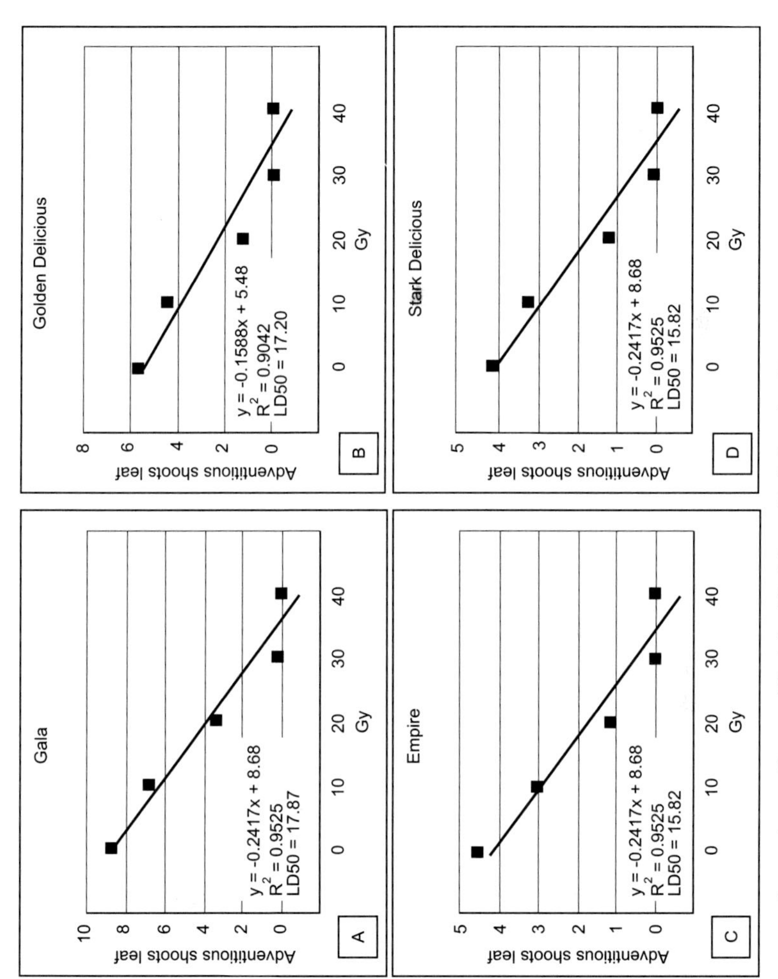

Fig. 2.5 Response of apple cultivars 'Gala', 'Golden Delicious', 'Empire' and 'Stark Delicious' leaves to γ-ray irradiation with 0, 10, 20, 30 and 40 Gy.

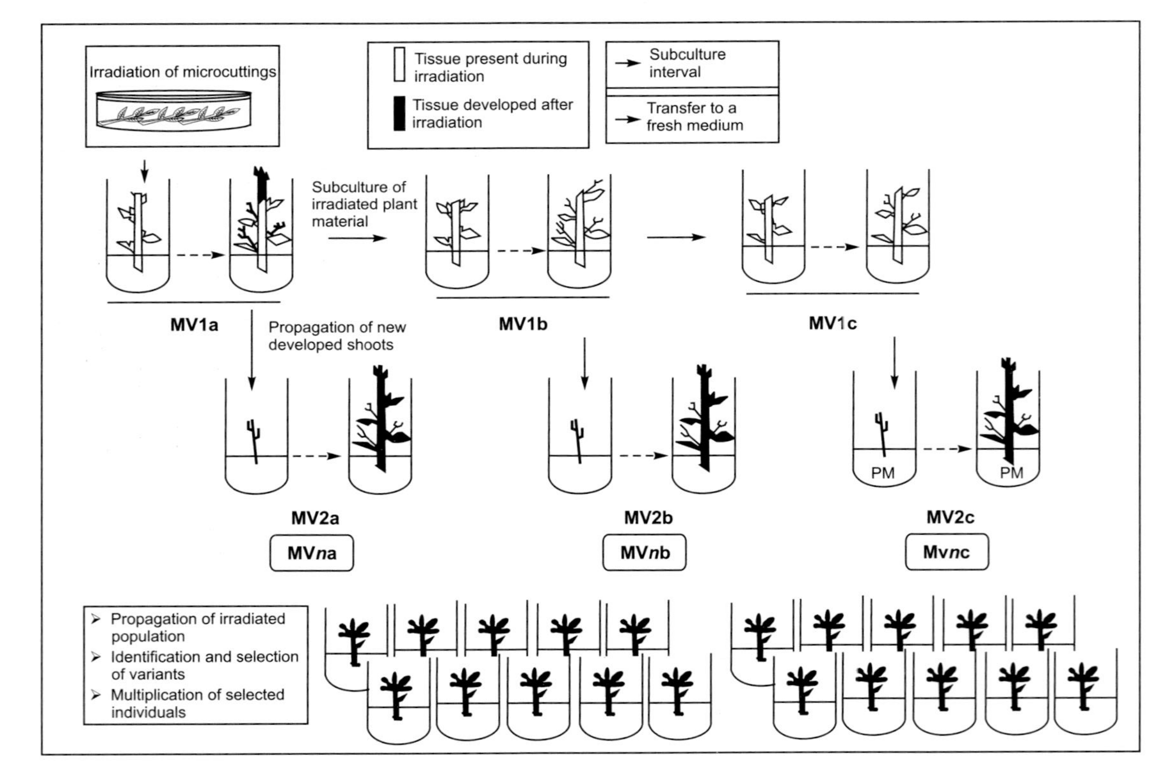

Fig. 2.6 Protocol for the manipulation of in vitro cultures of woody species after irradiation (modified from Predieri 2001).

between cultivars but also by physiological variations. These findings confirm that before executing the actual mutagenic treatment, it is advisable to perform a preliminary study on the specific sensitivity of the material to be used. Radiosensitivity has been studied on in vitro cultures of mazzard (*Prunus avium* L.) "F 12/1" treated with X-rays (Walther and Sauer 1985). The sensitivity found for shoot apices was higher than the one observed in basal parts, indicating that physiological factors can induce a difference in sensitivity even among the various axillary buds of a single microcutting. Differences in radiosensitivity were also observed among grapevine (*Vitis* spp. L.) rootstocks (Lima da Silva and Doazan 1995), and grapevine cultivars (Rosati et al., 1990). Shen et al., (1990) found large differences in γ-ray irradiated kiwifruit microcuttings (*Actinidia deliciosa* [A. Chev] Liang and Ferguson, cv. "Hayward"; *Actinidia chinensis* var. *hispida*, "Clone 4"). For banana (*Musa acuminata* Colla) shoot tips, radiosensitivity appeared to depend on the ploidy level of the treated genotypes. The LD50 were 20–25 Gy for the diploid (AA), 30–35 Gy for the triploid (AAA), and 35–40 Gy for the tetraploid (AAAA) (Novak et al., 1990). This positive correlation of tolerance to irradiation with the ploidy level was also observed in *Musa* spp. by Roux (1998). The possibility of selecting mutants following in vitro γ-ray irradiation of proliferating shoots has been shown on pear by Predieri et al., (1997) and Predieri and Zimmerman (2001) and also confirmed by RAPDs analysis (Schilirò et al., 2001). Treatments on explants already in tissue culture were able to yield reduced-size variants in pear (Predieri et al., 1997; Predieri 1998); a number of variants with increased sugar concentration or with extended shelf life were also identified (Predieri et al., 1998).

The LD50 recorded for the four different apple cultivars included in this study ranged from 15.82 to 17.87, suggesting evident similarity in radiosensitivity of the irradiated material, despite the differences in the regeneration rate of the non-irradiated leaves. The recorded LD50 is consistent with the one determined on apple rootstock M26 leaves (Predieri and Fasolo 1989). On the contrary, Pinet-Leblay et al., (1992) found that in the case of γ-ray irradiated pear leaves, the LD50 were genotype-dependent, varying from 20 to 50 Gy. Irradiation, followed by adventitious bud regeneration, has shown the potential to induce mutants with useful agronomic traits. Yang and Schmidt (1994) treated leaves of the cherry rootstock "209/1" (*Prunus cerasus* × *P. canescens*) with X-rays in vitro, and isolated and cloned a dwarf variant. The molecular analysis with RAPDs demonstrated to be a true mutant. Following γ-ray-induced variation in adventitious buds, Brunner and Keppl (1991) selected in the field a mutant for early fruit ripening in the apple cv. "Jonagold".

CONCLUSION

In vitro-induced mutagenesis with the use of physical mutagens appears to be a useful, ready-to-use technique for fruit improvement. Mutation-induction treatments performed on in vitro shoots or on plant tissues prior to regeneration have a potential for contributing to fruit breeding, allowing for single-trait changes while retaining the yield parameters. The first steps of such a protocol are easy and relatively cheap. However, only the use of accurate procedures and extensive field trials could allow the fulfillment of the great potential that this methodology promises. In vitro mutation induction results could be increased by the use of effective early screening methods for biotic agents and abiotic factors. Several screening methods have been developed regarding the resistance to pathogens (Leblay et al., 1993; Hammerschlag et al., 1995; Jain 1997; Masuda and Yoshioka 1997; Chevreau et al., 1998; Hammerschlag 2000); for tolerance to high pH (Dolcet-Sanjuan et al., 1992; Marino et al., 2000) or metals (Matsumoto and Yamaguchi 1990), and for selection of reduced vigor individuals (Lane et al., 1982; Scorza and Cordts 1989; Sarwar et al., 1998; Predieri and Govoni 1998).

If shared expertise on genetics, plant pathology, tissue culture physiology, and molecular biology is applied to the development of induced mutation procedures, this technique will prove to be a valuable contribution in making the most of the opportunities offered by in vitro culture for fruit improvement.

ACKNOWLEDGMENTS

Authors wish to thank Dr. A.E. Sztein for the revision of the manuscript.

REFERENCES

Ahloowalia B.S., 1998, In vitro techniques and mutagenesis for the improvement of vegetatively propagated plants. In: Jain S.M., Brar D.S. and Ahloowalia B.S. (Eds) Somaclonal Variation and Induced Mutations in Crop Improvement (pp 293–309). Kluwer Academic Publishers, Dordrecht.

Ahnström G., 1977, Radiobiology. In: Manual on Mutation Breeding. Second edition. Technical Reports Series No. 119. (pp 21–27). IAEA. Vienna.

Bhagwat B. and Duncan E.J., 1998, Mutation breeding of Highgate (*Musa acuminata*, AAA) for tolerance to *Fusarium oxysporum* f. sp. *cubense* using gamma irradiation. Euphytica **101**: 143–150.

Briggs R.W. and Constantin M.J., 1977, Radiation types and radiation sources. In: Manual on Mutation Breeding. Second edition. Technical Reports Series No. 119 (pp 7–20). IAEA, Vienna.

Briggs R.W. and Konzak C.F., 1977, Object and methods of treatment. In: Manual on Mutation Breeding. Second edition. Technical Reports Series No. 119 (pp 33–39). IAEA, Vienna.

Brunner H. and Keppl H., 1991, Radiation induced apple mutants of improved commercial value. In: Plant Mutation Breeding for Crop Improvement, Vol. 1 (pp 547–552). IAEA, Vienna.

Cassells A.C., 1998, In-vitro-induced mutations for disease resistance. In: Jain S.M., Brar D.S. and Ahloowalia B.S. (Eds) Somaclonal Variation and Induced Mutations in Crop Improvement (pp 367–378). Kluwer Academic Publishers, Dordrecht.

Chevreau E., Brisset M.N., Paulin J.P. and James, D.J., 1998, Fire blight resistance and genetic trueness-to-type of four somaclonal variants from the apple cultivar Greensleeves. Euphytica. **104**: 199–205.

D'Amato F., 1977, Cytogenetics of differentiation in tissue and cell cultures. In: Reinert J. and Bajaj Y.P.S. (Eds) Applied and Fundamental Aspects of Plant Cell, Tissue and Organ Culture (pp 343–357). Springer-Verlag, New York.

D'Amato F., 1992, Induced mutations in crop improvement: Basic and applied aspects. Agr. Med. 122 (special issue) 31–60.

Dolcet-Sanjuan R., Mok D.W.S. and Mok M.C., 1992, Characterization and in vitro selection for iron efficiency in *Pyrus* and *Cydonia.* In Vitro Cell. Dev. Biol. **28**: 25–29.

Donini B., 1982, Mutagenesis applied to improve fruit trees: techniques, methods and evaluation or radiation induced mutations. In: Induced Mutations in Vegetatively-Propagated Plants, II (pp 29–36). IAEA, Vienna.

Gaul H., 1977, Plant injury and lethality. In: Manual on Mutation Breeding. Second edition. In: Induced Mutations in Vegetatively Propagated Plants, II (pp 29–36). IAEA, Vienna.

Hammerschlag F.A., 2000, Resistant responses of peach somaclone 122-1 to *Xanthomonas campestris* pv. *pruni* and to *Pseudomonas syringae* pv. *syringae.* HortSci. **35**: 141–143.

Hammerschlag F.A., Ritchie D., Werner D., Hashmi G., Krusberg L., Meyer R. and Huettel R., 1995, In vitro selection of disease resistance in fruit trees. Acta Hort. **392**: 19–26.

Hartman J.B. and Vuylsteke D., 1999, Breeding for fungal resistance in *Musa.* In: Scarascia Mugnozza G.T., Porceddu E. and Pagnotta M.A. (Eds) Genetic and Breeding for Crop Quality and Resistance (pp 83–92). Kluwer Academic Publishers, Dordrecht.

Heinze B. and Schmidt J., 1995, Mutation work with somatic embryogenesis in woody plants. In: Jain S.M., Gupta K. and Newton J. (Eds) Somatic embryogenesis in woody plants, Vol. 1 (pp 379–398). Kluwer Academic Publishers, Dordrecht.

IAEA, 1998, Cellular Biology and Biotechnology Including Mutation Techniques for Creation of New Useful Banana Genotypes. Report Second Research Co-ordinated FAO/IAEA/BADC Meeting. Kuala Lampur, October 1997. IAEA, Vienna.

Jain S.M., 1997, Creation of variability by mutation and tissue culture for improving plants. Acta Hort. **447**: 69–77.

Janick J. and Moore J.N., 1975, Advances in fruit breeding. Purdue Univ. Press, West Lafayette.

Janick J. and Moore J.N., 1996, Fruit Breeding, Vol. I: Tree and Tropical Fruits. John Wiley and Sons, Inc. New York.

Kuksova V.B., Piven N.M. and Gleba Y.Y., 1997, Somaclonal variation and in vitro induced mutagenesis in grapevine. Plant Cell Tiss. Org. Cult. **49**: 17–27.

Lane W.D., Looney N.E. and Mage F., 1982, A selective tissue culture medium for growth of compact (dwarf) mutants of apple. Theor. Appl. Genet. **61**: 219–223.

Lapins K.O., 1983, Mutation breeding. In: Moore J.N. and Janick J. (Eds) Methods in Fruit Breeding (pp 74–99). Purdue Univ. Press, West Lafayette.

Leblay C., Chevreau E., Brisset M.N. and Paulin J.P., 1993, In vitro obtention and selection of pear mutants resistant to fire blight (*Erwinia amylovora*). Proceedings of FAO/IAEA Research Co-ordination Meeting "Induced-mutations and in vitro culture techniques for improving crop plant resistance to diseases". Grünbach, D., 7–11 October 1991. 27–36.

Lima da Silva A. and Doazan J.P., 1995, Gamma ray-mutagenesis on grapevine rootstocks cultivated in vitro. J. Int. Sci. de la Vigne et du Vin. **29**: 1–9.

Linsmaier E.M. and Skoog F., 1965, Organic growth requirements of tobacco tissue culture. Physiol. Plant. **18**: 100–127.

Mak C., Ho Y.W. and Tan Y.P., 1998, Micropropagation and mutation breeding techniques for the improvement of bananas. Acta Hort. **461**: 219–223.

Maluszynski M., Ahloowalia B.S. and Sigurbjörnsson B., 1995, Application of in vivo and in vitro mutation techniques for crop improvement. Euphytica **85**: 303–315.

Maluszynski M., Szarenko I. and Sigurbjörnsson B., 1996, Haploidy and mutation techniques. In: Jain S.M., Sopory S.K. and Veilleux R.E. (Eds) In vitro Haploid Production in Higher Plants, Vol. 1 (pp 67–930). Kluwer Academic Publishers, Dordrecht.

Marino G., Beghelli S, Rombolà A.D. and Cabrini L., 2000, In vitro performance at high culture pH and in vivo responses to Fe-deficiency of leaf-derived quince BA29 (*Cydonia oblonga*) somaclones regenerated at variable pH. J. Hort. Sci. Biotechnol. **75**: 433–440.

Masuda T. and Yoshioka T., 1997, In vitro selection of a mutant resistant to Alternaria blotch disease in "Indo" apple. Tech. News Inst. Rad. Breed. **56**: 1–2.

Matsumoto K. and Yamaguchi H., 1990, Selection of aluminium-tolerant variants from irradiated protocorm-like bodies in banana. Trop. Agric. **67**: 229–232.

Murashige T. and Skoog F.. 1962, A revised medium for rapid growth and bioassays with tobacco tissue cultures. Physiol. Plant. **15**: 241–251.

Novak F.J., 1991, In vitro mutation system for crop improvement. In: Plant Mutation Breeding for Crop Improvement, Vol. 2 (pp 327–342). IAEA, Vienna.

Novak F.J., Afza R., van Duren M. and Omar M.S., 1990, Mutation induction by gamma irradiation of in vitro cultured shoot-tips of banana and plantain (*Musa* cultivars). Tropic. Agric. **67**: 21–28.

Pinet-Leblay C., Turpin F.X. and Chevreau E., 1992, Effect of gamma and ultraviolet irradiation on adventitious regeneration from in vitro cultured pear leaves. Euphytica **62**: 225–233.

Predieri S., 1998, Compact pears obtained through in vitro mutagenesis. Acta Horti. **475**: 93–98.

Predieri S., 2001, Induced mutation and tissue culture in fruits. Plant Cell Tiss. Org. Cult. **64**: 185–210.

Predieri S. and Fasolo F., 1989, High-frequency shoot regeneration from leaves of the apple rootstock M26 (*Malus pumila* Mill.). Plant Cell Tiss. Org. Cult. **17**: 133–142.

Predieri S. and Gatti E., 2000, Effects of gamma radiation on plum (*Prunus salicina* Lindl.) ' Shiro'. Adv. Hort. Sci. **14**: 215–223.

Predieri S. and Govoni M., 1998, In vitro propagation of compact pear clones. Acta Horti. **475**: 127–132.

Predieri S. and Zimmerman R.H., 2001, Pear mutagenesis: in vitro treatment with gamma-rays and field selection for productivity and fruit traits. Euphytica **3**: 217–227.

Predieri S., Bertazza G. and Gennari F., 1998, Pear cv Bartlett mutants selection for high fruit quality: Analysis of soluble sugars and organic acids. Atti IV Giornate Scientifiche SOI, Sanremo, 1–3 Aprile 1998. 65–66.

Predieri S., Magli M. and Zimmerman R.H., 1997, Pear mutagenesis: in vitro treatment with gamma-rays and field selection for vegetative traits. Euphytica **93**: 227–237.

Predieri S., Rosati P. and Fornasini B., 1986, Mutagenesi sul pero in micropropagazione: messa a punto del metodo. Riv. Ortoflorofrutt. It. **70**: 369–379.

Rosati P., Silvestroni O., Intrieri C. and Murri G., 1990, Effects of in vitro gamma irradiation on two grapevine cultivars (*Vitis vinifera* L.) 5th Int. Symp. Grape Breed. S. Martin F.R.G. Sept. 1989. Vitis (special issue) 471–477.

Roux N.S., 1998, Improved methods to increase diversity in *Musa* using mutation and tissue culture techniques. In: Cellular Biology and Biotechnology Including Mutation Techniques for Creation of New Useful Banana Genotypes. Report Second Research Co-ordinated FAO/IAEA/BADC Meeting. Kuala Lumpur, October 1997 (pp 49–56). IAEA, Vienna.

Sanada T. and Amano E., 1998, Induced mutation in fruit trees. In: Somaclonal Variation and Induced Mutations in Crop Improvement. In: Jain S.M., Brar D.S. and Ahloowalia B.S. (Eds) Somaclonal Variation and Induced Mutations in Crop Improvement (pp 401–409). Kluwer Academic Publishers, Dordrecht.

Sarwar M., Skirvin R.M., Kushad M. and Norton M.A., 1998, Selecting dwarf apple (*Malus* × *domestica* Borkh.) trees in vitro: multiple cytokinin tolerance expressed among three strains of 'McIntosh' that differ in their growth habit under field conditions. Plant Cell Tiss. Org. Cult. **54**: 71–76.

Schilirò E., Predieri S. and Bertaccini A., 2001, Use of random amplified polymorphic DNA analysis to detect genetic variation in *Pyrus* species. Plant Mol. Biol. Rep. **19**: 271 a-h. http://www.nrc.ca/cisti/journals/ispmb/reporter.html.

Scorza R. and Cordts J.M., 1989, Differential sensitivity of "Compact Redhaven" and "Redhaven" peach shoot tips to B.A. in vitro. Hort. Sci. **24**: 334–336.

Shen X.S., Wan J.Z., Luo W.Y., and Ding X.L., 1990, Preliminary results of using in vitro axillary and adventitious buds in mutation breeding of Chinese goosberry. Euphytica **49**: 77–82.

Sigurbjörnsson B., 1977, Introduction: Mutations in plant breeding programmes. In: Manual on Mutation Breeding. Second edition. Technical Reports Series No. 119. (pp 1–6). IAEA, Vienna.

Spiegel-Roy P., 1990, Economic and agricultural impact of mutation breeding in fruit trees. Mutation Breeding Review **5**: 1–26.

Tan Y.P., Ho Y.W., Mak C. and Rusli I., 1993, "Fatom-1" an early flowering mutant derived from mutation induction of Grand nain, a cavendish banana. Mutation Breeding Newsletters **40**: 5–6.

Van Harten A.M., 1998, Mutation Breeding: Theory and Practical Applications. Cambridge Univ. Press, Cambridge.

Visser T., Verhaegh J.J. and De Vries D., 1971, Pre-selection of compact mutants induced by X-ray treatment in apple and pear. Euphytica **20**: 95–207.

Walther F. and Sauer A., 1985, Analysis of radiosensitivity—a basic requirement for in vitro somatic mutagenesis. I. *Prunus avium* L. Acta Hort. **169**: 97–104.

Wu F.F., Siddiqui S.H., Heinz D.J. and Ladds S.L., 1978, Evaluation of mathematical methods for predicting optimum dose of gamma radiation in sugar cane. Environ. Exp. Bot. **18**: 95–98.

Yang H. and Schmidt H., 1994, Selection of a mutant from adventitious shoots formed in X-ray treated cherry leaves and differentiation of standard and mutant with RAPDs. Euphytica **77**: 89–92.

3

Genetics Related to Doubled Haploid Induction in vitro

A.M. Torp, I. Bekesiova,
I.B. Holme, A.L. Hansen and S.B. Andersen*
The Royal Veterinary and Agricultural University, Plant Breeding, Thorvaldsensvej 40 DK-1871 Frederiksberg C, Denmark.

ABSTRACT

Frequent production of albino plants in cereal anther and microspore culture is an obstacle to the efficient use of these techniques aimed at producing homozygous plants for plant breeding. In a previous study, we have identified three QTLs on wheat chromosomes 2AL (*QGpp.kvl-2A*), 2BL (*QGpp.kvl-2B.1*) and 5BL (*QGpp.kvl-5B*) that affect the capacity to regenerate green plants from anther culture in a doubled haploid population (Ciano × Walter) of wheat. To confirm the presence of the QTL on chromosome 2BL, two doubled haploid lines (CW204 and CW219) from this population were selected to establish a F_2 population segregating for this QTL. Results verified the presence and location of the QTL in the F_2 population, but indicated that additional QTLs for green plant regeneration are also segregating in this population. Anther culture of the F_1-hybrids between CW204 and CW219 showed that the genetic ability to form albino plants were dominantly inherited relative to the formation of green plants. Segregation ratios of alleles from three microsatellite markers flanking the investigated QTL-area, showed no indication of selection for the allele favouring green plant formation among green anther culture regenerants from the F_1 hybrid. Overall, in this chapter, the dominance of the albino trait and lack of distorted segregation in favour of green type alleles has been discussed in relation to hypotheses for the mechanism behind albino plant formation in cereal anther and microspore culture.

INTRODUCTION

Many plant breeding programmes include a phase of inbreeding to make recombined material homozygous before selecting any new cultivars. For autogamous species, such inbreeding is traditionally obtained via repeated selfing in 5–6 generations. As an alternative to this process, plants with the haploid chromosome number may be produced from heterozygous material followed by chromosome doubling to obtain homozygous lines. Large-scale production of chromosome doubled

*Corresponding author: E-mail: sba@kvl.dk

haploid plants can save 3–4 generations in most programmes, because completely homozygous lines are obtained in the first generation.

Interspecific pollination followed by chromosome elimination is frequently used for haploid production in wheat and barley (Kasha and Kao 1970; Kisana et al., 1993). Techniques for large-scale doubled haploid production from microspores, however, may be more attractive, because these techniques have the potential for regeneration of larger numbers of haploid plants. Traditionally, such microspore-derived haploids in cereals are produced by means of anther culture. Recent progress in the development of isolated microspore culture has increased the response level from this technique, making isolated microspore culture a promising tool for the future production of doubled haploids in both barley and wheat (Hoekstra et al., 1992; Mejza et al., 1993; Jähne and Lörz 1995; Poulimatka et al., 1996; Touraev et al., 1996). In addition, new efficient methods for in vitro chromosome doubling have further improved the method for wheat (Hansen and Andersen 1998a, 1998b). However, the disadvantage in both the methods is that culture response is dependent on genotype (Foroughi-Wehr et al., 1982; Andersen et al., 1987; Larsen et al., 1991). This genotype dependence of the present culture technique limits the general use of the methods for doubled haploid production in breeding.

Results from cereal microspore culture can be divided into 3 components of response: the ability to form embryos; the ability to regenerate plants from embryos; and the ability for green plant formation. Formation of embryos as well as percentages of green plants obtained from anther culture are strongly affected by genotype of the donor plants, while plant regeneration ability is less affected by genes (Andersen et al., 1987; Tuvesson et al., 1989). Furthermore, embryo formation, regeneration, and green plant formation seem to be relatively independently inherited (Foroughi-Wehr et al., 1982; Andersen et al., 1987; Tuvesson et al., 1989; Torp et al., 2001). For embryo formation, the genotype dependency has been found to account for 25% to 60% of total variation in wheat and barley (Andersen et al., 1987; Tuvesson et al., 1989; Larsen et al., 1991; Holme et al., 1999).

Albino plants are frequently formed in high numbers in addition to green plants in many cereal anther and microspore cultures. The ability to form a high green plant percentage, i.e. the number of green plants divided by the total number of regenerants, has been found to show a considerable genetic component, accounting for 32% to 85% of the total variation of the trait in wheat and barley (Andersen et al., 1987; Tuvesson et al., 1989; Larsen et al., 1991; Zhou and Konzak 1992; Holme et al., 1999). A number of approaches have been used to locate the genes responsible for embryo formation and green plant formation in cereals.

Early research with anther culture response from reciprocal crosses between high and low responsive lines established the fact that the genes for embryo formation and green plant formation in wheat and barley are mainly chromosomally inherited (Tuvesson et al., 1989; Larsen et al., 1991; Zhou and Konzak 1992). In hexaploid wheat, the use of translocation lines (Henry and De Buyser 1985), substitution lines (Szakacs et al., 1988; Agache et al., 1989; Ghaemi et al., 1995), or nullisomic-tetrasomic and ditelosomic lines (De Buyser et al., 1992), enabled the identification of chromosomes or chromosome arms affecting embryo formation and green plant formation. Recently, molecular genetic markers have been used to locate genes for anther culture ability to specific positions on the chromosomes in barley (Manninen 2000), maize (Cowen et al., 1992; Murigneux et al., 1994; Beaumont et al., 1995; Dufour et al., 2001), rice (He et al., 1998; Yamagishi et al., 1998), and wheat (Torp et al., 2001).

Using genetic markers, deviation from the expected 1:1 segregation is frequently observed among anther and microspore culture-derived haploids. This segregation distortion has sometimes been observed to favour alleles from the parent with higher anther culture response (Tuvesson et al., 1991; Graner et al., 1991; Foissey and Delourme 1996). This may be explained by selection during culture of units with the chromosome segment from the responsive parent harbouring a gene affecting culture ability. Associations between areas with distorted segregation and anther culture ability have, therefore, been used as an effective tool to locate the genes controlling anther culture ability in several studies (Yamagishi et al., 1998; Manninen et al., 2000; Dufour et al., 2001). However, not all chromosomal areas displaying distorted segregation show association with anther culture ability, and not all chromosomal areas carrying genes for anther culture ability display distorted segregation. Only 2 out of 5 areas with distorted segregation were found to be associated with anther culture ability in a doubled haploid population of rice (Yamagishi et al., 1998), and only 1 out of 5 chromosomal areas with distorted segregation was found to be associated with anther culture ability in a doubled haploid population of barley (Manninen 2000). In both maize (Murigneux et al., 1994) and rice (He et al., 1998), distorted segregation was absent in mapped chromosomal areas associated with anther culture ability. In wheat, only one out of 3 QTL regions associated with green plant formation showed significantly distorted segregation in favour of the high responding parent (Torp et al., 2001). The existence of 1 set of genes with a distorted segregation and another set without segregation distortion indicate different modes of action of the genes involved in cereal anther culture ability.

Regeneration of chlorophyll-deficient (albino) plants is an interesting phenomenon associated with cereal in vitro culture. Most microspore-derived albino plants contain pro-plastid like structures blocked at different stages of their development into mature chloroplasts (Sun et al., 1974; Vaughn et al., 1980; Caredda et al., 1999, 2000). Molecular studies of albino plants derived from microspores of wheat, barley, and rice have revealed that many of these plants carry deletions or rearrangements in their plastid genome (Day and Ellis 1984, 1985; Harada et al., 1991; Hofinger et al., 2001). Plastid DNA (ptDNA) deletions have, thus, often been associated with formation of albino plants. However, the mechanism leading to expression of an albino phenotype is not always related to ptDNA deletions, as microspore-derived albino plants with apparently unaltered ptDNA genomes have been found in wheat, barley, and rice (Day and Ellis 1984; Dunford and Walden 1991; Harada et al., 1991; Hofinger et al., 2001). The albino plants from wheat (Hofinger et al., 2001) seemed to lack plastid ribosomes and showed characteristic deficiencies in plastid transcription and translation patterns. This could indicate that ptDNA deletions may be a secondary rather than a primary cause of albinism, at least in some cases (Hofinger et al., 2001). The various deficiencies in plastid development of albino plants could indicate that nuclear genes affecting albino plant formation exert their effect via interactions with events in plastid development. Alternatively, they may affect the relative survival rate of green and albino initials during in vitro culture. Identification of distinct QTLs affecting the tendency for green/albino plant formation may, thus, provide new tools for the study of these phenomena.

In a previous study, we have identified 3 QTLs for green plant percentage on wheat chromosomes 2AL (*QGpp.kvl-2A*), 2BL (*QGpp.kvl-2B.1*), and 5BL (*QGpp.kvl-5B*) in the cross Ciano × Walter (Torp et al., 2001). Amongst them, the QTL region on 2BL had the most profound influence on green plant percentage and was selected for further studies.

MATERIALS AND METHODS

Plant Material

Two DH lines, CW204 and CW219 from the original mapping population Ciano × Walter, were used as parents to establish a mapping population segregating for the QTL region on chromosome 2BL (*QGpp.kvl-2B.1*) affecting green plant formation (Torp et al., 2001). CW204 carries the green type allele from Ciano and was previously shown to produce 48% green plants, whereas CW219 carries the albino type allele from Walter and previously yielded approx. 29% green plants. The 2 parental lines were selected from the mapping population so that both parents carry the green type allele in the 2 other known QTL areas for green plant formation on chromosome 2AL and 5BL in Ciano × Walter (Torp et al., 2001). This selection of parental lines based on marker scores was used to obtain genetic segregation in the offspring only for the QTL on chromosome 2BL (*QGpp.kvl-2B.1*). The F_1 hybrid between CW204 and CW219 was bagged to self-pollinate in the greenhouse in order to produce the F_2 population.

Anther Culture

Ten plants from each of the parental lines, 10 F_1 hybrids and 103 F_2 plants were evaluated for anther culture response. Donor plants were sown in 12 cm plastic pots with Finnpeat (Pindstrup, Denmark) and raised in a growth chamber (12°C day and night, 16 h day length, 50–60% relative humidity, 400–450 $\mu E/m^2s$ light intensity). Pots were watered with a 0.025% full nutrient solution (Pioner^{TM}, Broeste, Denmark) whenever necessary. To obtain 10–12 spikes from each plant, as necessary for a reliable evaluation, peat was removed from the roots after harvest of the first set of spikes and the plants were re-potted in fresh peat so as to produce a second round of spikes. Methods used for selection and sterilization of spikes and conditions for anther culture were as described by Tuvesson et al., (1989). Anthers were plated on medium 190-2 (Wang and Hu Han 1984) modified so as to contain 9% maltose instead of sucrose, 1.5 mg/L 2,4-D, and 0.5 mg/L kinetin; and solidified with 0.35% Gelrite (Kelco). Cultures were heat shock treated for the first 3 days of culture at 32°C followed by incubation at 27–28°C until 35 days of culture. The embryos were transferred to a regeneration medium 190-2 (Wang and Hu Han 1984) after 35 days of culture and incubated for plant regeneration at 27–28°C. All in vitro cultures received continuous fluorescent light (60–80 $\mu E/m^2s$). Embryo formation (embryos/anther), regeneration (plants/embryo), and percentage of green regenerants were recorded on the basis of anthers cultured for each plant separately.

Marker Analyses

F_2 plants were typed with microsatellite markers (Röder et al., 1998) known to map in the QTL area (*QGpp.kvl-2B.1*) on chromosome 2BL. DNA was extracted from young leaf material using the CTAB method described by Tinker et al., (1993). PCR and subsequent gel analysis of microsatellites using silver stained poly-acrylamide gel was performed, as previously described (Torp et al., 2001). For each microsatellite marker, scores of individual plants were recorded as homozygous green (AA) type, homozygous albino type (aa), dominant green (A-), dominant albino (a-), or heterozygous (Aa), as described for the software programmes Joinmap and PlabQTL used for map construction and QTL analysis, respectively.

Map Construction and QTL Analyses

To improve the comparability between QTL positions in the present cross CW204 × CW219 and the previously studied Ciano × Walter population, a consensus linkage map of chromosome 2BL was constructed for the 2 mapping populations based on the principles outlined by Stam (1993). Joinmap 2.0 software (Stam and Van Ooijen 1995) was used for map construction and to calculate map distances with the Kosambi mapping function (Kosambi 1944).

To improve homogeneity and distribution of residuals and the frequency of green regenerants for each F_2 plant, arcsin $\sqrt{P}$ was transformed. QTL analysis for percentage of green plants among F_2 offsprings used interval mapping procedures of the software PlabQTL (Utz and Melchinger 1996). An empirical threshold value for LOD scores was estimated in PlabQTL by permutation analysis based on the principles described by Churchill and Doerge (1994). A LOD value of 2.12, corresponding to the 5% LOD threshold after 2000 permutations, was selected for declaring a QTL significant. A one-LOD support interval for the QTL position was estimated in PlabQTL.

Segregation Distortion

Three microsatellite markers flanking the QTL area (*QGpp.kvl-2B.1*) on chromosome 2BL were used to test for distorted segregation in a population of 283 green regenerants obtained from anther culture of the F_1 hybrid between CW204 × CW219. Conditions used for anther culture and microsatellite analyses were as described above. Test for deviations of alleles from the expected 1:1 segregation ratio for each microsatellite marker used Chi-square analyses.

RESULTS

This study of one particular genetic factor underlying anther culture ability comprised culture of anthers of 10–15 spikes from each of 103 F_2 plants, together with 10 plants from each parental line and their F_1 hybrid. From the 107 809 anthers cultured, 24 294 embryos were extracted, from which 3231 green and 6639 albino plants were regenerated. Thus, from the whole material, 3 green plants per hundred cultured anthers and an average of 32.7% green plants among regenerants were obtained.

As expected, based on previous results, the trait is quantitative in nature. The histogram (Fig. 3.1a) of the 103 F_2 plants grouped according to their percentage of green regenerants shows a continuous distribution from zero to hundred per cent green plants. From Fig. 3.1a, it is also clear that the 2 parental lines produced clearly different percentages of green plants in the experiment. The high-yielding parent CW204 produced an average of 60.0% green plants, whereas for the low yielding parental line CW219, only 20.2% of the regenerated plants turned green. Among the offspring there are F_2 plants, which produced either considerably higher or lower percentages of green regenerants compared with the parents. This phenomenon may be explained by both genetic (transgression) and environmental factors affecting the green plant percentage as well as due to sampling error, which will affect the measurements because the percentage of green plants has been estimated from a limited sample of regenerated plants from each line.

Overall dominance of the trait is clear from Fig. 3.1a as the F_1 hybrid yielded a low percentage of green plants (24.1%), which was only slightly higher than the one obtained for the low-yielding parent CW219. It indicates that genetic ability to form many green plants in the anther culture is recessively inherited relative to the formation of albino plants. The overall distribution in Fig. 3.1a is clearly skewed towards lower frequencies of green plants. This may, to some extent, be explained by the percentage scale used for scoring, which restricts the scores towards both extremes of the scale. A widely-used approach to overcome this type of scaling problem is to employ a hyperbolic transformation (arcus sine of the square root of frequency). Theoretically, this transformation should relieve the skewness and simultaneously provide a more uniform sampling error for the transformed score of each plant. At least a reduction of skewness is obtained with these data through the hyperbolic transformation (Fig. 3.1b). For these reasons, this transformation of data was used throughout the study.

The approximately 30 cM chromosomal region on the long arm of 2BL, associated with green plant percentage in the cross Ciano × Walter (*QGpp.kvl-2B.1*, Torp et al., 2001), was genotyped with 6 polymorphic microsatellite markers. A genetic map for the region studied was subsequently calculated in Joinmap. Unfortunately, one of the markers *Xgwm526* could not be included in this map, since the number of recombinations between *Xgwm526* and the other markers in the map was too high to calculate a meaningful map distance. In an attempt to solve this problem, a consensus map was calculated for the 2 mapping populations, CW204 × CW219 and Ciano × Walter, based on the principles outlined by Stam (1993). However, this did not solve the problems, as it was still not possible to position *Xgwm526* on the map. The consensus map (without *Xgwm526*) was, nevertheless, used for the QTL analysis, since it improves comparability between QTL positions in CW204 × CW219 and Ciano × Walter populations. The consensus map was scanned for QTLs for the green plant percentage using simple

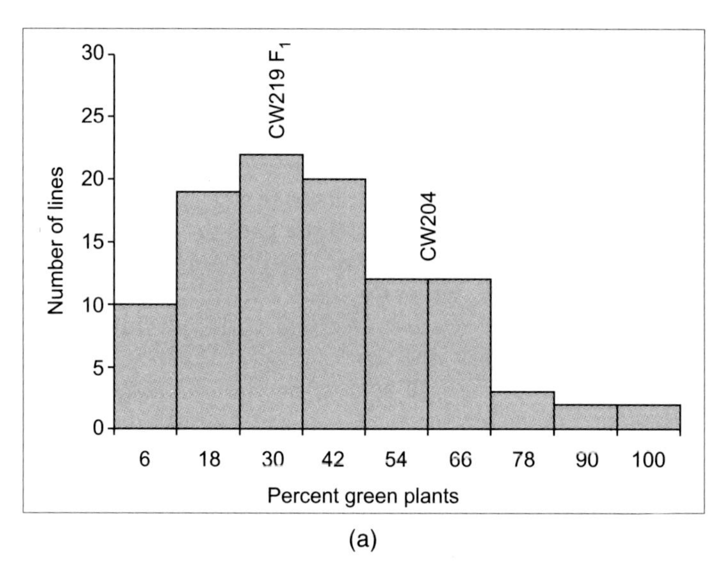

(a)

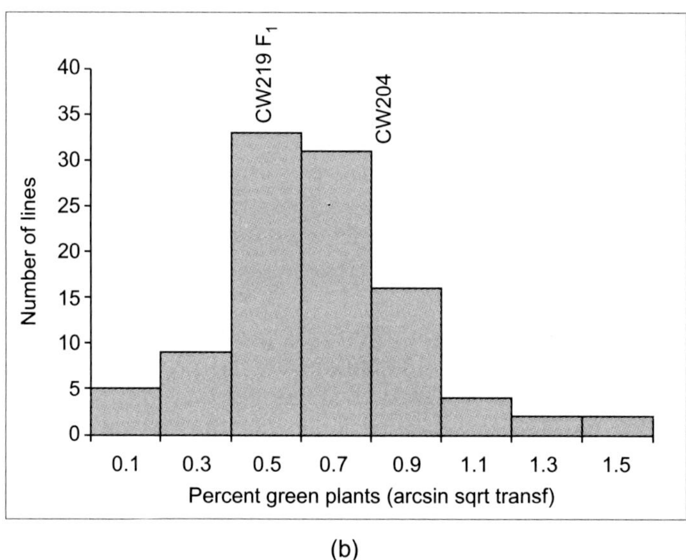

(b)

Fig. 3.1 Distribution of green plant percentage among 103 F_2 plants derived from the F_1 hybrid between CW204 and CW219. (a) percentage scale, (b) arcsin $\sqrt{P}$ transformed scale.

interval mapping. The expected QTL for the trait was indicated with a LOD-score of 2.15, which is significant at the 5% level, based on 2000 permutations of the data. A plot of LOD-scores for each possible position of the QTL along the chromosome segment (Fig. 3.2) shows that the most

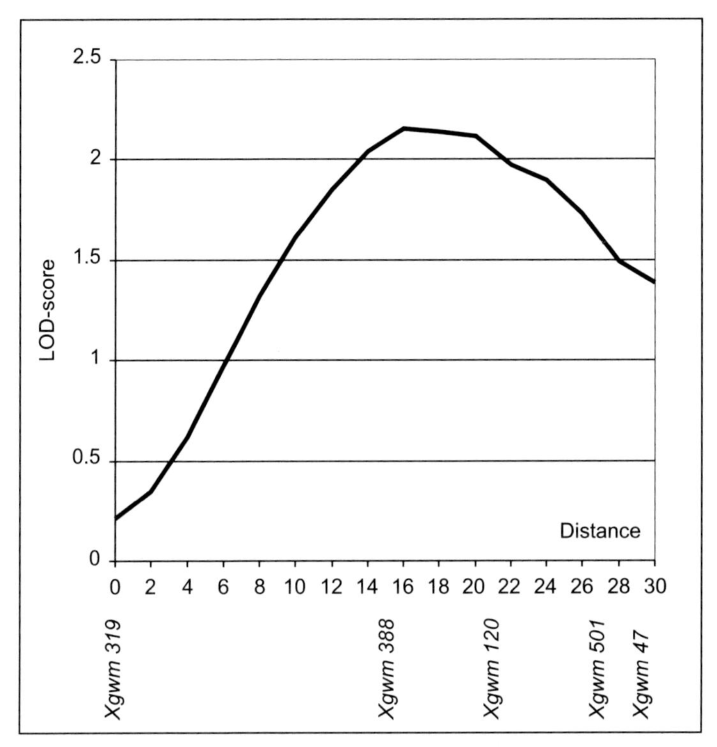

Fig. 3.2 QTL likelihood map for green plant percentage on chromosome 2BL in the F_2 population. The grey box represents the one-LOD support interval for the QTL location. The arrow indicates the estimated QTL position for the corresponding QTL in the Ciano x Walter population (Torp et al., 2001).

likely position of this QTL with the present set of data is approximately 16–18 cM from the marker *Xgwm319*, between the 2 markers *Xgwm388* and *Xgwm120*. This position is close to the position previously indicated for *QGpp.kvl-2B.1* in the cross Ciano × Walter (Torp et al., 2001), indicated with an arrow in Fig. 3.2. An approximate 95% confidence interval for the position of the QTL was calculated as the one-LOD support interval, which is delimited by the positions on either side of the estimated QTL location corresponding to a decrease in LOD-score of 1 LOD unit (Lander and Botstein 1989). Figure 3.2 shows that the ~95% confidence interval for position of the gene is still around 30 cM, even with the larger mapping population used in this study.

From the model fitted using interval mapping additive and dominance effects in the QTL locus studied were estimated (Table 3.1). As expected, the positive allele for green plant percentage was contributed by the high responding parent, CW204. It is also seen from the estimates that this QTL

Table 3.1 Estimated additive and dominance effects for the QTL on chromosome 2BL (arcsin $\sqrt{P}$ scale).

	Estimates
Mean	0.662 ± 0.042
Additive effect*	0.063 ± 0.042
Dominance effect	–0.124 ± 0.064

*green type allele from CW204

shows strong dominance of the albino-type allele over the green plant-favouring allele, consistent with the overall type of dominance for the trait. The negative sign of estimate of dominance contribution is due to the green plant frequency being the positive direction. These estimates of additive and dominance from the interval mapping based model may indicate over-dominance in the locus because the dominance contribution (–0.124) is almost twice as big as the additive effect (0.063). When standard errors of estimates are taken into consideration, however, the dominance contribution is not significantly larger than the additive one and over-dominance in this locus is not safe to conclude. The predicted values for the two parents CW204 and CW219 were calculated from the estimated additive effect, and back-transformed to the percentage scale. Based on the one QTL model, CW204 and CW219 were predicted to produce 43.9% and 31.8% green plants, respectively. These values do not correspond completely with the observed phenotypic values for the 2 parents—CW204 (60.0%) and CW219 (20.2%)—indicating that additional QTLs for green plant formation are segregating in this F_2 population.

Possible selection for the chromosomal segment harbouring the gene during the anther culture process was studied through marker analysis of 283 green regenerants from the F_1 hybrid between CW204 and CW219. Observed segregation of all 3 microsatellite markers were close to 1:1 as expected, provided no selection among gametes carrying the anther culture-favouring allele from CW204 takes place (Table 3.2). The microsatellite marker *Xgwm319*, situated at one end of the chromosomal region, had a segregation significantly different from the 1:1 ratio at the 0.05 per cent level. The scores, however, show that deviation from the expected ratio is because the anther culture unfavourable allele derived from CW219 is over represented. An over representation of the unfavourable CW219 allele is also observed for the 2 other markers studied (*Xgwm501*, *Xgwm526*), although for these 2 markers, deviation is far from significant in the Chi-square test.

DISCUSSION

Many experiments reported in the literature as well as the present study demonstrate the dual nature of genetics behind the response level in

Table 3.2 Segregation of three microsatellite markers flanking the QTL area among 283 green regenerants from the F_1 hybrid between CW204 and CW219.

Microsatellite marker	*No. of Green type CW204 allele*	*No. of Albino type CW219 Allele*	*Chi-square 1:1 segregation*
Xgwm319	124	159	4.33*
Xgwm501	130	153	1.87
Xgwm526	129	154	2.21

*significant at $p = 0.05$.

anther and microspore culture of cereals. Particularly, when it comes to the peculiar phenomenon of formation of albino plants in addition to green regenerants, there are now evidence both for existence of genes being selected during the culture as well as genetic loci in which selection between alleles do not occur (He et al., 1998; Yamagishi et al., 1998; Manninen 2000; Torp et al., 2001).

One simple explanation for differences between the lines in their ability to respond would be that some alleles in response genes affect the survival or developmental rate of plant initials during the culture. During culture of microspores from an F_1-hybrid between high and low responding parents, microspores carrying culture-favouring alleles may have a higher chance of developing into a green plant rather than microspores with culture non-favouring alleles. Selection of initials with culture-favouring alleles will lead to the well-known distorted segregation often observed among anther or microspore culture offspring. In such a scenario, the response genes may be said to have a gametophytic type of action mechanism because they exert their effect locally in individual initials corresponding to the allele carried in their genome. It is also evident that dominance would not be expected in the case of a gametophytic action mechanism because the 2 alleles in each locus will act independently in different microspores. Only in a situation with strong selection in vitro for individuals with culture-favouring alleles, an overall dominance may be expected. This is because the over-representation of one type among regenerants will create a picture of the trait, indistinguishable from true genetic dominance. This pseudo-dominance, however, will always be in the direction of the culture-favourable allele.

Molecular models that explain the mechanism of action of response genes, which show no distorted segregation during anther or microspore culture, must comprise what could be named a sporophytic type of action mechanism. If no selection takes place among genotypes in a mixture of microspores from an F_1 hybrid, the causal agent must affect all initials equally whether they carry culture-favouring alleles or not. This means that action of the gene is not localized to its host cell, but acts across cell boundaries. Even so, response genes may produce the causal agent either in haploid initials or in diploid parental tissues, if the product is exchanged uniformly among plant initials. Such a model with a sporophytic mode of action will easily explain the dominance of the trait in both directions. The observation of general strong dominance of the ability to form many albino plants in this and other studies (Quimio and Zapata 1990; Torp et al., 2001) indicates that genes affecting this trait have a predominantly sporophytic mode of action.

Original models proposed to explain the formation of albino plants in cereal anther cultures took their start from the phenomenon of maternal inheritance of plastids observed in most angiosperm species. The hypothesis says that the degradation of plastid genomes during pollen development is a natural phenomenon in some species, which do not pass chloroplasts to their offspring from the male partner (Day and Ellis 1984). This was well supported by the observation of large scale deletions in plastid DNA genomes of albino regenerants (Day and Ellis 1984, 1985; Harada et al., 1991; Hofinger et al., 2001). Support for these ideas can also be found from experiments with the unicellular algae *Chlamydomonas*, where plastid DNA from one of the parents (mt^- mating type) is selectively destroyed within 1 h after zygote formation (Kuroiwa et al., 1982; Nishimura et al., 1999). This selective destruction is probably due to the action of a nuclear-encoded nuclease (Gilham et al., 1987, 1991; Van-Winkle-Swift et al., 1994). The hypothesis, based on a general mechanism for destruction of male chloroplasts during gametophyte development, however, has been weakened by reports on relatively frequent albino plant formation in ovule/ovary cultures (Liu and Zhou 1984), as well as somatic cell cultures from cereal species (Cho et al., 1998; Yoshida et al., 1999). The phenomenon, however, may still be explained if the stress induced by in vitro culture is assumed to induce a normally-suppressed maternal inheritance type of pathway, leading to specific destruction of plastids.

A more general model, based on changes due to stress from the in vitro culture, suggests an association between the formation of albino plants and ongoing apoptosis that is programmed cell death (Caredda et al., 1999, 2000). Apoptosis was recently observed in both developing anthers and anthers subjected to in vitro culture (Wang et al., 1999). A mechanism based on apoptosis would predict that the nuclear genome conserve totipotency during culture, whereas the plastids may lose the ability to develop into chloroplasts due to ongoing DNA degradation (Caredda et al., 2000).

A more specific molecular model to explain the formation of albino plants is based on the observation of albino plants lacking chloroplast ribosomes (Hofinger et al., 2001). Plant chloroplasts have different 70S ribosomes from the cytoplasmic 80S eucaryotic ribosomes and the 78S ribosomes found in mitochondria. These 70S procaryote type of ribosomes are apparently considerably more sensitive to both temperature extremes as well as a number of different chemicals with antibiotic activity (Feierabend and Schrader-Reichhardt 1976; VanBogelen and Neidhardt 1990; Yoshida et al., 1998; Zubko and Day 1998). Ribosome formation in the plastids of rye, wheat, barley, and oat can be specifically inhibited by growth at an elevated temperature between 32°C and 34°C (Feierabend and Schrader-Reichhardt 1976; Feierabend and Mikus 1977), while ribosomes of maize and rice are sensitive to low temperatures (Hopkins and Elfman 1984; Yoshida et al., 1996). In *Brassica napus*, an irreversible change to albino phenotype has been demonstrated after external treatment with streptomycin-like antibiotics known to inactivate 70S ribosomes (Zubko and Day 1998). If all ribosomes in a pro-plastid are inactivated by the external agent, the organelle is prevented from recovering after removal of the agent, because it cannot perform plastid protein synthesis. Upon division of cells containing many ribosome-free plastids, some daughter cells will end up without functional plastids and give rise to cell lines that are unable to develop mature chloroplasts (Zubko and Day 1998). If the hypothesis on formation of cell lines with ribosome-free plastids is used to explain the formation of albino plants in anther and microspore culture of cereals, the ribosome-inactivating stress or chemical factor must be assumed to be generated by the culture conditions and by the anther/microspore complex.

Genetic differences between plant lines with high and low frequencies of green plants in anther and microspore cultures could then be due to either differences in ribosomal targets or in synthesis of chemical compounds in the cultures. Genetic changes in plastid-encoded ribosomal components to provide less sensitive plastid ribosomes lead to a clear maternal inheritance of the trait (Yurina et al., 1978), which is not observed in anther and microspore cultures (Tuvesson et al., 1989; Larsen et al., 1991; Zhou and Konzak 1992). Nuclear mutations, which result in a programmed loss of plastid ribosomes in the affected tissue, are known in both barley (*albostrians*, Knoth and Hagemann 1977) and maize (*iojap*, Shumway and Weier 1967; Thompson et al., 1983). However, both these mutations are recessive, which does not fit with the dominance observed for albino plant formation. At the same time, genetic changes in chromosomally-inherited plastid ribosomal components are expected to produce a gametophytic mode of action in anther culture, unless most of the plastid ribosomes in each microspore are formed before meiosis. This is because such protein components are expected to act locally in the host cell without exchange to neighbouring cells. In the case of albino plant formation, the strong dominance for high albino plant frequency observed, therefore, seems to exclude this mechanism. Finally, changes in synthetic pathways of low molecular weight chemical compounds capable of streptomycin like inactivation of plastid ribosomes could be imagined. Such mutations in nuclear genes leading to a production of reduced concentrations of the chemical agent, could thus explain the relatively rare occurrence of lines in wheat (Andersen et al., 1988; Holme et al., 1999) and barley (Knudsen et al., 1989) with high green plant regeneration capacity. The idea is also compatible with the dominance of high albino frequency in the cultures observed in this and other studies, because deleterious mutations in synthetic pathways are generally recessive to the wild type. This means that the heterozygous hybrid will produce the ribosome-inactivating compound and show a high frequency of albino plant formation.

CONCLUSION

The results obtained in this study confirmed the presence and location of a previously identified QTL for green plant regeneration on wheat chromosome 2BL in a new mapping population. In the present study, no selection of alleles from the 3 microsatellites markers was observed in favour of the highest responding parent in the QTL region investigated. Also, inheritance of the albino trait showed dominance over green plant formation in the F_1-hybrid. Both results strongly indicate that genes controlling albino formation have a predominantly sporophytic mode of action.

ACKNOWLEDGEMENTS

We thank H. Faarup, and B. K. Hansen for their technical assistance. The study was financially supported by the Royal Veterinary and Agricultural University, the Danish Research Academy and the Danish Cereal Network.

REFERENCES

Agache S., Bachelier B., De Buyser J., Henry Y., Snape J., 1989, Genetic analysis of anther culture response in wheat using aneuploid, chromosome substitution and translocation lines. Theor. Appl. Genet. **77:** 7–11.

Andersen S.B., Due I.K., Olesen A., 1987, The response of anther culture in a genetically wide material of winter wheat (*Triticum aestivum L.*). Plant Breeding **99**: 181–186.

Andersen S.B., Due I.K., Olesen A., 1988, Results with anther culture in some important Scandinavian varieties of winter wheat. Acta Agric. Scand. **38**: 289–292.

Beaumont V.H., Rocheford T.R., Widholm J.M., 1995, Mapping the anther culture response genes in maize (*Zea mays* L.). Genome **38:** 968–975.

Caredda S., Devaux P., Sangwan R.S., Clement C., 1999, Differential development of plastids during microspore embryogenesis in barley. Protoplasma **208**: 248–256.

Caredda S., Doncoeur C., Devaux P., Sangwan R.S., 2000, Plastid differentiation during androgenesis in albino and non-albino producing cultivars of barley (*Hordeum vulgare* L.). Sex. Plant Reprod. **13**: 95–104.

Cho M.J., Jiang W., Lemaux P.G., 1998, Transformation of recalcitrant barley cultivars through improvement of regenerability and decreased albinism. Plant Sci. **138**: 229–244.

Churchill G.A., Doerge R.W., 1994, Empirical threshold values for quantitative trait mapping. Genetics **138**: 963–971.

Cowen N.M., Johnson C.D., Armstrong K., Miller M., Woosley A., Pescitelli S., Shokut M., Belmar S., Petolino J.F., 1992, Mapping genes conditioning in vitro androgenesis in maize using RFLP analysis. Theor. Appl. Genet. **84**: 720–724.

Day A., Ellis T.H.N., 1984, Chloroplast DNA deletions associated with wheat plants regenerated from pollen. Possible basis for maternal inheritance of chloroplasts. Cell **39**: 359–368.

Day A., Ellis T.H.N., 1985, Deleted forms of plastid DNA in albino plants from cereal anther culture. Curr. Genet. **9:** 671–678.

De Buyser J., Hachemi-Rachedi S., Lemee M.L., Sejourne S., Marcotte J.L., Henry Y., 1992, Aneuploid analysis of anther culture response in wheat. Plant Breeding **109**: 339–342.

Dufour P., Johnsson C., Antoine-Michard S., Cheng R., Murigneux A., Beckert M., 2001, Segregation distortion at marker loci: variation during microspore embryogenesis in maize. Theor. Appl. Genet. **102**: 993–1001.

Dunford R., Walden R.M., 1991, Plastid genome structure and plastid-related transcipt levels in albino barley plants derived from anther culture. Curr. Genet. **20**: 339–347.

Feierabend J., Schrader-Reichhardt U., 1976, Biochemical differentiation of plastids and other organelles in rye leaves with a high-temperature-induced deficiency of plastid ribosomes. Planta **129:** 133–145.

Feierabend J., Mikus M., 1977, Occurrence of a high temperature sensitivity of chloroplast ribosome formation in several higher plants. Plant Physiol. **59**: 863–867.

Foisset N., Delourme R., 1996, Segregation distortion in androgenic plants In: In vitro Haploid Production in Higher Plants. Vol. 2. (Eds S.M. Jain, S.K. Sopory, R.E. Veilleux). pp. 189–201. (Kluwer Academic Publishers: The Netherlands).

Foroughi-Wehr B., Friedt W., Wenzel G., 1982, On the genetic improvement of androgenetic haploid formation in *Hordeum vulgare* L. Theor. Appl. Genet. **62**: 233–239.

Ghaemi M., Sarrafi A., Morris R., 1995, Reciprocal substitutions analysis of embryo induction and plant regeneration from anther culture in wheat (*Triticum aestivum* L.). Genome **38**: 158–165.

Gilham N.W., Boynton J.E., Johnson A.M., Burkhart B.D., 1987, Mating type linked mutations which disrupt the uniparental transmission of chloroplast genes in *Chlamydomonas*. Genetics 115, 677–684.

Gilham N.W., Boynton J.E., Harris E.H., 1991, Transmission of plastid genes. Cell Cult. Somatic Cell Genet. Plants **7A**: 55–92.

Graner A., Jahoor A., Schondelmaier J., Siedler H., Pillen K., Fishbeck G., Wenzel G., Hermann R.G., 1991, Construction of an RFLP map of barley. Theor. Appl. Genet. **81**: 250–256.

Hansen N.J.P., Andersen S.B., 1998a, In vitro chromosome doubling with colchicine during microspore culture in wheat (*Triticum aestivum* L.). Euphytica **102**: 101–108.

Hansen N.J.P., Andersen S.B., 1998b, Efficient production of doubled haploid wheat plants by in vitro treatment of microspores with trifluralin or APM. Plant Breeding **117**: 401–405.

Harada T., Sato T., Asaka D., Matsukawa I., 1991, Large-scale deletions of rice plastid DNA in anther culture. Theor. Appl. Genet. **81**: 157–161.

He P., Shen L., Lu C., Chen Y., Zhu L., 1998, Analysis of quantitative trait loci which contribute to anther culturability in rice (*Oryza sativa* L.). Mol. Breed. **4:** 165–172.

Henry Y., De Buyser J., 1985, Effect of the 1B/1R translocation on anther culture ability in wheat (*Triticum aestivum* L.). Plant Cell Rep. **4:** 307–310.

Hoekstra S., van Zijderveld M.H., Louwerse J.D., Heidekamp F., van der Mark F., 1992, Anther and microspore culture of *Hordeum vulgare* L. cv. Igri. Plant Sci. **86**: 89–96.

Hofinger B.J., Ankele E., Gülly Ch, Heberle-Bors E., Pfosser M.F., 2001, The involvement of the plastid genome in albino plant regeneration from microspores in wheat. In: Biotechnological Approaches for Utilisation of Gametic Cells. (Ed. B Bohanec). COST 824 Final Meeting, Bled, Slovenia, 1–5 July 2000, pp. 215–228.

Holme I.B., Olesen A., Hansen N.J.P., Andersen S.B., 1999, Anther and isolated microspore culture response of wheat, *Triticum aestivum* L., lines from northwestern and eastern Europe. Plant Breeding **118**: 111–117.

Hopkins W.G., Elfman B., 1984, Temperature-induced chloroplast ribosome deficiency in virescent maize. J Hered. **75**: 207–211.

Jähne A., Lörz H., 1995, Cereal microspore culture. Plant Sci. **109**: 1–12.

Kasha K.J., Kao K.N., 1970, High frequency haploid production in barley (*Hordeum vulgare* L.). Nature **225**: 874–876.

Kisana N.S., Nkongolo K.K., Quick J.S., Johnson D.L., 1993, Production of doubled haploids by anther culture and wheat × maize method in a wheat breeding programme. Plant Breeding **110**: 96–102.

Kosambi D.D., 1944, The estimation of map distances from recombination values. Ann. Eugen. **12**: 172–175.

Knoth R., Hagemann R., 1977, Struktur und Funktion der genetishen Information in den Plastiden. Biol. Zbl. **96**: 141–150.

Knudsen S., Due I.K., Andersen S.B., 1989, Components of response in barley anther culture. Plant Breeding **103**: 241–246.

Kuroiwa T., Kawano S., Nishibayashi S., 1982, Epifluorescent microscopic evidence for maternal inheritance of chloroplast DNA. Nature **298**: 481–483.

Lander E.S., Botstein D., 1989, Mapping Mendelian factors underlying quantitative traits using RFLP linkage maps. Genetics **121**: 185–199.

Larsen E.T., Tuvesson I.K.D., Andersen S.B., 1991, Nuclear genes affecting percentage of green plants in barley (*Hordeum vulgare* L.) anther culture. Theor. Appl. Genet. **82**: 417–420.

Liu Z., Zhou C., 1984, Investigation of ploidy and other characters of the gynogenic plants in *Oryza sativa* L. Acta Genetica Sinica **11**: 113–119.

Manninen O.M., 2000, Association between anther culture response and molecular markers on chromosomes 2H, 3H and 4H of barley (*Hordeum vulgare* L.). Theor. Appl. Genet. **100**: 57–62.

Mejza S.J., Morgant V., DiBona D.E., Wong J.R., 1993, Plant regeneration from isolated microspores of *Triticum aestivum*. Plant Cell Rep. **12**: 149–153.

Murigneux A., Bentolila S., Hardy T., Baud S., Guitton C., Jullien H., Ben Tahar S., Freyssinet G., Beckert M., 1994, Genotypic variation of quantitative trait loci controlling in vitro androgenesis in maize. Genome **37**: 970–976.

Nishimura Y., Misumi O., Matsunage S., Higashiyama T., Yokota A., Kuroiwa T., 1999, The active digestion of uniparental chloroplast DNA in a single zygote of *Chlamydomonas reinhardtii* is revealed by using the optical tweezer. Proc. Natl. Acad. Sci. USA **96**: 12577–12582.

Puolimatka M., Laine S., Pauk J., 1996, Effect of ovary co-cultivation and culture medium on embryogenesis of directly isolated microspores of wheat. Cereal Res. Comm. **24**: 393–400.

Quimio C.A., Zapata F.J., 1990, Diallel analysis of callus induction and green-plant regeneration in rice anther culture. Crop Sci. **30**: 188–192.

Röder M.S., Korzun V., Wendehake K., Plaschke J., Tixier M.H., Leroy P., Ganal M.W., 1998, A microsatellite map of wheat. Genetics **149**: 2007–2023.

Shumway L.K., Weier T.E., 1967, The chloroplast structure of iojap maize. Amer. J. Biol. **54**: 773–780.

Stam P., 1993, Construction of integrated genetic linkage maps by means of a new computer package: Join Map. Plant J. **3**: 739–744.

Stam P., Van Ooijen J.W., 1995, Joinmap (tm) version 2.0: Software for the calculation of genetic linkage maps. CPRO-DLO, Wageningen.

Szakács É., Kovács G., Pauk J., Barnabás B., 1988, Substitution analysis of callus induction and plant regeneration from anther culture in wheat (*Triticum aestivum* L). Plant Cell Rep. **7**: 127–129.

Sun C.S., Wang C.C., Chu C.C., 1974, The ultrastucture of plastids in the albino pollen-plants of rice. Scientia Sinica **17**: 793–802.

Thompson D., Walbot V., Coe E.H., 1983, Plastid development in iojap- and chloroplast mutator-affected maize plants. Amer. J. Bot. **70**: 940–950.

Tinker N.A., Fortin M.G., Mather D.E., 1993, Random amplified polymorphic DNA and pedigree relationships in spring barley. Theor. Appl. Genet. **85**: 976–984.

Torp A.M., Hansen A.L., Andersen S.B., 2001, Chromosomal regions associated with green plant regeneration in wheat (*Triticum aestivum* L.) anther culture. Euphytica **119**: 377–387.

Touraev A., Indrianto A., Wratschko I., Vicente O., Heberle-Bors E., 1996, Efficient microspore embryogenesis in wheat (*Triticum aestivum* L.) induced by starvation at high temperature. Sex. Plant Reprod. **9**: 209–215.

Tuvesson I.K.D., Pedersen S., Andersen S.B., 1989, Nuclear genes affecting albinism in wheat (*Triticum aestivum* L.) anther culture. Theor. Appl. Genet. **78**: 879–883.

Tuvesson I.K.D., Pedersen S., Olesen A., Andersen S.B., 1991, An effect of the 1BL/1RS chromosome on albino frequency in wheat (*Triticum aestivum* L.) anther culture. J Genet. Breed. **45**: 345–348.

Utz H.F., Melchinger A.E., 1996, PLABQTL: A program for composite interval mapping of QTL. Journal of Agricultural Genomics 2: http://www.cabi-publishing.org/jag.

VanBogelen R.A., Neidhardt F.C., 1990, Ribosomes as sensors of heat and cold shock in *Escherichia coli*. Proc. Natl. Acad. Sci. USA **87:** 5589–5593.

VanWinkle-Swift K., Hoffman R., Shi L., Parker S., 1994, A suppressor of a mating-type limited zygotic lethal allele also suppresses uniparental chloroplast gene transmission in *Chlamydomonas monoica*. Genetics **136**: 867–877.

Vaughn K.C., DeBonte L.R., Wilson K.G., 1980, Organelle alteration as a mechanism for maternal inheritance. Science **208**: 196–198.

Wang M., Hoekstra S., van Bergen S., Lamers G.E.M., Oppedijk B.J., van der Heijden M.W., de Priester W., Schilperoort R.A., 1999, Apoptosis in developing anthers and the role of ABA in this process during androgenesis in *Hordeum vulgare* L. Plant Mol. Biol. **39**: 489–501.

Wang X., Han H., 1984, The effect of potato II medium for triticale anther culture. Plant Sci. Lett. **36**: 237–239.

Yamagishi, M., Otani M., Higashi M., Fukuta Y., Fukoi K., Yano M., Shimada T., 1998, Chromosomal regions controlling anther culturability in rice (*Oryza sativa* L.). Euphytica **103**: 227–234.

Yoshida R., Kanno A., Sato T., Kameya T., 1996, Cool-temperature-induced chlorosis in rice plants. Plant Physiol. **110**: 997–1005.

Yoshida R., Sato T., Kanno A., Kameya T., 1998, Streptomycin mimics the cool temperature response in rice plants. J. Exp. Bot. **49**: 221–227.

Yoshida S., Kasai Y., Watanabe K., Fujino M., 1999, Proline stimulates albino regeneration and seed-derived rice callus under high osmosis. J. Plant Physiol. **155**: 107–109.

Yurina N.P., Odintsova M.S., Maliga P., 1978, An altered chloroplast ribosomal protein in a streptomycin resistant tobacco mutant. Theor. Appl. Genet. **52**: 125–128.

Zhou H., Konzak C.F., 1992, Genetic control of green plant regeneration from anther culture of wheat. Genome **35**: 957–961.

Zubko M.K., Day A., 1998, Stable albinism induced without mutagenesis: a model for ribosome-free plastid inheritance. The Plant Journal **15**: 265–271.

4

Stability of Novel Gene Expression in Transgenic Conifers: An Issue of Concern?

Christian Walter
New Zealand Forest Research Institute Ltd.,
Sala Street, Rotorua, New Zealand

ABSTRACT

Biotechnology and in particular genetic engineering, has in the last two decades of the twentieth century, demonstrated significant benefit to various areas of science and commercial endeavours, including medicine and agriculture. More recently, modern biotechnology protocols such as genetic fingerprinting and genetic transformation have successfully been developed for trees, including conifers.

This chapter discusses the emerging role of genetic engineering in both understanding the fundamental biological pathways in forest trees, and in the commercial applications of conifer species. The focus is on aspects of interest for future forest biotechnology, in particular the stability of novel genes introduced into forest trees. The potential causes and impacts of gene silencing will be assessed and discussed in the context of forest tree genetic engineering.

Possible strategies to avoid silencing and expression instability will also be deliberated upon and initial results on transgene stability in some forest tree species will be provided.

INTRODUCTION

Over the past 20 years, genetic engineering technologies have contributed significantly towards the improvement of agricultural crops. Plants with engineered resistance against herbicides or insects are used in commercial operations worldwide (James 2000). We are becoming increasingly aware that this technology has the potential to add to the quality and yield of agricultural products, and to solve a range of environmental problems in both developed and also developing countries (Krattiger 1998). New types of genetically-engineered products for human consumption hold the promise to significantly contribute to human health and welfare (Shintani and DellaPenna 1998). The production of "Golden Rice", through genetic

E-mail: christian.walter@forestresearch.co.nz

engineering, is just one example to illustrate this point (Guerinot 2000). Golden rice can provide millions of people in India with essential doses of vitamin A and iron, thereby preventing serious diseases.

Further, genetic engineering offers the option to reduce the use of agrochemicals, leading to a more environmentally-acceptable agriculture that is truly sustainable and affordable to third world countries (www.isaaa.org/). A good example is genetically-engineered insect resistance in agricultural or forest tree species. This has the potential to significantly reduce the use of insecticides on these crops (Tabashnik 1997; Krattiger, 1997).

In forestry, a further beneficial effect should be considered. The world demand for forest products, including pulp and paper and high quality timber for building purposes and furniture production is steadily increasing (Zhu et al., 1999). At the same time, harvesting of trees from natural rainforests is already showing devastating effects on natural ecosystems and the global climate. Plantation forestry offers the potential to supply the required primary products in an ecologically sustainable manner, and forest biotechnology holds the promise to significantly add to the sustain-ability and environmental acceptability. Intensively and sustainably managed plantation forests, therefore, provide the option for the world population to leave natural forests alone (Fenning and Gershenzon, 2002).

The development of molecular biology platforms, including genetic engineering, has taken longer for forestry than agricultural crops. This is mainly due to additional challenges with species that have proven difficult to propagate in tissue culture (Becwar 1993; Ahuja 2000) and that are characterized by a long rotation time and long breeding cycles (sometimes 8–10 years). Further, forestry by tradition is different from modern agriculture in that in many countries, the deployment of clonally-propagated material is rather the exception than the rule. In many countries, forestry is based on natural regeneration and selective felling. For example, of the 417.6 million ha. forests in Canada, 1 million ha. are harvested each year. Half of this area is left to natural regeneration, while the other half is planted (Anon 2001). In some countries, however, clonal plantation forestry is practised to a greater extent and the total estate is growing (Kanowski 1997; Menzies and Aimers-Halliday 1997). It is important to note that the successful deployment of genetically-engineered material depends on a clonal propagation and plantation concept in order to fully recover its benefits.

Conventional tree improvement programmes aiming at the production of superior germplasm have been applied over the last 50 years and have traditionally made use of the identification of superior traits, followed by selective breeding (Carson et al., 1989; Jayawickrama and Carson 2000). Propagation technologies such as embryogenesis or fascicle cuttings were developed in order to provide superior planting stock for commercial plantations (Menzies and Aimers-Halliday 1997). Molecular biology research over the last 10 years has added techniques for quality assurance such as marker-aided selection (MAS) and genetic fingerprinting (Wilcox et al., 2001). More recently, significant progress in developing genetic engineering protocols has been made and these are now available for most major forest tree species of scientific and commercial importance (Jouanin et al., 1993; Walter et al., 1998a; Ahuja 2000; Bishop-Hurley et al., 2001).

In this chapter, genetic engineering of conifers is discussed briefly. It focusses on a specific aspect of conifer genetic engineering, and is related to the fidelity of gene expression over the life span of transgenic trees. This is an important consideration for scientists and commercial users of tree genetic engineering alike. Sound scientific evaluation of gene function and exploitation of novel traits in conifers depend on the continued and appropriate expression of introduced genes, and transgenic trees will only enter the commercial scale successfully if variation of gene expression can be confidently controlled.

GENETIC ENGINEERING OF TREES

The ability of *Agrobacterium tumefaciens*, a common soil bacterium, to genetically transform a wide range of host plants has revolutionized plant molecular biology. It can be said that this bacterium has opened the door to genetic engineering in plants. In the natural environment, it induces crown galls on plants by transferring a portion of its genome (the T-DNA or transferred DNA) into host cells (Van Larebeke et al., 1975). The T-DNA is eventually integrated into the plant cells genome and expressed, leading to the initiation of undifferentiated growth and the formation of a crown gall. Modified strains of *Agrobacterium*, where the T-DNA is replaced by a gene of interest, can be used to transfer any DNA into a plant cell, followed by its integration into the genome.

On the basis of this technology, many plant species have been modified with novel genes and transgenic plants regenerated (Zupan et al., 2000; Newell 2000; Tzfira et al., 2000).

In forestry, *Agrobacterium* has been used to transform many tree species, for instance poplar (Parsons et al., 1986; Fillatti et al., 1987), *Eucalyptus* spp (MacRae and van Staden, 1999) *Picea* spp (Klimaszewska et al., 2001), *Larix decidua* (Huang et al., 1991) and *Pinus strobus* (Levee et al., 1999). *Pinus radiata* has also been transformed and genetically-modified plants regenerated (Holland et al., 1997; Charity et al., 2001).

As an alternative to *Agrobacterium* and for species that were originally not amenable to this transformation technology, artificial gene transfer methods such as Biolistics® have been developed. This technique introduces pure DNA into cells by physical means, so that DNA integration into the genome can take place (Klein et al., 1987). Biolistic® techniques have also successfully been applied to conifer transformation, including regeneration of transgenic plants (Charest et al., 1993; Ellis et al., 1993; Clapham et al., 2000; Bishop-Hurley et al., 2001; Find et al., 2001; Walter et al., 1998b and c). This technology is briefly discussed in the following chapter.

RESULTS AND DISCUSSION

Genetic Engineering of *Pinus radiata*

A biolistic transformation protocol was developed in order to bombard embryogenic suspension cells of various conifers, including *Pinus radiata, P. taeda, Abies nordmaniana* and *Picea abies* (Walter et al., 1998b&c; Find et al., 2001). An antibiotic selection gene (*npt*II) provided the basis for selection of transgenic cells. Expression of the *uid*A reporter gene, co-transferred with *npt*II, was analyzed in the putatively-transformed tissue lines—both histochemically and fluorometrically. Histochemical expression has routinely been used to confirm successful transformation of embryogenic material at an early stage (Figure 4.1). Individual lines of the species transformed displayed differences in expression intensity in

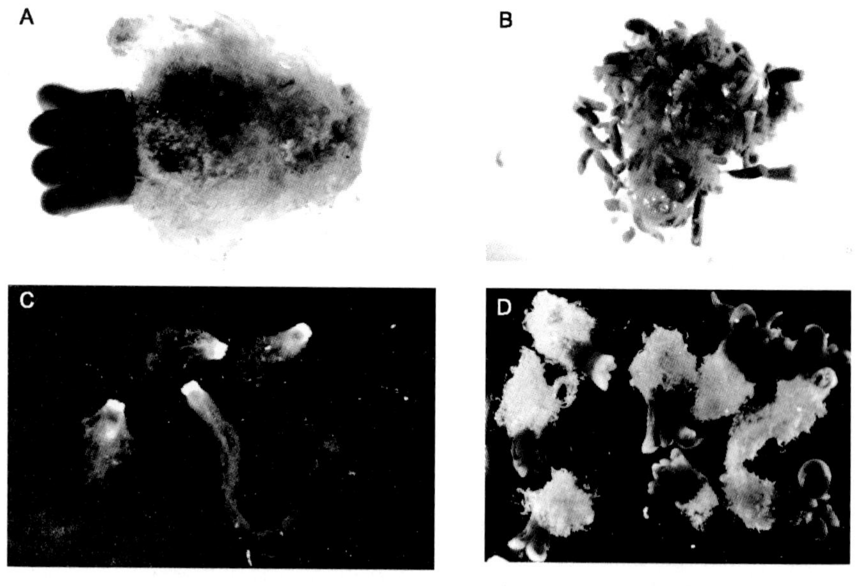

Fig. 4.1 Histochemical analysis of conifer embryogenic tissue. Transgenic embryos of *Pinus taeda* (A), *Picea abies* (B), *Pinus radiata* (C) and *Abies nordmaniana* (D), expressing the *uid*A gene. Note the different staining patterns and intensities.

histochemical assays and this variability was confirmed in fluorometric assays.

*Uid*A expression was also confirmed to be dependent on the developmental stage of embryogenic tissue and regenerating plantlets. For example, very early in embryo development, the intensity of the blue stain was relatively high in *P. radiata* and it remained intense during the "bullet" stage of somatic embryos. However, once the cotyledons were formed, *uid*A expressed strongly in the hypocotyl but only weakly in the cotyledons, particularly towards their distal ends. In the transformed plants, blue stain was observed in the guard cells of needle stomata and the vascular tissue of stems. No *uid*A expression was observed in non-transformed tissue or plants. Conversely, when *Picea abies* was transformed with the same constructs (Bishop-Hurley et al., 2001), *uid*A expression was usually high in developing embryos. Transgenic plants regenerated, displaying blue staining at various levels. In some transclones, these levels were constant over a period of time (for example S9-1), in others strong variation was observed (for example S20-2) (Table 4.1).

Table 4.1 Expression of *uid*A in transgenic plants of *Picea abies*. Histochemical assay. Expression levels are indicated from low (+) to very high (+++++). T1–T8: Times when samples were taken, in one-weekly intervals. NT: Not Tested.

Transclone	*T1*	*T2*	*T3*	*T4*	*T5*	*T6*	*T7*	*T8*
AS11-1	++++	NT	+++++	NT	++++	NT	++++	+++++
S9-1	++++	++++	++++	+++++	+++++	+++++	+++++	+++++
S9-4	+++++	+++	+++++	+++++	+++++	+++	+++	+++++
S20-1	+++	+++++	+++	++++	+++	++++	++++	+++++
S20-2	++++	+++	++	+++	NT	+++++	+++	+++

Putative transgenic tissue and trees were analyzed by Southern hybridization, PCR and *npt*II ELISA (Walter et al., 1998 a&b) (Figure 4.2). They are currently being analyzed in a field trial to obtain data on long-term gene expression and their performance is compared to non-transformed control trees.

Gene Expression Patterns in Transgenic Conifers

In transgenic *Pinus radiata* tissue, the expression of introduced genes varies with regard to individual independently-transformed lines. Fluorometric studies of transgenic tissue (Figure 4.3) confirmed two times to 1,300 times higher *uid*A activity compared to non-transformed controls. Variation of gene expression in different transclones may be resulting from

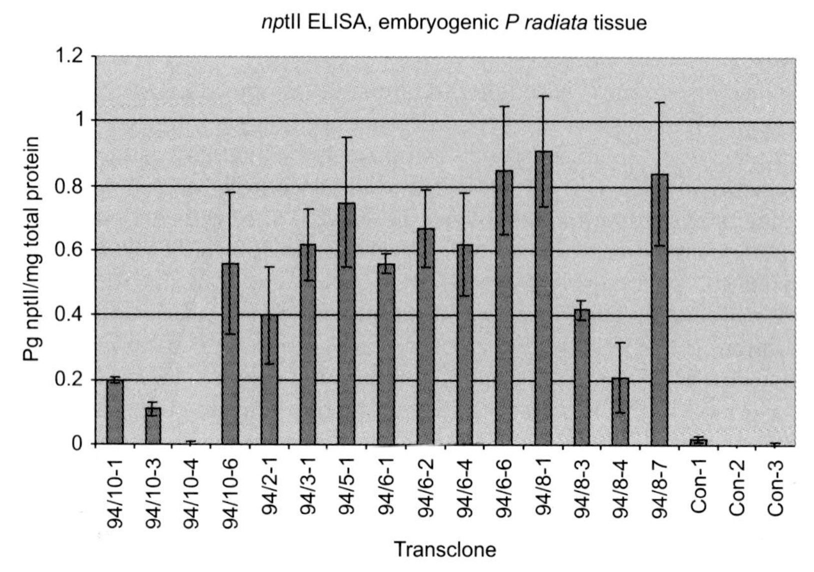

Fig. 4.2 ELISA analysis of *npt*II expression in *Pinus radiata* tissue. Con-1, 2 and 3 are non-transformed controls. All other samples are from transclones. All samples are the same genotype of *P. radiata*.

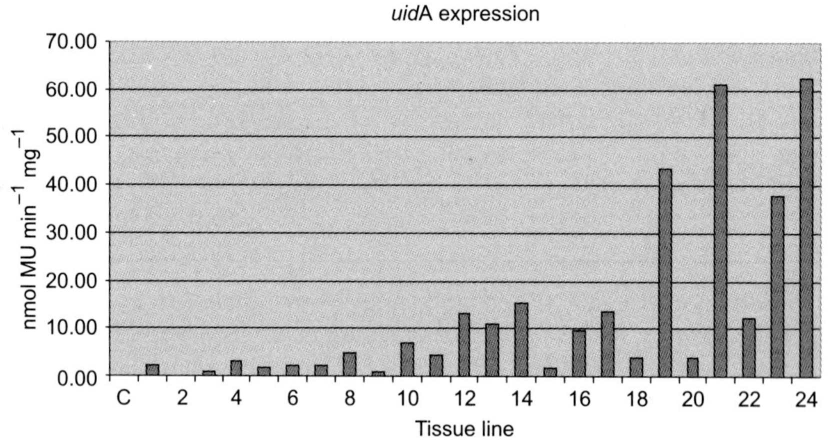

Fig. 4.3 Fluorometric analysis of *Pinus radiata* transclones expressing the *uid*A gene. C: Non-transformed control tissue; 1–24: Transformed lines. All lines are from the same *P. radiata* genotype.

copy number effects of the introduced genes, co-suppression or position effects or gene silencing (Walter et al. 1992; Meyer 1995; Matzke and

Matzke 1995; Kuai and Morris 1996; Fladung 1999; Kumar and Fladung 2000). Further, promoters may change gene expression characteristics with time and depending on the physical location of the respective cell within the plant. A specific promoter may not be active in, for example roots, but highly active in needles.

Interestingly, the level of *uid*A expression in transgenic material, as measured by fluorometric assays (Figure 4.3) could not be correlated to the number of copies integrated into the *Pinus radiata* genome (data not shown). Copy numbers varied between very low and probably more than 100. Further, tandem arrangements and insertion of fragmented copies were found on a regular basis (Walter 1998b).

Promoters Influencing Gene Expression Levels

Foreign genes and promoters can be recognized by the plant cell and their expression influenced or, in some cases, even be switched off (Walter et al., 1992). Furthermore, promoters which have been well characterized in their organism of origin do not necessarily show the same expression characteristics over a period of time in transgenic plants (Charest et al., 1993; Walter et al., 1994). Transferred genes and heterologous promoters are potentially subject to specific silencing phenomena in their new host, and a number of such mechanisms have been discussed (Matzke and Matzke 1995; Kuai et al., 1996). If genetic engineering is to be used in conifer plantation forestry, foreign gene expression needs to be tightly controllable and predictable. One possible solution to the problem could be the use of endogenous conifer promoters. These can be analyzed for their expression characteristics in the plants of origin, in different organs and at different ages/different sites, and can potentially be better predicted with regard to their behaviour in transgenic plant material. It is also hypothesized that endogenous promoters are less subject to gene silencing phenomena (Walter 1998a). However, further research is required to test this hypothesis.

Many foreign genes for proposed transfer into forest trees would not have to be expressed through the life-span of the tree, but rather, during a specific developmental stage, in a specific tissue or a given time period. A toxin gene to fight insect attack, for instance (Van der Salm et al., 1994), is only required when insects are actually present. Also, herbicide resistance genes need only be expressed for a limited number of years, in the nursery and in the first years following plantation establishment. Expression of some genes over the whole life span of the tree could pose environmental risks, such as the development of resistance in insects against a toxin produced by the tree (Tabashnik 1994; Alstad and Andow 1995). Consequently, research needs to be concentrated at the isolation and characterization of different promoter sequences, such as the copper inducible promoter, which can be switched on by the application of copper (Mett et al., 1993).

Although much is known about promoter activity in angiosperms, relatively little is known in conifers about levels of gene expression controlled by a specific promoter (Bekkaoui et al., 1990; Ellis et al., 1991; Walter et al., 1994).

To address this problem in conifers in more detail, eight different heterologous promoter constructs were fused upstream relative to a *uid*A reporter gene. The constructs were bombarded into embryogenic tissue of four conifer species and the expression of the *uid*A gene monitored three days after the transformation event. Expression characteristics were similar for all four conifers but differed widely with regard to the promoter construct used (Walter, unpublished data). Surprisingly, the 35S promoter (Franck et al., 1980), which is regarded as a very strong constitutive promoter in angiosperms, showed the lowest expression levels in all four species. Conversely, the maize ubiquitin promoter (Cornejo et al., 1993), which is strongly expressed in monocotyledons (but not in dicotyledons), showed very high expression levels in these experiments. It is also important to note that the two Emu-related (Last et al., 1991) promoter constructs used expressed at relatively high levels in *P. radiata* (Walter et al , 1994). One of them (pEmu) has proved to be highly active in monocotyledons; the other one (4ocsd35s) in dicotyledons, but not vice versa. Further, when an additional start codon with a Kozak consensus site in-frame with the *uid*A start codon was added close to it, a significant increase in *uid*A expression in conifers was detected. The Kozak consensus sequence has been shown to act as a translational enhancer since it improves the binding of ribosomes to mRNA (Kozak, 1989).

The information derived from this study helps us to understand promoter function in conifers and to design the appropriate promoter/ gene constructs for a particular application. However, the stable expression of genes driven by these promoters and the long-term characteristics of gene expression must still be evaluated.

Gene Expression in Conifers: Stable or Silenced?

While the data on transgene stability in annual plants and some non-coniferous trees is now available, not much is known on the stability of transgenes in conifers. Indications are that individual conifer transclones show a wide variety of gene expression levels using an identical construct (D. Ellis, pers. comm. and C. Walter, unpublished). Silencing effects have also been observed (A. Wagner, pers. comm.) but few studies have actually assessed this in greater detail.

A small field trial of genetically-transformed *Pinus radiata* plants was established in Rotorua, New Zealand in 1998. This trial was the first using transgenic radiata pines and the genes integrated included the *npt*II gene for the selection of transclones and the *uid*A reporter gene. The latter was under the control of the double CaMV 35S promoter and a Kozak consensus site was placed around the ATG start codon of the *uid*A gene. The vector used (pCW 122) was fully described in Walter et al., (1994).

Plants in this trial were analyzed with regard to continued expression of the *npt*II gene, which was under the control of the single 35S promoter, and including a Kozak consensus sequence around the ATG start codon. *npt*II expression was quantified by *npt*II ELISA (Agdia Patho-Screen kit, according to manufacturers, instructions) conducted on freshly-grown needles of various transgenic and non-transgenic control plants in the field. Measurements were taken at 5–6 weekly intervals during the year 2001, including winter, spring and the beginning of summer in New Zealand. The results of this experiment (Table 4.2) indicate high variation of *npt*II expression over a period of time in the same plant. This could be due to changes in gene expression, changes in post-transcriptional degradation of the protein, or to a chimeric transformation event where some cells (or parts of the plant) are transformed while others are not.

Table 4.2 *NPT*II ELISA with needles from transgenic *Pinus radiata* plants in a contained field test. Results are expressed as pg *NPT*II/mg total protein. T1–T5: Samples taken at 6-weekly intervals. Plant 137 is a non-transformed control plant.

Transclone	*T1*	*T2*	*T3*	*T4*	*T5*
49	2.76	NT	NT	1.27	2.4
52	1.96	NT	NT	0.54	1.51
67	NT	NT	2.06	2.84	2.1
68	NT	2.49	NT	2.63	3.09
137	0.06	0.01	0	0	0
147	7.09	2.88	3.22	2.64	4.03

Outlook

There is evidence which indicates that variation in gene expression is often correlated with high complexity of transgene integration. Transformation techniques that result in multi-copy or fragmented copy integrations usually show a higher degree of variation or even silencing rather than compared to those where single and complete integrations can be expected. The latter is generally true for *Agrobacterium*-mediated transformed material, and in many cases, it is desirable to use *Agrobacterium* as the transforming agent. However, silencing and more complex gene integrations were also observed in *Agrobacterium*-mediated transformed plants.

Conifer gene technology is still in its infancy. Many goals have been achieved in a relatively short time, including transformation with genes that may have an economic impact on forestry such as insect and herbicide resistance and those related to lignin biosynthesis. However, many other aspects of conifer gene expression and transgene expression in particular are still not very well understood. Gene silencing effects have been observed for annual plants and many non-coniferous trees, and indications are that silencing will also affect conifer gene expression. Strategies must be developed to avoid or control variation of gene expression if genetic engineering is to play a major role in conifer forestry. To this end, gene transfer technologies must be optimized so that the goal of precise and targeted integration will be achieved in the long term. In addition, promoters and genetic elements that can stabilize gene expression need to be assessed in the conifer background in greater detail.

Conifer genetic engineering has the potential to make transformational contributions to forestry once the issues such as silencing and control of gene expression are fully understood and under control. Once this is achieved, genetic engineering will contribute to value creation in forestry and provide an option for sustainable and socially and environmentally-friendly forestry for future years.

ACKNOWLEDGEMENTS

The author wishes to thank several colleagues for providing unpublished data for this contribution and for reviewing the manuscript: Tomoko Pearson, Lynette Grace, Jens Find, Lyn Holland, Ute Armbruster, Judy Moody and Armin Wagner. Part of this work was funded by the New Zealand Foundation for Research, Science and Technology (FRST) and by GEENZ.

REFERENCES

Ahuja M.R., 2000, Genetic engineering in forest trees: State of the art and future perspectives. In: S.M. Jain and S.C. Minocha (Eds). Molecular Biology of Woody Plants, Vol. 2, 485–511. Kluwer, Netherlands.

Alstad D.N., Andow D.A., 1995, Managing the evolution of insect resistance to transgenic plants. Science **268**: 1894–1896.

Anon, 2001, The state of Canada's forests. Sustainable forestry: a reality in Canada. ISBN 0-662-30746-1.

Bekkaoui F., Datla R.S.S., Pilon M., Tautorus T.E., Crosby W., Dunstan D.I., 1990, The effects of promoter on transient expression in conifer cell lines. Theoretical and Applied Genetics **79**: 353–359.

Becwar M.R., 1993, Conifer somatic embryogenesis and clonal forestry, In: M.R. Ahuja and W.J. Libby, (Eds.). Clonal Forestry I Genetics and Biotechnology **1**: 200–203., Springer-Verlag, Berlin, Heidelberg.

Bishop-Hurley S.L.R., Zabkievicz J., Grace L.J., Gardner R.C., Walter C., 2001, Conifer genetic engineering: transgenic *Pinus radiata* (D. Don) and *Picea abies* (Karst) plants are resistant to the herbicide Buster. Plant Cell Reports, **20**: 235–243.

Carson M.J., Burdon R.D., Carson S.D., Firth A., Shelbourne C.J.A., Vincent T.G., 1989, Realising genetic gains in production forests. In: Proceedings IUFRO working parties on Douglas fir, Lodgepole pine, *Sitka* and *Abies* spp. Breeding Genetic Resources. Session: Genetic gains in production forests. Olympia, Washington.

Charest P.J., Calero N., Lachance D., Datla R.S.S., Duchesne L.C., Tsang E.W.T., 1993, Microprojectile-DNA delivery in conifer species: factors affecting assessment of transient gene expression using the β-glucuronidase reporter gene. Plant Cell Reports **12**: 189–193.

Charity J.A., Holland L., Donaldson S.S., Grace L.J., Walter C., 2001, *Agrobacterium*-mediated transformation of *Pinus radiata* organogenic tissue using vacuum infiltration. Plant Cell, Tissue and Organ Culture.

Cornejo M.J., Luth D., Blankenship K.M., Anderson O.D., Blechl A.E., 1993, Activity of a maize ubiquitin promoter in transgenic rice. Plant Molecular Biology **23**: 567–581.

Clapham D., Demel P., Elfstrand M., Koop H.U., Sabala I., van Arnold S., 2000, Gene transfer by particle bombardment of embryogenic cultures of *Picea abies* and the production of transgenic plantlets. Scandinavian Journal of Forest Research **15**: 151–160.

Ellis D.D., McCabe D., Russell D., Martinell B., McCown B.H., 1991, Expression of inducible angiosperm promoters in a gymnosperm, *Picea glauca* (white spruce). Plant Molecular Biology **17**: 19–27.

Ellis D.D., McCabe D.E., McInnis S., Ramachandran R., Russell D.R., Wallace K.M. Martinell B.J., Roberts D.R., Raffa K.F., McCown B.H., 1993, Stable transformation of *Picea glauca* by particle acceleration. Bio/Technology **11**: 84–89.

Fenning, T.M. and Gershenzon, J. (2002). Where will the wood come from? Plantation forests and the role of biotechnology. *Trends in Biotechnology* **20(7)**, 291–296.

Fillatti J.J., Sellmer J., McCown B., Haissig B., Comai L., 1987, *Agrobacterium*-mediated transformation and regeneration of *Populus*. Molecular and General Genetics **206**: 192–199.

Find J., Grace L., Walter C., 2001, Stable transformation of nordmanns fir (*Abies nordmanniana*) by particle bombardment and regeneration of transgenic plants. Poster presentation at Wood Breeding and Biotechnology Conference, Bordeaux, France, 10–14 June 2001.

Fladung M., 1999, Gene stability in transgenic aspen (Populus). I. Flanking DNA sequences and T-DNA structure. Mol. Gen. Genet. 260 **6**: 574–581.

Franck, A., Guilley, H., Jonard G., Richards K., Richards and Hirth L., 1980, Nucleotide Sequence of Cauliflower Mosaic Virus DNA. Cell **21**: 285–294.

Guerinot M.L., 2000, Perspectives: plant biology. The green revolution strikes gold. Science, **287**: 241–243.

Holland L., Gemmell J.E., Charity J.A., Walter C., 1997, Foreign gene transfer into *Pinus radiata* cotyledons by *Agrobacterium tumefaciens*. NZ J. For. Sci. **27**: 289–304.

Huang Y., Diner A.M., Karnosky D.F., 1991, *Agrobacterium rhizogenes* mediated genetic transformation and regeneration of a conifer: *Larix decidua*. In Vitro Cell Developmental Biology **27**: 201–207.

James C., 2000, Global Status of Commercialised Transgenic Crops: 2000. ISAAA Briefs No. 21. ISAAA: Ithaca, NY.

Jayawickrama K.J.S., Carson M.J., 2000, A breeding strategy for the New Zealand Radiata Pine Breeding Cooperative. *Silvae Genetica 49*,2.

Jouanin L., Brasileiro A.C.M., Leple J.C., Pilate G., Cornu D., 1993, Genetic transformation: A short review of methods and their applications, results and perspectives for forest trees. Ann. Sci. For. **50**: 325–336.

Kanowski P.J., 1997, Afforestation and plantation forestry. Special Paper for the XI World Forestry Congress 13–22 October 1997. Working paper 1997/6.

Klein T.M., Wolf E.D., Wu R., Sanford J.C., 1987, High-velocity microprojectiles for delivering nucleic acids into living cells. Nature **327**: 70–73.

Klimaszewska K., Lachance D., Pelletier G., Lelu A.M., Seguin A., 2001, Regeneration of transgenic *Picea glauca, P. mariana* and *P. abies* after cocultivation of embryogenic tissue with *Agrobacterium tumefaciens*. In Vitro Cell Dev. Biol. (in press).

Kozak M., 1989, The scanning model for translation: An update. Journal of Cell Biology **109**: 229–241.

Krattiger A.F., 1997, Insect resistance in crops: a case study of Bacillus thuringiensis (Bt.) and its transfer to developing countries.

Krattiger A.F., 1998, The Importance of Ag-Biotechnology for Global Prosperity. ISAAA Briefs No. 6. ISAAA: Ithaca, NY.

Kuai B., Morris P., 1996, Screening for stable transformants and stability of glucuronidase gene expression on suspension cultured cells of tall fescue (*Festuca arundinacea*). Plant Cell Reports **15**: 804–808.

Kumar S., Fladung M., 2000, Determination of transgene repeat formation and promoter methylation in transgenic plants. Biotechniques **28**(6): 1128–1134.

Last D.I., Brettell R.I.S., Chamberlain A.M., Chaudhury A.M., Larking P.J., Marsh E.L., Peacock W.J., Dennis E.S., 1991, pEmu: An improved promoter for gene expression in cereal cells. Theoretical and Applied Genetics **81**: 581–588.

Levée V., Garin E., Klimaszewska K., Seguin A., 1999, Stable genetic transformation of white pine (*Pinus strobus* L.) after cocultivation of embryogenic tissues with *Agrobacterium tumefaciens*. Molecular Breed. **5**: 429–440.

Matzke M.A., Matzke A.J.M., 1995, How and why do plants inactivate homologous (trans)genes? Plant Physiology **107**: 679–685.

MacRae S., van Staden J., 1999, Transgenic Eucalyptus. In Y.P.S. Bajaj (Ed.) Biotechnology in Agriculture and Forestry 44. Springer Berlin, Heidelberg.

Menzies M.I., Aimers-Halliday J., 1997, Propagation options for clonal forestry with radiata pine. Genetics of Radiata Pine. Proceedings IUFRO Conference, Rotorua, New Zealand. FRI Bulletin No. **203**: 256–263.

Mett V.L., Lochhead L.P., Reynolds P.H.S., 1993, Copper controllable gene expression system for whole plants. Proceedings of the National Academy of Science, USA **90**: 4567–71.

Meyer P., 1995, Variation of transgene expression in plants. Euphytica **85:** 359–366.

Newell C.A., 2000, Plant transformation technology. Developments and applications. Mol. Biotechnology **16**(1): 53–65.

Parsons T.J., Sinkar V.P., Stettler R.F., Nester E.W., Gordon M.P., 1986, Transformation of poplar by *Agrobacterium tumefaciens*. Biotechnology **4**: 533–536.

Shintani D. and DellaPenna D., 1998. Elevating the vitamin E contents of plants through metabolic engineering. Science **282**: 2098–2100.

Tabashnik B.E., 1994, Evolution of resistance to *Bacillus thuringiensis*. Annual Review of Entomology **39**: 47–79.

Tabashnik B.E. 1997, Seeking the root of insect resistance to transgenic plants. Proceedings of the National Academy of Science of the USA **94**: 3488–3490.

Tzfira T., Rhee Y., Chen M.H., Kunik T., Citovsky V., 2000, Nucleic acid transport in plant-microbe interactions: the molecules that walk through cell walls. Annual Review of Microbiology **54**: 187–219.

Van Larebeke N., Genetello C., Schell J., Schilperoort R.A., Hermans A.K., Van Montagu M., Hernalsteens J.P., 1975, Acquisition of tumor-inducing ability by non-oncogenic Agrobacteria as a result of plasmid transfer. Nature **255**: 742–743.

Van der Salm T., Bosch D., Honée G., Feng L., Munsterman E., Bakker P., Stiekema W.J., Visser B., 1994, Insect resistance of transgenic plants that express modified *Bacillus thuringiensis cryIA (b)* and *cryIC* genes: a resistance management strategy. Plant Molecular Biology **26**: 51–59.

Walter C., Broer I., Hillemann D., Pühler A., 1992, High frequency, heat treatment-induced inactivation of the phosphinothricin resistance gene in transgenic single cell suspension cultures of *Medicago sativa*. Molecular and General Genetics **235**: 189–196.

Walter C., Smith D.R., Connett M.B., Grace L.J., White D.W.R., 1994, A biolistic approach for the transfer and expression of a *gus* reporter gene in embryogenic cultures of *Pinus radiata*. Plant Cell Reports **14**: 69–74.

Walter C., Carson S.D., Menzies M.I., Richardson T., Carson M., 1998a, Review: Application of biotechnology to forestry-molecular biology of conifers. World Journal of Microbiology and Biotechnology **14**: 321–330.

Walter C., Grace L.J., Wagner A., Walden A.R., White D.W.R., Donaldson S.S., Hinton H.H., Gardner R.C., Smith D.R., 1998b, Stable transformation and regeneration of transgenic plants of *Pinus radiata* D. Don. Plant Cell Reports **17**: 460–468.

Walter C., Grace L.J., Donaldson S.S., Moody J., Gemmell J.E., van der Maas S., Kwaalen H., Loenneborg A., 1998c, An efficient biolistic transformation protocol for *Picea abies* (L.) Karst embryogenic tissue and regeneration of transgenic plants. Canadian Journal of Forest Research 29 (10): 1539–1546.

Wilcox P.L., Carson S.D., Richardson T.E., Ball R.D., Horgan G.P., Carter P., 2001, Benefit-cost analysis of marker-based selection in seed orchard production populations of Pinus radiata. Accepted by Canadian Journal of Forest Science.

Zhu, S. *et al.* (1999) Global forest products consumption, production, trade and prices: global forest products model projections to 2010. FAO Report. See under *Forest products outlook study at:* http://www.fao.org/forestry/foris/index.jsp?lang id=1&geo id=42&start id=2711.

Zupan J., Muith T.R., Draper O., Zambryski P., 2000. The transfer of DNA from *Agrobacterium tumefaciens* into plants: A feast of fundamental insights. The Plant Journal **23**(1): 11–28.

Generation of Transgenic Creeping Bentgrass (*Agrostis palustris* Huds.) Plants from Mature Seed-derived Highly Regenerative Tissues

Myeong-Je Cho*, Khanh Van Le, Dorothy Okamoto, Yong-Bum Kim, Hae-Woon Choi, Peggy G. Lemaux

Department of Plant and Microbial Biology, University of California, Berkeley, CA 94720, USA

ABSTRACT

Mature seeds of creeping bentgrass (*Agrostis palustris* Huds. cv. Putter) were placed on medium containing 2,4-D (4.5 or 9.0 μM), BAP (0, 0.44 or 2.2 μM) and cupric sulfate (0.1 or 5.0 μM) under dim-light conditions to induce and proli-ferate highly regenerative, compact, green tissues. These highly regenerative tissues were transformed with a mixture of three plasmids containing genes for hygromycin phosphotransferase (*hpt*), phosphinothricin acetyltransferase (*bar*) and β-glucuronidase (*uidA; gus*) at a molar ratio of 1:1:1. Of 296 individual explants bombarded, fifteen independent transgenic lines (5.1%) were obtained after an 8- to 16-week selection period for hygromycin resistance with 30 to 100 mg/L of hygromycin B; regenerability of transgenic lines was 93%. The presence and stable integration of transgene(s) in plants were confirmed by PCR and DNA blot hybridization. The co-expression frequency of all three transgenes (*hpt/bar/uidA*) in T_0 plants was 29%; for two transgenes (*hpt/bar* or *hpt/uidA*), coexpression frequencies were 40–43%.

INTRODUCTION

Creeping bentgrass (*Agrostis palustris* Huds.) is a popular cool-season, perennial turfgrass widely used on golf courses and tennis greens (Miller 1984). Therefore, the ability to genetically engineer this crop-in order to improve its performance and pest-resistance qualities or to enhance its alternative uses is - of great importance. Successful transformation of creeping bentgrass has recently been reported using either embryogenic callus (Zhong et al., 1993; Hartman et al., 1994), protoplasts (Lee et al.,

*Corresponding author: E-mail: mjcho@nature.berkeley.edu

1996; Asano et al., 1998) or suspension cultures (Xiao and Ha, 1997). However, these approaches involve laborious steps that are difficult to reproduce in terms of initiation and maintenance of cultures.

We have recently established a very efficient in vitro system to proliferate highly regenerative, green tissues derived from immature scutellar tissues of barley (Cho et al., 1998; Lemaux et al., 1999), wheat (Cho et al., 1999a), maize (Cho et al., 1999a) and sorghum (Cho and Lemaux, 2001), and from mature seed-derived embryogenic callus tissues of oat (Cho et al., 1999b), rice (Cho et al., 2004), tall and red fescues (Cho et al., 2000b), Kentucky bluegrass (Ha et al., 2001) and orchardgrass (Cho et al., 2001). These tissues can be used to generate large numbers of shoots and can be maintained for more than a year with minimal losses in regenerability. Such tissues have been used for successful transformation of previously-recalcitrant cultivars of, for example, barley, wheat, Kentucky bluegrass and maize (Cho et al., 1998; Cho et al., 1999a; Ha et al., 2001). In addition, DNA methylation analyses showed that barley plants regenerated from the highly regenerative tissues incurred fewer methylation polymorphisms and had better agronomic performance than those from embryogenic callus tissues (Zhang et al., 1999; Bregitzer et al., 2002).

In this present study, we describe a highly efficient transformation system for creeping bentgrass via microprojectile bombardment of the highly regenerative, green tissues derived from mature seed-derived embryogenic callus followed by a report on the molecular and biochemical characterization of T_0 plants from these tissues.

MATERIALS AND METHODS

Plant Material and Culture of Explants

Mature seeds of creeping bentgrass (cv. Putter), were surface-sterilized for 20 min. in 20% (v/v) bleach (5.25% sodium hypochlorite), followed by 3 washes in sterile water. The seeds were placed on 3 different Murashige and Skoog (1962)-based callus-induction media: (1) D′ medium containing 9.0 μM 2,4-D and 0.1 μM copper; (2) D′BC2 medium containing 9.0 μM 2,4-D, 0.44 μM BAP and 5.0 μM copper; and (3) DBC3 medium containing 4.5 μM 2,4-D, 2.2 μM BAP and 5.0 μM copper (Cho et al., 1998). Five to 7 days after initiation, germinating shoots and roots from mature seeds of Putter were completely removed by manual excision. After three weeks of incubation at 24±1°C under dim light conditions (approximately 10 to 30 $\mu E\ m^{-2}\ s^{-1}$, 16 h-light), the highest-quality, compact, green tissues were selected and subcultured at 3- to 4-week intervals in the same medium.

Plasmids

Plasmids, pAHC15, pAHC20 and pAct1IHPT-4, were used for transformation (Fig. 5.1). pAHC15 and pAHC20 (Christensen and Quail 1996) contain *uidA* (β-glucuronidase, *gus*) and *bar*, respectively, each under control of the maize ubiqutin *ubi*1 promoter and intron (*ubi*1/*ubi*1I) and *nos*. pActlIHPT-4 (Cho et al., 1998) contains the hygromycin phosphotransferase (*hpt*)-coding sequence under control of the rice actin1 promoter (*act*1), its intron (*act*1I) and the *nos* 3' terminator.

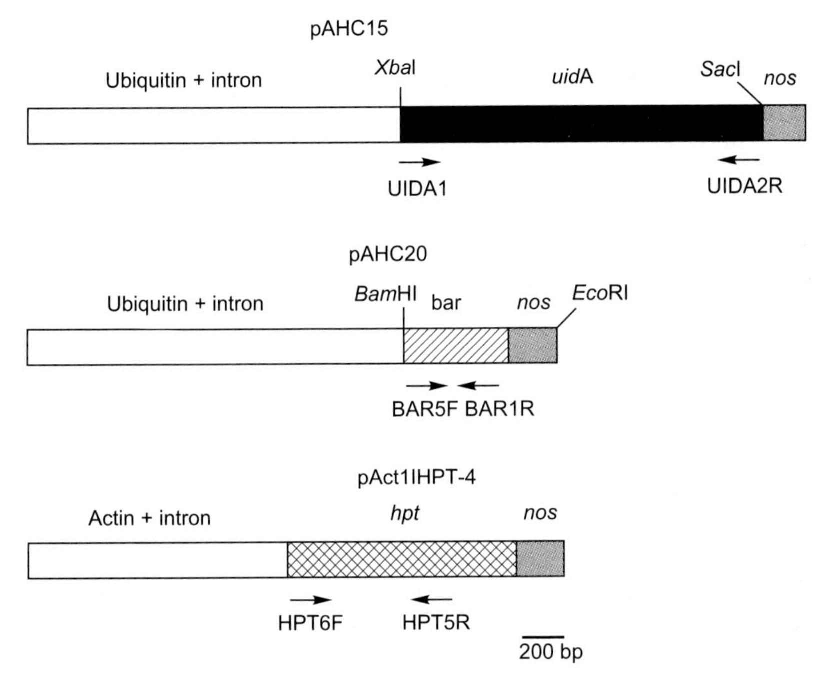

Fig. 5.1 Schematic diagram of plasmids used for creeping bentgrass transformation. The positions of PCR primer sets for each plasmid are indicated by arrows and the locations of enzyme restriction sites used for DNA blot hybridization analysis are indicated.

Particle Bombardment and Stable Transformation

Approximately, 4-to 5-month-old highly regenerative cultures were used for bombardment. The tissues (approximately 30 pieces, 3–4 mm) were transferred for osmotic pre-treatment to DBC3 medium containing equimolar amounts of mannitol and sorbitol so as to give a final concentration of 0.4 M. Four hours after treatment with the osmoticum, the tissues were bombarded in the manner described previously (Lemaux et al., 1996). Gold particles (1.0 μm) were coated with 25 μg of a 1:1:1 molar ratio of a mixture of pAct1IHPT-4, pAHC20 and pAHC15, followed by bombardment using a PDS-1000 He biolistic device at 900 psi. Sixteen to 18 hr after bombardment, tissues were moved to DBC3 medium without osmoticum, supplemented with 30 mg/L hygromycin B and grown at 24±1°C under dim light. From the second-round selection onward, tissues were subcultured at 3- to 4-week intervals and maintained on the same medium containing 100 mg/L hygromycin B. Following the identification of adequate quantities of highly regenerative green tissues on each medium, the tissues were plated on phytohormone-free BCI-DM$^-$ medium (Wan and Lemaux 1994) without a selective agent and exposed to higher-intensity light (approximately 45–55 μE m^{-2} s^{-1}). After four weeks, the regenerated shoots were transferred to soil and grown in the greenhouse.

Histochemical GUS Assay and Herbicide Application

Plant tissues from each putatively transformed line were tested for GUS activity by histochemical staining (Jefferson et al., 1987). To determine the herbicide sensitivity of transgenic plants, a section of leaf blade was painted once using a cotton swab with Basta solution [a 0.25% solution (v/v) of Basta™ plus 0.1% Tween 20] or entire leaf surfaces of a whole plant were sprayed twice with Basta solution. Plants were scored 1 week after herbicide application.

Genomic DNA Isolation, Polymerase Chain Reaction (PCR) and DNA Blot Hybridization

To test for the presence of *hpt, uidA* and *bar* in genomic DNA of putatively-transformed lines, 500 ng of genomic DNA, isolated from leaf tissues, was amplified by PCR using three primer sets (Fig. 5.1), HPT6F (5′-AAGCCTG AACTCACCGCGACG-3′) plus HPT5R (5′-AAGACCAATGCGGAGCATA TAC-3′) (Cho et al., 1998), UIDA1 (5′-agcggccgcaTTACGTCCTGTAGAAA CC-3′) plus UIDA2R (5′-agagctcTCATTGTTTGCCTCCCTG-3') (Cho et al., 1998) and BAR5F (5′-CATCGAGACAAGCACGGTCAACTTC-3′) plus BAR1R (5′-ATATCCGAGCGCCTCGTGCATGCG-3′) (Lemaux et al., 1996), respectively. Amplifications were performed in a 25-μl reaction with *Taq* DNA polymerase (Promega, Madison, WI) as described (Cho et al., 1998).

For DNA hybridization analysis, 10 μg of total genomic DNA from the leaf tissue of each line was digested with *Xba*I/*Sac*I and *Bam*HI/*Eco*RI for detection of *uidA* and *bar* from pAHC15 and pAHC20 transformants, respectively, separated on a 0.8% agarose gel, transferred to Zeta-Probe GT membrane and hybridized with a radiolabeled *uidA*-or *bar*-specific probe, following the manufacturer's instructions. The *uidA*-containing 1.9-kb *Xba*I/*Sac*I fragment from pDhGN-2 (M.-J. Cho, unpublished) and the *bar*-containing 0.6-kb *Pst*I fragment from pAHC20 were purified using a QIAEX gel extraction kit (QIAGEN, Chatsworth, CA) and labeled with $\alpha - {}^{32}P$-dCTP using random primers.

RESULTS AND DISCUSSION

Establishment of in vitro Culturing System

To establish a highly efficient in vitro system for culturing and regenerating bentgrass tissue, three different media, D′, D′BC2 and DBC3, were tested for callus induction. Non-regenerable callus tissues were produced in many cases on D′ medium. A higher callus-induction frequency and larger numbers of embryogenic structures were observed on D′BC2 medium rather than on DBC3. The use of DBC3 at the initial callus-induction step resulted in high rates of seed germination and low frequencies of callus induction (data not shown). However, DBC3 was optimal for maintaining highly regenerative, green tissues (Fig. 5.2a) derived from embryogenic callus initiated on D′BC2 medium. The frequency of shoot regeneration was increased 1.7- and 2.6-fold with D′BC2 and DBC3 medium, respectively, compared to D′ medium (Table 5.1).

Table 5.1 Regenerability of creeping bentgrass tissues on three different media

Culture medium	*Ingredients*			*Regenerability*[a] (*No. of shoots/tissue piece*)
	2,4-D (μM)	*BAP (μM)*	*$CuSO_4$ (μM)*	
D′	9.0	0.0	0.1	35.7±4.3
D′BC2	9.0	0.44	5.0	62.1±14.7
DBC3[b]	4.5	2.2	5.0	92.5±26.7

[a]Ten to 17 pieces of tissue (~5 to 7 mm in diameter) from each treatment were transferred to regeneration medium and the shoots were counted after 21 days. Values represent means ± SD of four replicates of each treatment.
[b]D′BC2 for initiation and DBC3 for maintenance

Bombardment and Selection of Hygromycin-resistant Clones

Four-to 5-month-old highly regenerative tissues (Fig. 5.2A) containing multiple light-green shoot meristem-like structures were used for bombardment. In barley, the expression of a gene associated with the maintenance of the meristematic state in barley meristematic tissue, a

Fig. 5.2 Production of transgenic creeping bentgrass lines and functional expression of transgenes in transgenic plants.

(A) Highly regenerative, green tissues of Putter induced from mature seeds initiated on D′BC2 medium and maintained on DBC3 medium.

(B) Putter tissues on third round of selection with 100 mg/L hygromycin B.

(C) Plantlets regenerated from highly-regenerative tissues of a transgenic Putter line.

(D) Transgenic plants two months after transfer to soil from Magenta box.

(E) GUS activity in leaf tissues of transgenic Putter (right) and non-transgenic (left) plants.

(F) Basta-resistant transgenic Putter lines (right) and Basta-sensitive (left) lines.

knotted 1 homologue, was studied in tissue derived from the excised shoot apex and from highly regenerative tissues (Zhang et al., 1998). The pattern was similar, although not identical, in these two tissues (Zhang et al., 1998), suggesting that they had physiological and developmental similarities (Lemaux et al., 1999). These highly regenerative tissues are believed to have a high percentage of cells that are capable of sustained cell division, accounting for the fact that they can be maintained in culture for more than two years with minimal loss in regenerability. Similar results were obtained with highly regenerative, green tissues of barley (Cho et al., 1998), oat (Cho et al, 1999b), wheat (Cho et al., 1999a), tall and red fescues (Cho et al., 2000), Kentucky bluegrass (Ha et al., 2001) and orchardgrass (Cho et al., 2001).

During selection on hygromycin, most tissues gradually turned brown or white. In general, green, hygromycin-resistant tissues were observed at the third- or fourth-round selection (Fig. 5.2B). Putatively transformed lines were proliferated on the same medium until there was sufficient material for regeneration. Hygromycin-resistant highly regenerative green tissues were regenerated (Fig. 5.2C) and plantlets were then transferred to soil and grown in the greenhouse (Fig. 5.2D). Using this protocol, 15 independent transgenic lines were obtained from 296 pieces of tissue, giving a 5.1% transformation frequency (Table 5.2). Of these fifteen hygromycin-resistant lines, 93% (14/15) were regenerable, similar to the frequency previously observed for transgenic oat (100%) (Cho et al., 1999b), transgenic tall (100%) and red (82%) fescue (Cho et al., 2000) and orchardgrass (91%) (Cho et al., 2001a).

Molecular and Biochemical Analysis of Transgenic Plants

GUS activity was detected in putative transgenic leaf tissues (Fig. 5.2E) and callus tissues (data not shown). Of the 15 independent hygromycin-resistant lines examined, 6 were positive for GUS activity, giving a 40% coexpression efficiency (Table 5.2). Strong *uidA* expression was detected in leaf tissue in CBTrans-3, -5, -8, -9, -12 and -13; GUS expression was not observed in the negative control (data not shown); 6 out of 7 containing *hpt* and *uidA* expressed GUS. Six out of 14 transgenic lines tested were Basta-resistant (Fig. 5.2F), giving a 43% coexpression efficiency with *hpt* (Table 5.2); 6 out of 9 containing *hpt* and *bar* were Basta-resistant. Coexpression frequency of the three transgenes *(hpt/bar/uidA)* was 29% (4/14); data on CBTrans-8 was not included because it was not regenerable.

Presence and integration of the introduced *uidA* and *bar* genes in genomic DNA of T_0 plants were analyzed using PCR (Table 5.2) and DNA blot hybridization (Fig. 5.3). Of the fifteen T_0 lines analyzed, the presence of a 1.9-kb *uidA* fragment in seven T_0 lines, CBTrans-3, -4, -5, -8, -9, -12 and

Table 5.2 Analysis of transgenic creeping bentgrass callus and plants

Plasmids used for transformation	*Transgenic lines*	*PCR (Callus/Plants)*			*Transgene expression (Plants)*	
		hpt	*bar*	*uidA*	*Herbicide resistance*	*GUS*
pAct1IHPT-4	CBtrans-1	+	–	–	–	–
+ pAHC20	CBtrans-2	+	–	–	–	–
+ pAHC15	CBtrans-3	+	+	+	–	+
	CBtrans-4	+	+	+	–	–
	CBtrans-5	+	+	+	+	+
	CBtrans-6	+	–	–	–	–
	CBtrans-7	+	+	–	+	–
	CBtrans-8[a]	+	+	+	n.d.[c]	+
	Cbtrans-9	+	+	+	+	+
	CBtrans-10	+	+	–	+	–
	CBtrans-11	+	+	–	–	–
	CBtrans-12	+	+	+	+	+
	CBtrans-13	+	+	+	+	+
	CBtrans-14	+	–	–	–	–
	CBtrans-15	+	–	–	–	–
	15/296 = 5.1%[b]	15/15 = 100%	10/15 = 67%	7/15 = 47%	6/14 = 43%	6/15 = 40%

[a]Non-regenerable line
[b]Transformation frequency: 15 independent transgenic lines were obtained from bombardment of 296 green tissue pieces
[c]not determined

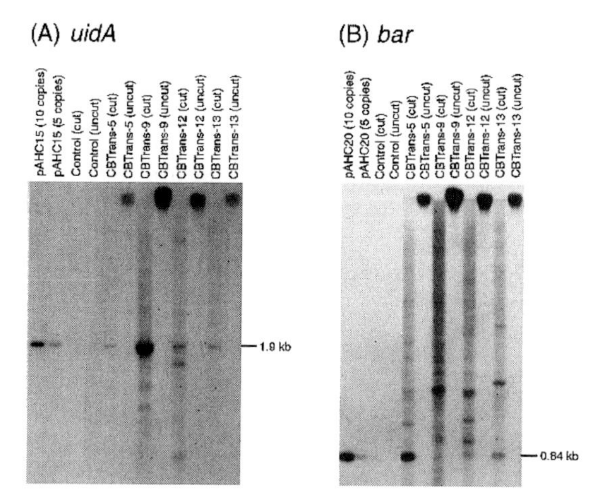

Fig. 5.3 DNA blot hybridization analysis of transgenic plants
Ten µg of genomic DNA per lane, isolated from leaf tissues of non-transformed control and transformed plants, was digested with *Xba*I/*Sac*I and *Bam*HI/*Eco*RI to release *uidA* and *bar*, respectively. Gel blots were hybridized with radiolabeled *uidA* (A) or *bar* (B) probe. Lanes designated 5 and 10 copies represent 5 and 10 copies of *uidA* (A) and *bar* (B) per creeping bentgrass genome.

-13, was confirmed by PCR amplification. A 0.34-kb fragment in ten T_0 lines, CBTrans-3, -4, -5, -7, -8, -9, -10, -11, -12 and -13, confirmed the presence of *bar*. Integration of *uidA* and *bar* into genomic DNA was further confirmed by DNA hybridization analysis on uncut DNA samples and both genes were present in undigested high molecular weight genomic DNA (Fig. 5.3). In addition, genomic DNA from CBTrans-5, -9, -12 and -13, when transformed with pAHC15 and positive for GUS activity and by PCR, produced the expected 1.9-kb-*uidA* fragment after digestion with *Xba*I and *Sac*I (Fig. 5.3a). All four transgenic lines (CBTrans-5, -9, -12 and -13) having Basta resistance produced the 0.84-kb *bar* fragment after digestion with *Bam*HI and *Eco*RI (Fig. 5.3b). Non-transformed plants did not contain DNA, which hybridized with either the *uidA* or *bar* probes.

In conclusion, we established an efficient and reproducible system for generating transformed creeping bentgrass (cv. Putter) plants using highly regenerative, green tissues derived from mature seed-derived embryogenic callus.

Acknowledgments

The authors thank Dr Christopher L. Schardl (University of Kentucky, Lexington, KY) for Putter seeds and Dr P. Quail (Plant Gene Expression Center, USDA-ARS, Albany, CA) for pAHC20, and pAHC15. M.-J. Cho, D. Okamoto, Y.-B. Kim and H.W. Choi were supported by funds from the BioSTAR Program of the University of California and Ventria, Bioscience, Sacramento, CA. P.G. Lemaux was supported by the USDA Cooperative Extension Service through the University of California.

REFERENCES

Asano Y., Ito Y., Fukami M., Sugiura K., Fujiie A., 1998, Herbicide-resistant transgenic creeping bentgrass plants obtained by electroporation using an altered buffer. Plant Cell Rep. **17**: 963–967.

Bregitzer P., Zhang S., Cho M.-J., Lemaux P.G., 2002, Reduced somaclonal variation in barley is associated with culturing highly differentiated, meristematic tissues. Crop Sci. 42: 1303-1308.

Cho M.-J., Buchanan B.B., Lemaux P.G. 1999a, Development of transformation systems for monocotyledonous crop species and production of foreign proteins in transgenic barley and wheat seeds. In: Application of Transformation Technology in Plant Breeding. 30th Anniversary Korean Breeding Soc. pp. 39–50.

Cho M.-J. Jiang W., Lemaux P.G., 1999b, High-frequency transformation of oat via microprojectile bombardment of seed-derived highly regenerative cultures. Plant Sci. **148**: 9–117.

Cho M.-J., Choi H.W., Lemaux P.G., 2001, Transformed T_0 orchardgrass (*Dactylis glomerata* L.) plants produced from highly regenerative tissues derived from mature seeds. Plant Cell Rep. **20**: 318–324.

Cho M.-J., Ha C.D., Lemaux P.G., 2000b, Production of transgenic tall fescue and red fescue plants by particle bombardment of mature seed-derived highly regenerative tissues. Plant Cell Rep. **19**: 1084–1089.

Cho M.-J., Jiang W., Lemaux P.G., 1998, Transformation of recalcitrant barley cultivars through improvement of regenerability and decreased albinism. Plant Sci. **138**: 229–244.

Cho M.-J., Lemaux P.G., 2001, An efficient system for transformation and plant regeneration of sorghum (*Sorghum bicolor* L.) using highly regenerative, green tissues. In Vitro Cell. Devel. Biol. 37(3II): 38A.

Cho M.-J., Yano H., Okamoto D., Kim H.-K., Jung H.-R., Newcomb K., Le V.K., Yoo H.S., Langham R., Buchanan B.B., Lemaux P.G. 2004, Stable transformation of rice (*Oryza sativa* L.) via microprojectile bombardment of highly regenerative, green tissues derived from mature seed. Plant Cell Rep. (in press).

Christensen A.H., Quail P.H., 1996, Ubiquitin promoter-based vectors for high-level expression of selectable and/or screenable marker genes in monocotyledonous plants. Transgenic Res. **5**: 1–6.

Ha C.D., Lemaux P.G., Cho M.-J., 2001, Stable transformation of a recalcitrant Kentucky bluegrass (*Poa pratensis* L.) cultivar using mature seed-derived highly regenerative tissues. In Vitro Cell. Dev. Biol. Plant **37**: 6–11.

Hartman C.L., Lee L., Day P.R., Tumer N.E., 1994, Herbicide resistant turfgrass (*Agrostis palustris* Huds.) by bioloistic transformation. Bio/Technology **12**: 919–92.

Jefferson R.A., Kavanagh T.A., Bevan M.W., 1987, GUS fusions: β-glucuronidase as a sensitive and versatile gene fusion marker in higher plants. EMBO J **6**: 3901–3907.

Lee L., Laramore C.L., Day P.R., Tumer N.E., 1996, Transformation and regeneration of creeping bentgrass (*Agrostis palustris* Hurls.) protoplasts. Crop Sci. **36**: 401–406.

Lemaux P.G., Cho M.-J., Louwerse J., Williams R., Wan Y., 1996, Bombardment-mediated transformation methods for barley, Bio-Rad. Bulletin 2007, pp. 1–6.

Lemaux P.G., Cho M.-J., Zhang S., Bregitzer P., 1999, Transgenic cereals: *Hordeum vulgare* (barley). In: Vasil I.K. (Ed.). Molecular Improvement of Cereal Crops. Kluwer Academic Pub., pp. 255–316.

Miller D.A., 1984, Forage Crops. McGraw-Hill, NY, pp. 396–409.

Murashige T., Skoog F., 1962, A revised medium for rapid growth and bioassays with tobacco tissue cultures. Physiol. Plant **15**: 473–497.

Wan Y., Lemaux P.G., 1994, Generation of large numbers of independently transformed fertile barley plants. Plant Physiol. **104**: 37–48.

Xiao L., Ha S.B., 1997, Efficient selection and regeneration of creeping bentgrass transformants following particle bombardment. Plant Cell Rep. **16**: 874–878.

Zhang S., Williams-Carrier R., Jackson D., Lemaux P.G., 1998, Expression of CDC2Zm and KNOTTED1 during in vitro axillary shoot meristem proliferation and adventitious shoot meristem formation in maize (*Zea mays* L.) and barley (*Hordeum vulgare* L.). Planta **204**: 542–549.

Zhang S., Zhang S., Cho M.-J. Bregitzer P, Lemaux P.G., 1999, Comparative analysis of genomic DNA methylation status and field performance of plants derived from embryogenic calli and shoot meristematic cultures. In: Altman A., Ziv M., Izhar S. (Eds.) Plant Biotechnology and In Vitro Biology in the 21st Century. Kluwer, Dordrecht, Netherlands, pp. 263–267.

Zhong H., Bolyard M.G., Srinivasan C., Sticklen M.B., 1993, Transgenic plants of turfgrass (*Agrostis palustris* Huds.) from microprojectile bombardment of embryogenic callus. Plant Cell Rep. **13**: 1–6.

6

Chromosomal and Molecular Analysis of Somaclonal Variation in Certain *Allium* Species

Bong-Bo Seo, Geum-Sook Do and Jyotsna Devi*

Department of Biology, Kyungpook National University, Taegu 702-701, Republic of Korea

**Department of Plant Breeding & Genetics, Assam Agricultural University, Jorhat, India*

ABSTRACT

Somaclonal variant obtained through tissue culture is easy access of genetic source improvement such as quality, productivity and tolerance to disease. Some promising auto- and aneuploidy levels regenerated as variants have been identified through chromosomal analysis in the certain *Allium* species. In this chapter, we describe the combined data obtained so far from tissue culture in some *Allium* species. Main contents are composed of callus induction, cytological analysis of callus cell, plant regeneration, and genetic variation in regenerated plants involving cytological analysis, C-banding analysis, isozyme and total protein analysis, morphological analysis, and molecular analysis by FISH technique.

INTRODUCTION

Onion, garlic, and leek, all varieties of common vegetables, are used as food or spices in most countries of the world. In early classifications of angiosperms, *Allium* and related genera were placed in the Liliaceae family. In the more recent and competent taxonomic treatment of the monocotyledons, they are recognized as the distinct family Alliaceae close to the Amaryllidaceae (Dahlgren et al., 1985). *Allium* is a large genus, probably of more than 500 species, which are perennial, mostly bulbous plants. This genus is widely distributed over the warm temperate zones of the northern hemisphere.

The development of many cultivars of *Allium* has proceeded by mass selection and hybridization. More recently, commercial production of hybrid seeds has become possible due to the discovery of male sterility in onion. However, a high degree of hybrid sterility and limited genetic recombination ability are major limitations to using interspecific hybridization for onion improvement. Therefore, there has been strong interest in developing in vitro technology in order to facilitate unconventional breeding of commercially-valuable *Allium* species by utilizing culture-induced genetic variants. In 1981, Larkin and Scowcroft coined a general term 'somaclonal variation' for plant variants derived from any form of cell or tissue cultures.

Somaclonal variation arising in tissue culture has been looked upon as a novel source of mutants for crop improvement. The variations found in the callus cells originate either from pre-existing variations in the somatic cells of the explant or from new mutations generated during growth in culture. An alternative explanation is that the tissue culture phase imposes stress, which may enhance spontaneous mutation rate or induce instability such as chromosome breakage and DNA transposition (Muller et al., 1990; Cecchini et al., 1992). It has also been suggested that genetic instability may be associated with the fraction of repeated sequences of DNA present in the plant genome (Breiman et al., 1987; Johnson et al., 1987; Bebeli et al., 1990; Karp et al., 1992). Somaclonal variation may also be due to molecular changes caused by mitotic crossing over (MCO) in regenerated plants, leading to single gene mutations. Minute changes in the structure of chromosomes can alter the expression and genetic transmission of specific genes. Further recombination, or chromosome breakage, at any preferential region of a particular chromosome(s) affects the genome in a disproportionate high frequency, resulting in altered phenotypic expressions (Evans and Sharp 1986). Thus, a whole range of genetic changes from point mutations, amplifications, deletions of DNA sequences to gross changes in chromosome number and structure may occur during in vitro culture.

An analysis of these genetic variations in the regenerated plants is necessary in order to exploit the somaclonal variants for crop improvement. The cytological characteristics commonly noticed in somaclonal variation are either structural or numerical variations of the chromosome, such as polyploidy, aneuploidy, and chromosome rearrangements. These variations are due to the chromosomal reorganization at various levels and errors of mitosis and meiosis. An analysis of C-banded karyotypes also gives a good picture of structural variations in the chromosomes. Chromosomal changes are a clear indication of somaclonal variation (Lee and Phillips 1988).

Fluorescence in situ hybridization (FISH) has been used in many plants to identify the chromosomes accurately, using species-specific repetitive sequences, ribosomal genes, and even unique sequences (Mukai et al., 1990; Maluszynska and Heslop-Harrison 1993; Jiang and Gill 1994). The ribosomal genes are of great value for karyotype analysis because of their universal occurrence. They are also useful for comparative studies of genome organization. Many authors have reported on the physical mapping of the repeated DNA sequences of the 5S and 18S-5.8S-26S rRNA gene families, using FISH in many plant species (Castilho and Heslop-Harrison 1995; Linares et al., 1996; Mukai et al., 1990; Song and Gustafson 1993). Appels et al., (1988) successfully collected information on physical location of the 5S rRNA loci in wheat, rye and barley. Using in situ hybridization (ISH), it was found that 5S rRNA loci are located distal to the secondary constrictions on chromosome 1B of wheat and 1R of rye and on non-nucleolar chromosome in barley. Another additional 5S rRNA locus in rye was located on 5R (Reddy and Appels 1989). Furthermore, a multicolour FISH technique is a useful tool for simultaneous detection of two or more sequences with different colours in the same cells (Ried et al., 1992: Mukai et al., 1993).

In this chapter, the results of various experiments conducted on the callus and regenerated plants of different *Allium* species utilizing cytological, C-banding and FISH technique have been presented and discussed.

DESCRIPTION OF *ALLIUM* SPECIES

Although many species are highly responsive in tissue culture procedure, most of them are not suitable for cytogenetic studies. *Allium* species are suitable for studying karyological variability of tissue cultures because of a small number of chromosomes with clear morphology and length. At the same time, it appears possible to derive long-term cultures with a high regenerative ability, which makes it possible to study in detail the karyological changes in the course of the whole tissue culture cycle, i.e. initial explant, dedifferentiation, short- or long-term culture, regeneration, regenerant (Novak et al., 1986). The different species used for the study of somaclonal variation are described below:

***Allium sativum* L.:** Garlic (*Allium sativum* L.), an economically important crop, is an obligate apomictic plant vegetatively propagated by cloves. It is impossible to breed this plant just by conventional crossing method because of early degeneration of flowers. It is a diploid (2n = 16) with two pairs of satellite chromosomes.

***Allium cyaneum* R.:** It is a plant found in mountainous regions at an altitude above 1000 m. Karyotype analysis of this species has revealed somatic chromosomes (2n = 16), including one pair of satellite chromosome plus 0-5 B chromosomes. Some terminal C-band markers were observed on a few chromosomes (Seo et al., 1989).

***Allium tuberosum* Rottl.:** This species is widely cultivated in South Asia, including Korea, Japan and China. The leaves of this non-bulbous species have a characteristic flavour of genus *Allium*, alliin, and are consumed in the form of a vegetable. Both cultivated and wild accessions of it have the same chromosome number, but minor phenotypic differences exist between them. *A. tuberosum* (2n = 4x = 32) is an autotetraploid, consisting of four sets of seven metacentric and one submetacentric chromosome with satellite. Seo (1977) reported that this species has a few C-bands in the satellite and terminal regions of some chromosomes.

***Allium wakegi* Araki.:** A vegetatively propagated crop, this species is grown in Korea, China, Japan and other southeastern Asia. Highly preferred as green salad onion, it has great potentiality as breeding material due to its good texture and delicate flavour. However, it becomes unsuitable for crossing programme due to its complete sterility, attributed to its hybrid origin. *A. wakegi* is an amphihaploid (2n = xy = 16), as ascertained by irregular meiotic configuration in PMC (Iwasa 1964) and unique C-banding pattern of each chromosome (Seo and Kim 1975). Tashiro (1984) proposed that *A. wakegi* must be a product of the interspecific hybridization between *A. ascalonicum* and *A. fistulosum.*

***Allium senescens* var. *minor*:** A wild species reported from Korea, this crop has potentiality to be used in breeding programme to improve the cultivated onion due to its favourable characters such as a long flowering period, winter hardiness and resistance to bacterial and fungal diseases. Cytogenetic analysis of this diploid (2n = 16) plant shows clear C-bands with constant inter-bands in the chromosomes (Kim et al., 1990a).

***Allium victorialis* var. *platyphyllum*:** This species occurs in subalpine coniferous or mountain broad-leaved forests. Normal chromosome compositions of it is diploid (2n = 16), consisting of a pair of subtelocentric with satellite, a pair of submetacentric, and six pairs of metacentric chromosomes (Kim et al., 1990).

***Allium fistulosum*:** This species, possessing high levels of resistance to several important pest and diseases, is of primary importance in Asia, i.e. China, Korea and Japan. Elsewhere, it is rarely grown. It is a diploid (2n = 16), possessing significantly larger satellites than most species of the section Cepa.

CALLUS INDUCTION

Generally, somaclonal variations are observed among plants regenerated through somatic organogenesis from the callus cells. Relatively low levels of variations are observed in plants regenerated through the tissue culture system, which avoids callus phase. Therefore, induction and successful establishment of callus is necessary to exploit the potential of somaclonal variation.

In our laboratory, callus was initiated in *Allium* species from bulb, cloves or flower buds. The bulbs were washed thoroughly, surface sterilized by immersing the entire bulbs in 70% (v/v) ethanol for 4–5 min and in 5 % (w/v) sodium hypochlorite solution for 20 min with vigorous agitation, followed by washing three times in sterile distilled water. Each bulb was dissected into 5–6 mm pieces. The basal sections included shoot apex, middle sections leaf sheath bases and the upper sections included only leaf blades. All these explants were inoculated into various media.

Nair and Seo (1992) cultured the basal sections, leaf sheath bases and leaf blades of *Allium senescens* var. *minor* and noted that all the explants enlarged in size but further growth was observed only in basal explants. MS medium (Murashige and Skoog 1962) with 1 mg/l of 2,4-D and kinetin-stimulated callus from 60% of the explants. Similar result was observed in B5 medium (Gamborg et al., 1968) also. Callus initiation on basal bulb explants of *A. victorialis* var. *platyphyllum* was visible after 60 days of culture in BDS medium (Dunstan and Short 1977), containing 2 mg/l 2, 4-D and 1 mg/l BA (Seo et al., 1995, 1996). Lee et al. (1998) cultured the basal sections of bulb of *A. cyaneum* in MS and B5 medium with different hormone combinations and achieved maximum frequency of callus formation on MS medium supplemented with 1 mg/l 2,4-D and BAP. Callus initiation was below 30% in B5 medium.

The cloves of *A. sativum* were sterilized by dipping in 70% ethanol and 5% sodium hypochlorite for 10 min. each. After removing the scales and leaf sheaths, inner most leaflets were cut horizontally into discs of 2–5 mm length and transferred to MS medium supplemented with 2 mg/l 2,4-D, NAA, kinetin and 15% coconut milk. The leaflet discs proliferated into sufficient callus for subculturing after 120 days of initial culture (Kim and Seo, 1991).

Callus was obtained from flower buds of *A. tuberosum* (Do et al., 1999) and *A. senescens* var. *minor* (Nair, 1993). Inflorescences were sterilized in a manner similar to the bulbs and individual flower buds with 2–3 mm long pedicels were removed aseptically from the closed spathe and inoculated into MS, B5 or BDS medium with different hormone concentrations. Callus were initiated from the nectar regions of the petals after about 21 days (Fig. 6.1).

Callus formation from protoplasts of *A. wakegi* was reported for the first time from our laboratory (Seo et al., 1994). The protoplasts were isolated from in vitro grown leaves and inoculated to 100 μL drops of KM8P (Kao and Michayluk 1975) liquid medium with 0.1 M mannitol, 0.4 M glucose, and hormones in petri dishes sealed with parafilm and incubated in the dark at 25°C. Fresh media with less concentration of mannitol were added at 10 days' interval for one month. After two to three months of culture, the mini-calli obtained were transferred to MS medium for further development of callus.

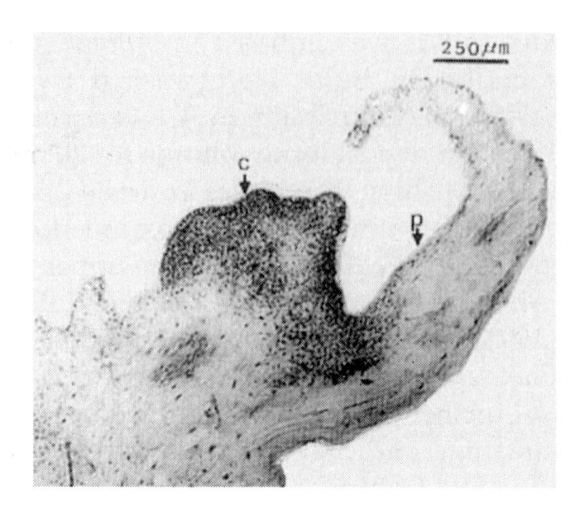

Fig. 6.1 Histological analysis of flower buds of *A. senescens* var. *minor* showing callus initiation from the nectar regions. p—petal, c—callus.

CYTOLOGICAL ANALYSIS OF CALLUS

The genetic instability of the callus induced from bulbs and flower buds of *A. senescens* var. *minor* with respect to culture medium and culture age was investigated by Nair (1993). Occurrence of micronuclei, binucleate, trinucleate, hexanucleate cells and spindle abnormalities were observed and the frequency of such variations varied slightly, according to the culture medium. Anaphase bridges were frequently observed, accompanied with or without laggards. Dicentric chromosomes, telomeric adhesion, ring chromosome, chromatid deletion, and centromere breakage were observed in cells isolated from the callus (Fig. 6.2). Kim (1990) analyzed the genetic variations among callus cells of *A. wakegi* and observed low mitotic indices (1.25–2.12%) and several mitotic abnormalities. Kim also observed different frequencies of euploid and aneuploid cells among whom, diploid and tetraploid cells were most frequent. The structurally-rearranged chromosomes were also observed among the various ploids of callus cells.

Karyotypical analysis of callus of *A. sativum* was done by Kim and Seo (1991) at every 30 days' interval and the frequency of various ploidy levels observed is shown in Fig. 6.3. At 180 days, the frequency of diploid cells was the lowest. The frequency of hypodiploid cells increased with culture age. The hyperdiploid cells increased at first and then decreased from 150-day-old callus. The number of tetraploid cells increased till 180 days but decreased after that period. The various metaphase chromosome complements observed in callus cells are shown in Fig. 6.4.

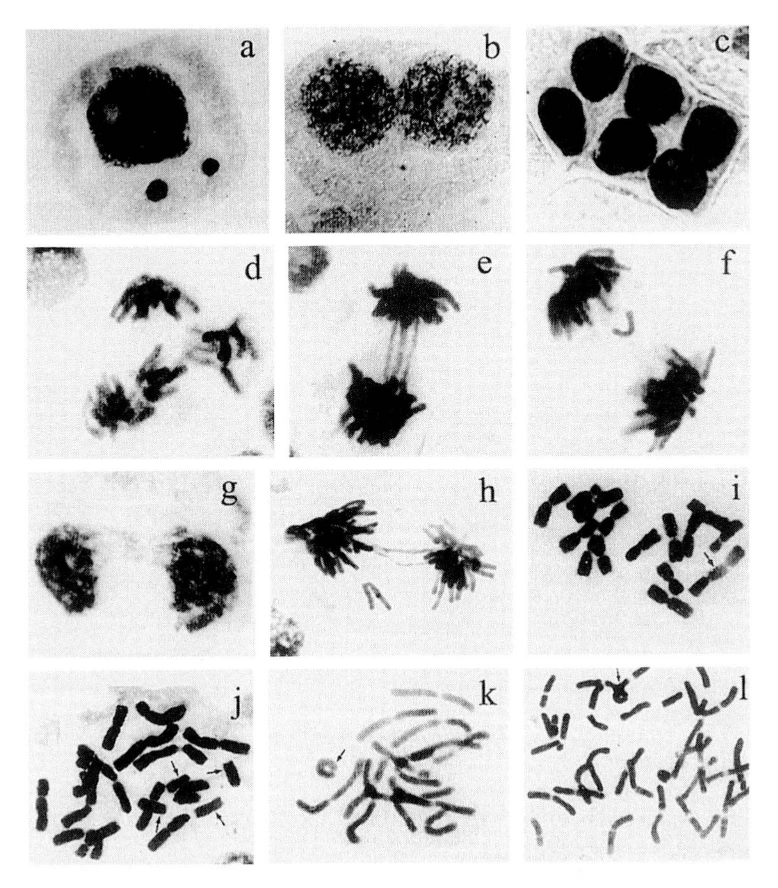

Fig. 6.2 Mitotic abnormalities observed in the callus cells. (a) Micronuclei, 2 in number. (b) Binucleate cell. (c) Hexanucleate cell. (d) Spindle abnormalities. (e) Anaphase bridge, 3 in number. (f) Laggard without bridge. (g) Chromatin bridge. (h) Laggard along with anaphase bridge. (i) Dicentric chromosome. (j) Centromere breakage. (k) Ring chromosome. (l) Telomeric adhesion.

Calli obtained from basal disc explants of *A. victorialis* var. *platyphyllum* were grown in three kinds of nutrient media (MS, BDS, and B5) and the frequencies of mitotic index and the chromosomal aberrations were analyzed (Seo et al., 1995). The mitotic index varied from 0.55% to 1.01% with respect to culture media and age. The mitotic irregularities like micro-, bi- and multi-nuclei, chromosome bridge and laggards were noted in each type of calli. The chromosome number variations observed in metaphase stage were identified as aneuploid and tetraploid. Structural

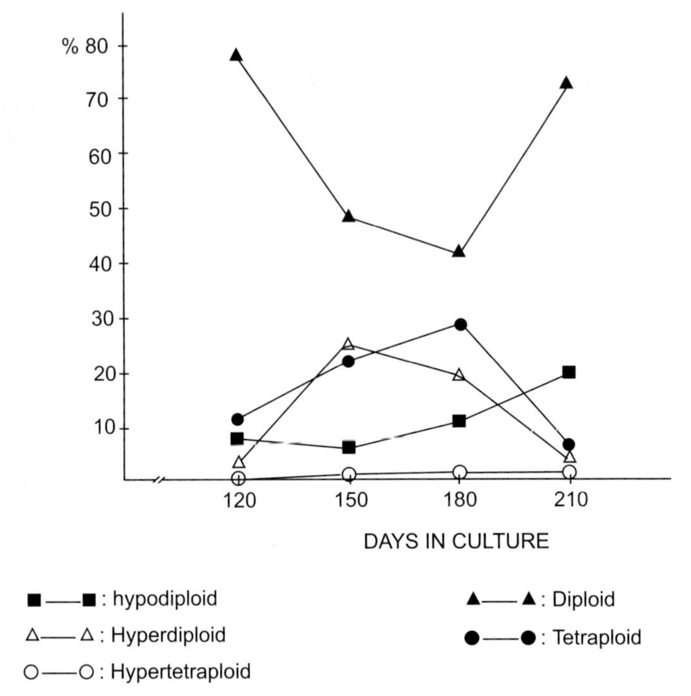

Fig. 6.3 Frequency of various ploidy levels in the callus cells of *A. sativum.*

variations such as dicentric chromosomes, centromere breakage and small chromosomes were observed.

An impressively heterogeneous cell populations were observed in the callus derived from various species of *Allium.* According to Roy (1980), the difference in the chromosomal behaviour of the callus was due to the genetic constitution of each species. The different ploidy level observed in the calli maintained in different medium might be because of the variation in the nutrient levels of the medium. Novak (1974) suggested the origin of polyploid state in callus to endomitosis or endoreduplication, directly stimulated by the conditions of the in vitro culture. The exogenous auxin and cytokinin in the culture medium can be supposed to stimulate the division of polyploid cells (Novak 1974). Chromosome variations were also observed due to the higher dose of nutrients like calcium chloride in potato and levels of EDTA in *Haploppapus* cells (Sree Ramalu, 1984). Abraham et al. (1992) reported cytological abonormalities induced by magnesium sulphate in callus cultures of *Vicia faba.* Baylis and Gould (1974) proposed the scheme for aneuploid generation as tetraploidization, followed by segregation and rearrangement, while Novak (1981) related the origin of aneuploids directly to the presence of micronuclei, chromosome bridge and laggard in callus cells. In *A. victorialis* var. *platyphyllum,*

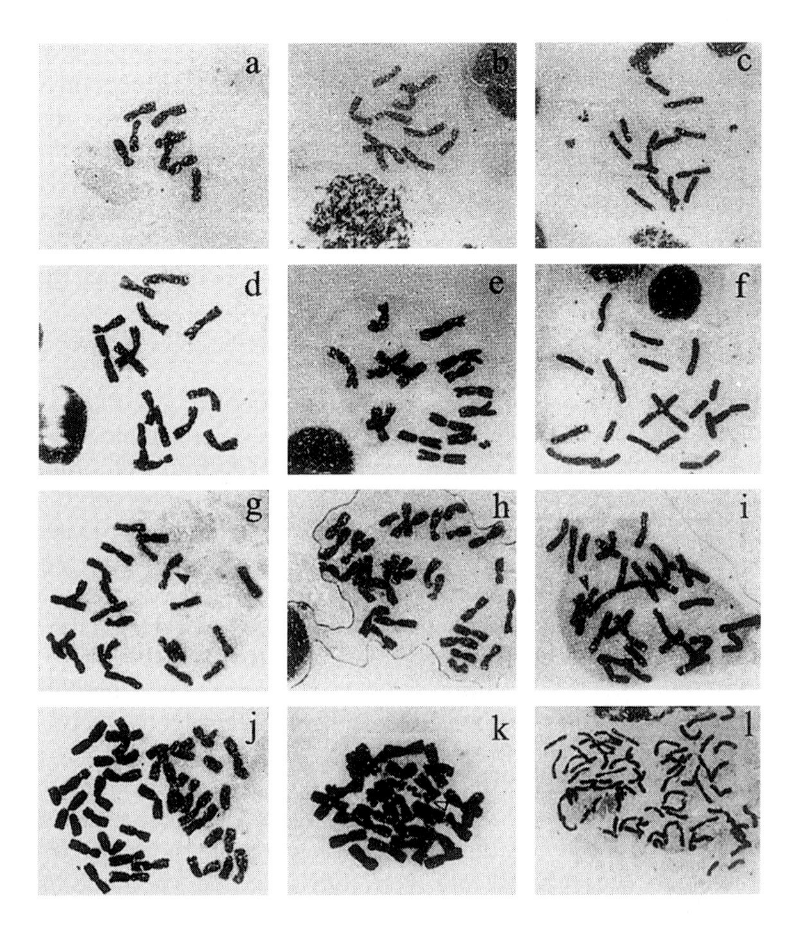

Fig. 6.4 The metaphase plates with various chromosome numbers observed in callus cells. (a) Haploid, 2n = 8. (b) Hypodiploid, 2n = 10. (c) Hypodiploid, 2n = 13. (d) Hypodiploid, 2n = 15. (e) Normal diploid, 2n = 16. (f) Hyperdiploid, 2n = 17. (g) Hyperdiploid, 2n = 24. (h) Hyperdiploid, 2n = 30. (i) Hyperdiploid, 2n = 31. (j) Tetraploid, 2n = 32. (k) Hypertetraploid, 2n = 35. (l) Octoploid, 2n = 64.

the chromosome number variations among callus cells were correlated with the mitotic abnormalities (Seo et al., 1995).

PLANT REGENERATION

Organogenesis of callus leading to regeneration of complete plants is highly desirable to exploit the benefit of somaclonal variation. In our laboratory, successful regeneration of plants from callus was achieved for different species of *Allium*.

Kim (1990) observed that when callus of *A. wakegi* was transferred to a regeneration medium, only limited karyotypes tolerated the process. Amphidiploid cells as well as other cells with varied chromosome composition regenerated into plants. Many shoots were produced about 2–3 weeks after transferring callus and roots appeared a few days after the shoots were transplanted to a rooting medium. When transplanted to the soil, 200 plants survived, which produced 1314 bulbs after the first growth term. Many of the regenerants were aneuploid (2n = 26, 27, 28 or 30) but except for one aneuploid (2n = 28), the others did not survive in soil.

Callus of *A. senescens* var. *minor* was subcultured in MS, BDS, and B5 medium supplemented with BA (4.4 μM) alone or in combination with 2,4-D (4.5 μM) or NAA (5.4 μM) and kept under continuous illumination for shoot regeneration (Nair 1993). Maximum shoot regeneration was obtained from the callus maintained in MS medium with 2,4-D and BA. The shoots regenerated from the callus maintained in BDS medium supplemented with 2,4-D and BA showed minimum survival rate when transferred to the soil. Only a few shoots were regenerated from the callus maintained in B5 medium supplemented with BA. The shoots obtained from different media were transferred to respective basal medium for rooting. The rooted plantlets were transferred to vermiculite, kept under high humidity for one month and transferred to soil. After one year, out of 840 plants, 646 survived in the soil.

Seo et al. (1996) observed that the period needed for plant regeneration of *A. victorialis* var. *platyphyllum* was relatively long as compared to other *Allium* species. The induced calli were inoculated on MS medium with different hormone combinations. After 5 to 6 weeks, the basal portion of the callus formed primordia, which developed into green healthy shoot buds under continuous illumination and roots appeared within two weeks of transfer to basal medium without hormones (Fig. 6.5). The mode of regeneration from callus depended on the hormone used (Table 6.1). The roots developed directly from the callus in MS basal medium supplemented with kinetin alone.

GENETIC VARIATION IN REGENERATED PLANTS

Evaluation of callus-derived regenerants is important to recover the tissue culture induced variants in regenerated plants. The karyotype stability pattern of regenerated mixoploid plants should be investigated in order to utilize the potential variants for crop improvement. Various techniques may be employed to study the genetic make-up of these plants.

Fig. 6.5 Regeneration of plants via callus from basal bulb segments of *A. victorialis*. var platyphyllum (a) Callus formed on BDS medium supplemented with 2 mg/l 2,4-D and 1 mg/l BA. (b) Shoot primordia formation after 240 days. (c) Direct root regeneration from callus. (d) Regenerated plantlets.

Cytological Analysis

Seo et al. (1996) studied the karyotypic variation from the root-tips of callus-derived regenerated *A. victorialis* var. *platyphyllum* plants. The karyotypic variants were classified into diploid, hypodiploid, mixoploid and hyperdiploid, according to the chromosome number. The highest frequency of chromosomal aberration among regenerants was aneuploid. The hypodiploid having one arm deleted chromosome following

Table 6.1 Response of *A. victorialis* var. *platyphyllum* callus to shoot regeneration in MS media supplemented with growth hormones

Plant growth regulators (mg/l)					*% of shoot regenerated*
Auxin	*Cytokinin*				
NAA	*BA*	*zeatin*	*kinetin*	*2ip*	
0.2	2				45.7 (16/35)
0.2		2			76.7 (23/30)
0.2			2		–
0.2				2	–
	2				22.7 (5/22)
		2			35.3 (6/17)
			2		a
				2	–

Numbers in parentheses indicate total regenerated shoots/number of callus.
a: root regeneration only

centromere breakage were also observed. Although translocation or inversion could not be identified by orcein-staining method, dicentric chromosome, centromere breakage and chromosome fragment were frequently observed in the regenerants.

The karyotypic stability pattern of regenerated mixoploid plantlets of *A. sativum* was investigated by Kim and Seo (1991). In their study, the mixoploid plants were stabilized into diploid or tetraploid condition after a 2- or 3-month period, depending on the frequency of diploid or tetraploid cells present in the plantlet. They presumed that only a part of various chromosomal aberrations in callus may be shown in the regenerated plants. The cell division may be in an abnormal manner during the early stages of growth and differentiation, but tend to be stabilized either as diploid or tetraploid as they grow. After transferring to the soil, chromosomal pattern did not exhibit any changes.

C-banding Analysis

C-banding techniques for plant chromosomes are sensitive tools for karyotyping and genome identification. Chromosome breakage, a frequent mutational event in tissue culture system and other physical events were identified by the distribution patterns of constitutive heterochromatin in plant chromosomes (Ashmore and Gould 1981; Bebeli et al., 1990; Karp et al., 1992) and C-bands are usually equated with heterochromatin distribution in the chromosomes (Sumner 1990). C-banding patterns representing heterochromatin have been widely used to identify homologous chromosomes (Kalkaman 1984). C-banding techniques were utilized in our laboratory to analyze the genetic variants regenerated from tissue cultures of *Allium* species.

For C-band formation, the root tips were pretreated with 0.002 M 8-hydroxyquinoline for 6 hours, macerated in 45% acetic acid for 2–3 hours at room temperature, hydrolyzed for 15 sec. in 1 N-HCl at 60°C, and squashed in 45% acetic acid. After being air dried overnight, the coverglasses were detached by immersing in 10% acetic acid bottom side up and dried for one day. The dried slides were immersed in 8% (w/v) aqueous solution of barium hydroxide for 7 min. at room temperature, rinsed in distilled water, incubated in 2XSSC at 60°C for 30 min., and washed twice with distilled water. The slides were stained in a Giemsa solution for 30 min. at room temperature, rinsed in distilled water, dried in desiccators, and mounted with Canada balsam through the process using dehydration by butanol.

Giemsa C-banded karyotypes were analysed in 200 regenerated plants of *A. wakegi* (Kim et al., 1990b). About 92.5% of the regenerated plants were amphihaploid, and the C-banding patterns of these plants were almost the same as that of the common species. Five per cent of the plants were tetraploid. The karyotypic rearrangements judged from C-banding pattern were more frequently observed in amphidiploid regenerants rather than in amphihaploid ones. The C-banded metaphases and their idiograms for the five amphidiploid plants and one hypoamphidiploid one with 2n = 28 are shown in Figs. 6.6 and 6.7. The regenerant AW 2B-01 showed exactly doubled C-banded karyotype of amphihaploid cultivar. The karyotype of AW 02-18 differed from that of AW 2B-01in that a part of non-banded chromosomal arm was added to one of the long arm of chromosome o. The variant AW 02-22 revealed considerably large satellites in both chromosome f but no satellite was found in chromosome n. Terminal bands were deleted in variant AW 09-15 from 3 chromosomes, i.e. e, h, and k. Double thick bands appeared in telomeric part of the long arm in one of the chromosome l, in variant AW 10-12. The variant AW 13-09 was hypoamphidiploid, 2n = 28, and in C-banding pattern, one chromosome of each pairs c, d, f, and k was lost, and one satellite in chromosome n pair was deleted (Figs. 6.6, 6.7).

Giemsa C-banding analysis of in vitro regenerated plants in *A. senescens* var. *minor* revealed a duplication of chromosome 3 and origin of a new karyotype (Nair and Seo 1995). Somatic connections between the new satellite chromosome and a third chromosome suggested a hypothesis for the origin of the new karyotype. This new genome had adaptive advantages under in vitro condition, resulting in an increased rate of regeneration from the callus. The culture medium from which the new karyotype evolved contained BA. As BA is a derivative of purine, there is a possibility of its integration into DNA molecule (Gould 1986).

Lee and Seo (1997) showed that all chromosomes of *A. wakegi* cultivar and AW 02-18 regenerant had clearly visible C-bands. The karyotype of *A. wakegi* cultivar consisted of two subtelocentric chromosomes with satellite

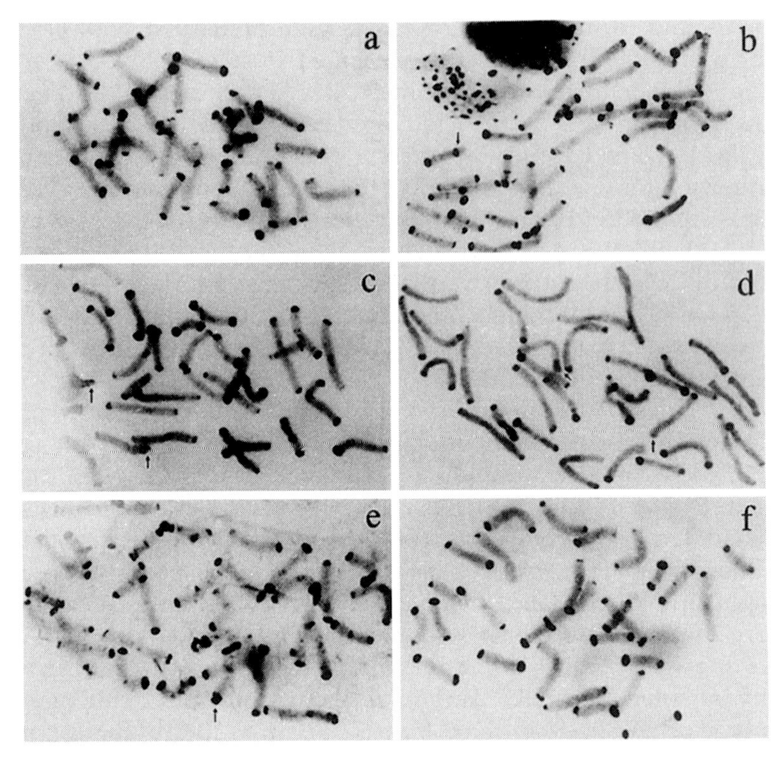

Fig. 6.6 C-banded metaphase plates of amphidiploid variants regenerated from tissue culture. (a) AW 2B-01. (b) AW 02-18. (c) AW 02-22. (d) AW 09-15. (e) AW 10-12 and a hypoamphidiploid (f) AW 13-09. Arrows indicate the rearranged chromosomes.

and fourteen metacentric chromosomes. Strongly stained bands were observed in the telomeric part of both short and long arm of all chromosomes, while weakly-stained bands were observed in centromeric regions. Chromosome 14 had a considerably large satellite, while chromosome 6 showed barely detectable satellite in the short arm.

Isozyme and Total Protein Analysis

Isozymes, direct gene products, have several advantages in studying the variations in a single gene level because the alleles at most isozyme loci are usually co-dominant and rarely exhibit epistatic interactions (Allicchio et al., 1987). Isozyme analysis was carried out in 12 variants regenerated from the tissue culture of *A. wakegi,* utilizing 5 different enzymes (Kim 1990). The isozyme patterns showed considerable variations between variants (Fig. 6.8a,b,c,d,e). ADH (alcohol dehydrogenase) showed only one band with slight migration differences between

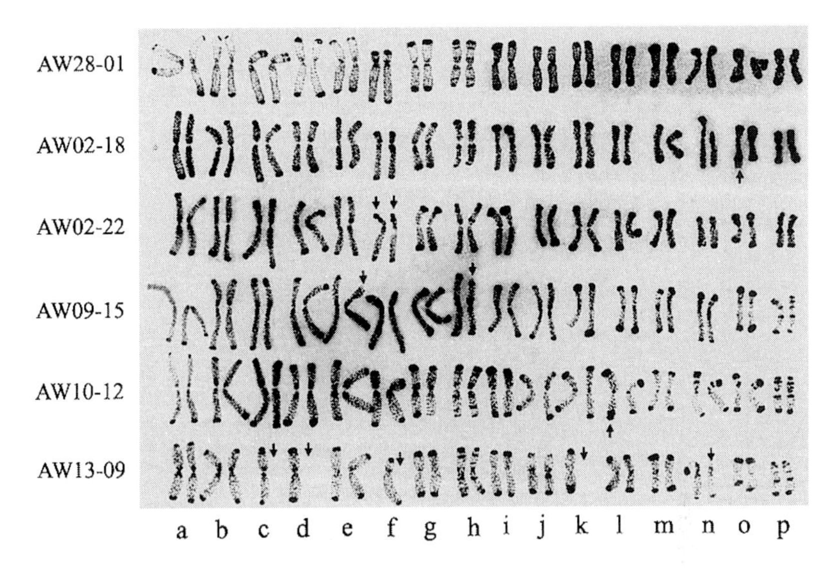

Fig. 6.7 Digrammatic representation of C-banded karyotypes of amphidiploid and hypoamphidiploid plants. Arrows indicate the rearranged or lost chromosomes.

variants, but intensities of the band—an indicative of enzyme activity—were quite different between the variants (Fig. 6.8a). The IDH (isocitrate dehydrogenase) patterns differed in migration among variants 2, 3, 4, 5, and 12 (Fig. 6.8b). Three zones of activity were recorded in MDH (malate dehydrogenase). The variant AW 10-28 (lane 2) gave no band in zone 1, but other variants showed variations in the intensity and mobility of bands (Fig. 6.8c). Similarly, PGM (phosphoglucomutase) isozyme patterns showed significant variation in band intensity and mobility (Fig. 6.8d). PGI (phosphogluco-isomerase) showed two zones of activity. The bands in zone 1 showed little difference in both intensity, but bands in zone 2 displayed considerable differences in both intensity and mobility. AW 13-02 (lane 5) and AW 09-15 (lane 10) showed a faint band in zone 2 (Fig. 6.8e). The results implied that a significant change in gene structure and gene expression had occurred during in vitro culture and regeneration of *A. wakegi*. The isozyme variations observed in regenerants might have resulted from DNA rearrangement, responding to a shock caused by tissue culture (Allicchio et al., 1987). The change in isozyme patterns observed seemed to be reasonable, since the variants had also exhibited considerable chromosomal variations. Kim (1990) further used 10 of these variants for the analysis of total protein patterns by native PAGE and SDS-PAGE. No significant difference was observed in the total protein patterns but the SDS-PAGE patterns showed a little difference between variants.

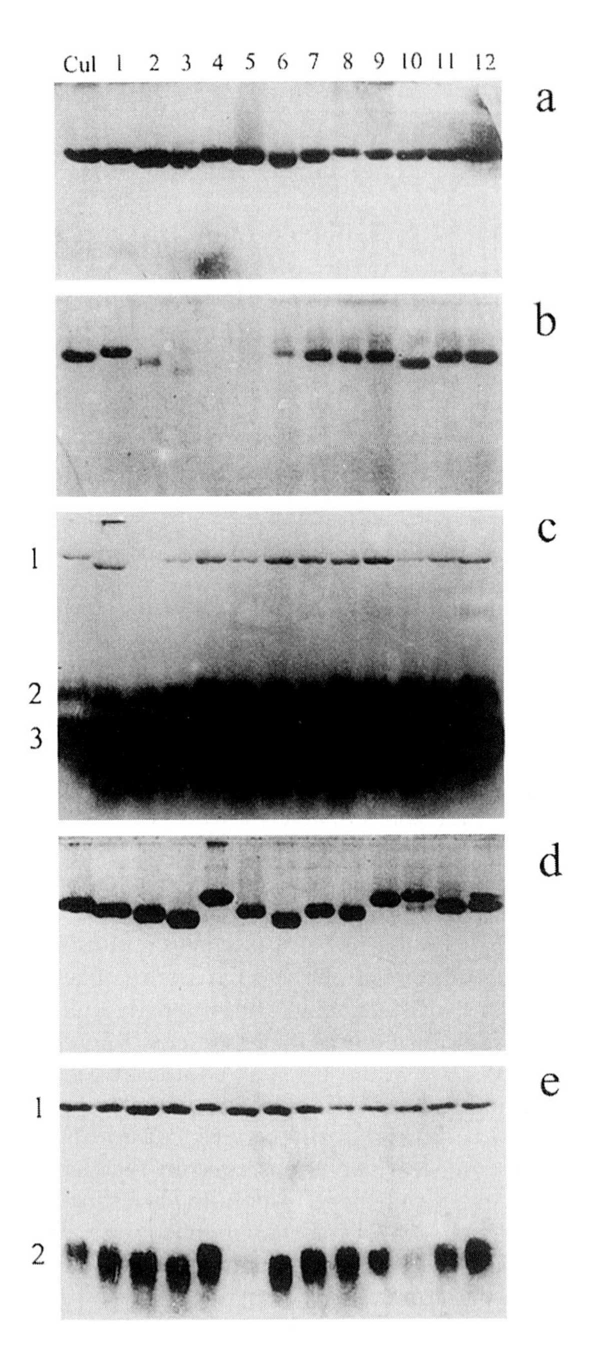

Fig. 6.8 The five isozymic patterns of the twelve regenerated cultivars. (a) ADH. (b) IDH. (c) MDH. (d) PGM. (e) PGI.

Morphological Analysis

Morphological analysis was carried out in selected regenerated amphidiploid plants of *A. wakegi* by Kim (1990). Amphidiploid plants differed in growth, size of leaf and bulb, propagation rate and other cytotaxonomic factors from the normal cultivar. However, some of the amphidiploid variants showed poor adaptability on cultivation. During an investigation on the growth patterns, AW 02-18 appeared to have distinct merit in growth, plant morphology, flowering, formation of pollen grain and meiotic configuration, while other variants showed some defective characters. The bulb size of AW 02-18 was almost twice the common cultivar, but the hypoamphidiploid exhibited deficient bulb size (Fig. 6.9). Although this variant showed complete sterility and lack of seed setting upon selfing, the possibility to restore fertility when crossed with other species is expected, considering the normal meiosis and normal pollen grain formation. Thus, this variant holds potentiality for improvement of this crop.

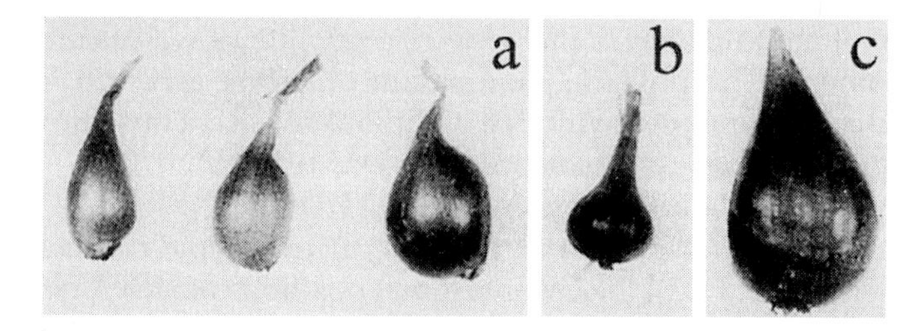

Fig. 6.9 Comparison of average size of bulb among various ploidy level. (a) Amphihaploid. (b) Hypoamphidiploid. (c) Amphidiploid.

Molecular Analysis by FISH Technique

Although C-banding has facilitated the identification of separate chromosomes in plants, the interpretation of a karyotype is often difficult when only a few terminal C-banded chromosomes are observed. Examination of the chromosomal distribution of 5S and 18S-26S ribosomal genes that are useful in identifying the types of genomic changes that might occur during in vitro culture (Maluszynska and Heslop-Harrison 1993). In plants, 5S and 18S-26S ribosomal RNA genes are present in many hundreds of tandemly-repeated units at one or more pairs of loci within the genome. The use of fluorescence in situ hybridization (FISH) technique allows the visualization of these gene families or repeated sequences in plant chromosomes. Investigations were done in our laboratory to analyse the chromosomal location of 5S and 18S-26S rRNA genes by FISH technique in callus derived regenerated variants of *Allium* species. The FISH protocols were carried out using the method of Mukai et al., (1993).

An accurate physical map showing the localization of 5S and 18S-26S rRNA genes had been constructed (Lee and Seo 1997) by bi-colour FISH in *A. wakegi* cultivar and its regenerant (AW 02-18). A rhodamine labelled 5S rDNA and a biotin-labelled 18S-26S rDNA were used as probes. The signals of 5S rDNA were detected on the intercalary region of short arm in chromosome 9 (one region) and 15 (two regions). The signals of 18S-26S rDNA were detected on the terminal region of short arm of chromosome 10 and at same regions of chromosome 6 and 14 including, the satellite. ISH allowed the identification of homologous chromosomes of chromosome 6, 9, 10, 14, and 15 in AW 02–18 regenerant (Lee and Seo 1997).

Chromosomal location of 5S and 18S-26S rRNA genes were analyzed in callus-derived plants of *A. cyaneum* (Lee et al., 1998). Using digoxigenin-labelled 5S rRNA and biotin-labelled 18S-26S rRNA gene probes, they compared the FISH patterns of regenerated autotetraploid plants with the diploid wild type. The 5S rRNA gene sites were detected on the interstitial region of both arms of chromosome 7 and on short arm of chromosome 4. The 18S-26S rRNA gene sites were detected on the terminal region of the short arm of chromosome 5, including the satellite as well as on a part of chromosome B. The physical localization of rRNA genes in tetraploid regenerants corresponded with that of diploid species. Thus, the results of FISH suggested that tetraploid regenerants originated from the exact doubling of normal diploids.

In a karyotype analysis of At30 (2n = 30), an aneuploid regenerant of *A. tuberosum* (2n = 4x = 32), the chromosomal positions of rRNA genes were physically mapped and compared with normal (Do et al., 1999). Both normal *A. tuberosum* and At30 exhibited two sets of 5S rDNA sites, one on the proximal position of the short arm of chromosome 3, and the other on the intercalary region on the long arm of chromosome 6. There was one 18S-5.8S-26S rDNA site in the secondary constriction, including flanking short chromosomal segments of the satellite and terminal regions on the short arm of chromosome 8 in wild type *A. tuberosum.* However, At30 showed only 3 labelled chromosome 8, indicating that this was one of the lost chromosomes of At30. Do et al. (2000) further conducted karyotype analysis by FISH to identify the lost chromosomes in three aneuploid somaclonal variants of *A. tuberosum*. Chromosome compositions of these variants were confirmed as being fixed lines during an experiment conducted over two years of greenhouse cultivation. The 5S rRNA gene signals in all the variants as well as the wild type were detected as two sets, one in the intercalary region of the short arm of chromosome 3, and the other in the long arm of chromosome 6. One 18S-5.8S-26S rRNA gene site was located on the terminal region of the short arm of chromosome 8, including the secondary constriction and satellite (Fig. 6.10). The three lost

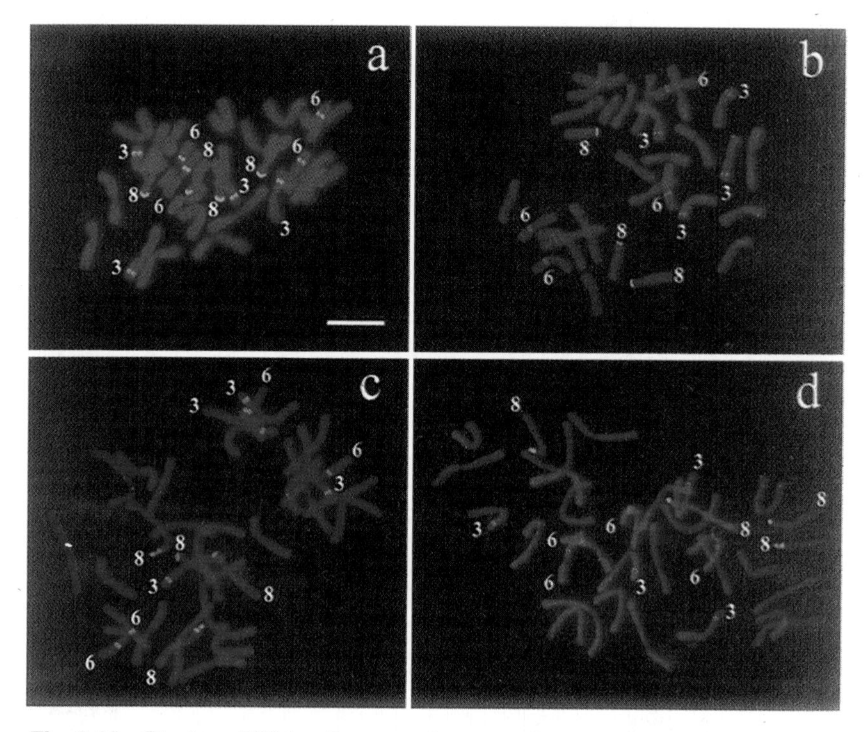

Fig. 6.10 Bicolour FISH patterns on the metaphases of (a) At29. (b) At30. (c) At31. d. Wild type in *A. tuberosum* using both 5S and 18S-5.8S-26S rDNA probes (bar = 10 mm). Digoxigenin-labelled 5S rDNA probe detected with antidigoxigenin-rhodamine conjugate (red) and biotin-labeled 18S-5.8S-26S rDNA probe detected with avidin-FITC conjugate (green). Numbers indicate corresponding chromosomes showing rDNA hybridization sites.

chromosomes of At29 variant (2n = 29) were all chromosome 2, the two for At30 (2n = 30) were chromosomes 7 and 8, and At31 (2n = 31) was missing one of chromosome 2 (Fig. 6.11).

CONCLUSION

Somaclonal variations, a novel source of variability, have found widespread application in plant breeding programmes. The techniques are relatively simple as compared to those of recombinant DNA since plants can be directly transferred to the fields and evaluated.

It has been observed that *Allium* callus is easily established from vegetative as well as reproductive parts and variation exists among

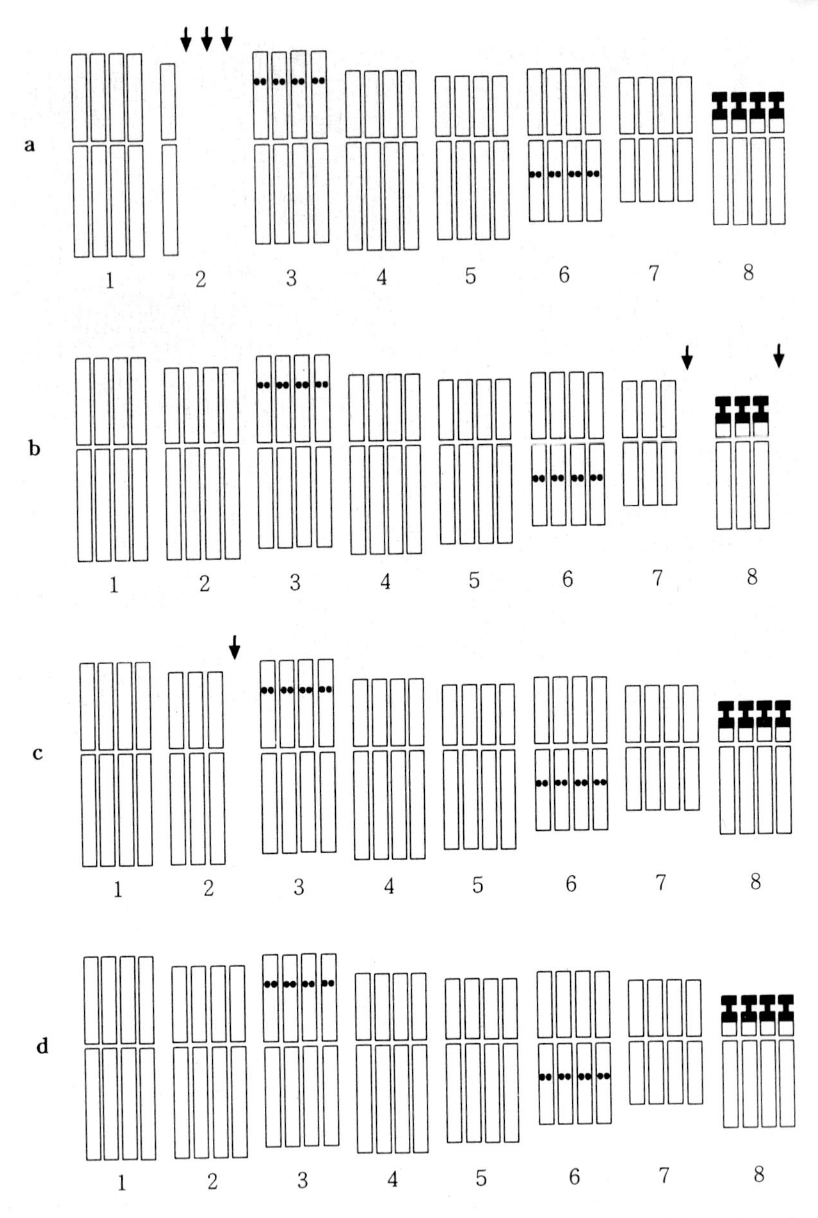

Fig. 6.11 Idiogram showing chromosomal localization of 5S and 18S-5.8S-26S rDNAs in somaclones, (a) At29, (b) At30, (c) At31, and (d) Wild type of *A. tuberosum* observed by FISH. Circles indicate 5S rDNA and rectangles indicate 18S-5.8S-26S rDNA loci. The arrows indicate the lost chromosomes that are res-ponsible for the abnormal karyotype.

regenerants. These variants would be important for *Allium* improvement, especially to improve the quality, productivity and tolerance to diseases. However, techniques are required for proper identification of the desirable somaclones. As *Allium* is a convenient material for cytogenetic studies, cytological and C-banding techniques could be easily employed to analyse the karyotypes in callus cells and regenerated plants. FISH has been found to be an efficient method for identifying the chromosome composition of regenerants by demonstrating the number and physical location of 5S and 18S-26S rRNA gene loci.

In the present study, some promising autotetraploids and aneuploids regenerated as variants have been identified through chromosomal analysis. The results of these investigations indicate the potential for genetic manipulation of cultivated *Allium* species using tissue culture.

REFERENCES

Abraham, S., Nair, A.S. and Nair, R.B., 1992, Cytological abnormalities induced by magnesium sulphate in callus cultures of *Vicia faba*. Cytologia **57**: 373–375.

Allicchio, R., Antonioli, C., Graziani, L., Rencarati, R. and Vannini, C., 1887, Isozyme variation in leaf callus regenerated plants of *Solanum tuberosum*. Plant Sci. **53**: 81–86.

Appels, R., Gerlach, W.L., Dennis, E.S., Swift, H. and Peacock W.J., 1988, Molecular and chromosomal organization of DNA sequences coding for the ribosomal RNAs in cereals. Chromosoma **78**: 293–311.

Ashmore, S.E. and Gould, A.R., 1981, Karyotypic evolution in a tumor derived plant tissue culture analysed by Giemsa C-banding. Protoplasma **106**: 297–308.

Bayliss, M.W. and Gould, A.R., 1974, Studies on the growth in culture of plant cells. XVIII. Nuclear cytology of *Acer pseudoplatanus* suspension cultures. J. Exp. Bot. **25**: 772–783.

Bebeli, P., Karp, A., and Kaltskies, P.J., 1990, Somaclonal variation from cultured immature embryos of sister lines of rye differing in heterochromatin content. Genome **33**: 177–183.

Breiman, A. Roten-Abarbanell, D., Karp, A., and Shaskin, H., 1987, Heritable somaclonal variation in wild barley (*Hordeum spontaneum*). Theor. Appl. Genet. **74**: 104–112.

Castilho, A. and Heslop-Harrison, J.S., 1995, Physical mapping of 5S and 18S-26S rDNA and repetitive DNA sequences in *Aegilops umbellulata*. Genome **38**: 91–96.

Cecchini, E., Natali, L., Cavallini, A. and Durante, M., 1992, DNA variation in regenerated plants of pea (*Pisum sativum*). Theor. Appl. Genet. **84**: 874–879.

Dahlgren, R.M.T., Clifford, H.T. and Yeo, P.F., 1985, The families of monocotyledon. Springer-Verlag, Berlin, 193. pp.

Do, G.S., Seo, B.B., Ko, J.M., Lee, S.H., Pak, J.H., Kim, I.S. and Song, S.D., 1999, Analysis of somaclonal variation through tissue culture and chromosomal localization of rDNA sites by fluorescent in situ hybridization in wild *Allium tuberosum* and a regenerated variant. Plant Cell, Tissue Organ Culture **57**: 113–119.

Do, G.S., Seo, B.B., Pak, J.H., Kim, I.S. and Song, S.D., 2000, Karyotypes of three somaclonal variants and wild plants of *Allium tuberosum* by bicolor FISH. J. Plant Biol. **43**: 143–148.

Dunstan, D.I. and Short, K.C., 1977, Improved growth of tissue cultures of onion (*Allium cepa*). Physiol. Plant. **41**: 70–72.

Evans, D.A., and Sharp, W.R., 1986, Somaclonal and gametoclonal variation. In: D.A. Evans et al. (Eds) Handbook of Plant Cell Culture. Vol. 4. MacMillan Publishing Company, New York, pp. 97–132.

Gamborg, O.I., Millar, R.A. and Ojima, K., 1968, Nutrient requirements of suspension cultures of soybean root cells. Exp. Cell Res. **50**: 151–158.

Gould, A.R. 1986, Factors controlling generation of variability in vitro. In: I.K. Vasil (Ed.) Cell Culture and Somatic Cell Genetics. Vol. 3. Academic Press, Inc. Orlando, Florida., pp. 549–564.

Iwasa, S., 1964, Cytogenetic studies in the Wakegi, *Allium fistulosum* var. *caespitosum*. J. Fac. Agric., Kyushu Univ. **13**: 165–177.

Jiang, J. and Gill, B.S., 1994, New 18S-26S ribosomal RNA gene loci: chromosomal landmarks for the evolution of polyploid wheats. Chromosoma **103**: 179–185.

Johnson, S.S., Phillips, R.L. and Rines, H.W., 1987, Possible role of heterochromatin in chromosome breakage induced by tissue culture in oats (*Avena sativa* L.). Genome **29**: 439–446.

Kalkaman, E.R., 1984, Analysis of the C-banded karyotypes of *Allium*. Genetika **65**: 141–148.

Kao, K.N. and Michayluk, M.R., 1975, Nutrient requirement for growth of *Vicia hajastana* cells and protoplasts at a very low population density in liquid media. Planta **126**: 105–110.

Karp, A., Owen, P.G., Steele, S.H., Bebeli, P.J. and Kaltskies, P.J. 1992, Variation in the telomeric heterochromatin in somaclones of rye. Genome **35**: 590–593.

Kim, H.H. 1990. Plant regeneration and genetic variations from tissue culture of *Allium wakegi*. Ph.D. thesis. Kyungpook National University, Taegu, Korea.

Kim, Y.K. and Seo, B.B., 1991, Karyotypic variation in callus and regenerated plant of *Allium sativum*. Korean J. Genetics **13**: 147–155.

Kim, H.H., Kim, J.W. and Seo, B.B., 1990a, C-banded karyotypes of *Allium victorialis, A. senescens* var. *minor*. Korean J. Genetics **12**: 55–61.

Kim, H.H., Seo, B.B. and Lee, J.Y., 1990b, Chromosomal study on the genetic variants regenerated from tissue cultures of *Allium wakegi*. Korean J. Genetics **12**: 104–112.

Larkin, P.J. and Scowcroft, W.R. 1981. Somaclonal variation—Novel source of variability from cell cultures for plant improvement. Theor. Appl. Genet. **60**: 197–214.

Lee, M. and Phillips, R.L., 1988, The chromosomal basis of somaclonal variation. Ann. Rev. Plant Physiol. and Plant Mol. Biol. **39**: 413–437.

Lee, S.H. and Seo, B.B. 1997. Chromosomal localization of 5S and 18-26S rRNA genes using fluorescence in situ hybridization in *Allium wakegi*. Korean J. Genetics **19**: 19–26.

Lee, S.H., Ryu, J.A., Do, G.S., Seo, B.B., Pak, J.H., Kim, I.S. and Song, S.D., 1998, Chromosome analysis by fluorescence in situ hybridization of callus derived regenerants in *Allium cyaneum* R. Plant Cell Rep. **18**: 209–213.

Linares, C., Gonzales, J., Ferrer, E. and Forminaya, A., 1996, The use of double fluorescence in situ hybridization to physically map the positions of 5S rDNA genes in relation to the chromosomal location of 18S-26S rDNA and a C genome specific DNA sequence in the genus *Avena*. Genome **39**: 535–542.

Maluszynska, J. and Heslop-Harrison, J.S., 1993, Localization of tandemly-repeated DNA sequences in *Arabidopsis thaliana*. Plant J. **1**: 159–166.

Mukai, Y., Endo, T.R. and Gill, B.S., 1990, Physical mapping of the 5S rRNA multigene family in common wheat. J. Hered. **81**: 290–295.

Mukai, Y., Nakahara, Y. and Yamamoto, M. 1993. Simultaneous discrimination of the three genomes in hexaploid wheat by multicolor fluorescence in situ hybridization using total genomic and highly repeated DNA probes. Genome **35**: 489–495.

Muller, E.P., Brown, T.H., Hartke, S. and Lorz, H., 1990, DNA variation in tissue-culture derived rice plants. Theor. Appl. Genet. **80**: 673–679.

Murashige, T. and Skoog, F., 1962, A revised medium for rapid growth and bioassays with tobacco tissue cultures. Physiol. Plant. **15**: 473–497.

Novak, F.J. 1974. The changes of karyotype in callus cultures of *Allium sativum* L. Cytologia **27**: 45–54.

Novak, F.J., 1981, Chromosomal characteristics of long-term callus cultures of *Allium sativum* L. Cytologia **46**: 371–379.

Novak, F.J., Havel L. and Dolezel J., 1986, *Allium.* In: D.A. Evans et al. (Eds.) Handbook of Plant Cell Culture. Vol. 4. MacMillan Publishing Company, New York, pp. 419–456.

Nair, A.S., 1993, The chromosomal origin of somaclonal variation in *Allium senescens* var. *minor.* Ph.D. thesis. Kyungpook National University, Taegu, Korea.

Nair, A.S. and Seo, B.B., 1992, Callus and plantlet regeneration from bulb explants in *Allium senescens* var. *minor.* Korean J. Plant Tissue Culture. **19**: 89–92.

Nair, A.S. and Seo, B.B., 1995, Hormonal effects on chromosomal variability among the regenerated plants in *Allium senescens* var. *minor*. Indian J. Exp. Biol. **33**: 533–536.

Reddy, P. and Appels, R., 1989, A second locus for the 5S rRNA multigene family in *Secale* L.: sequence divergence in two lineages of the family. Genome **32**: 465–467.

Ried, T., Baldini, A., Rand, T.C. and Ward, D.C., 1992, Simultaneous visualization of seven different DNA probes by in situ hybridization using combinational fluorescence and digital imaging microscopy. Proc. Natl. Acad. Sci. USA **89**: 1388–1392.

Roy, S.C., 1980, Chromosomal variation in the callus tissues *Allium tuberosum* and *A. cepa*. Protoplasma **102**: 171–176.

Seo, B.B. 1977, Cytogenetic studies of some tetraploids in *Allium*. Korean J. Bot. **20**: 71–76.

Seo, B.B. and Kim, H.H. 1975, Karyotypic analysis based on heterochromatin distribution in *Allium fistulosum* and *A. ascalonicum*. Korean J. Bot. **18**: 92–100.

Seo, B.B., Kim, H.H. and Kim, J.H., 1989, Giemsa C-banded karyotypes and their relationship of four diploid taxa in *Allium*. Korean J. Bot. **32**: 173–180.

Seo, B.B., Lee, S.H. and Nair, A.S., 1994, Callus formation from protoplast cultures of *Allium wakegi*. Bionature **14**: 23–28.

Seo, B.B., Lee, S.H., Pak, J.H., Kim, I.S. and Song, S.D., 1995, Karyological analysis of somaclonal variation in callus cells of *Allium victorialis* var. *platyphyllum*. J. Plant Biol. **38**: 321–328.

Seo, B.B., Lee, S.H., Pak, J.H., Kim, I.S. and Song, S.D., 1996, Karyological variation of callus-derived regenerants in *Allium victorialis* var. *platyphyllum*. J. Plant Biol. **39**: 231–235.

Song, Y.C. and Gustafson, J.P., 1993, Physical mapping of the 5S DNA complex in rice (*Oryza sativa*). Genome **36**: 658–661.

Sree Ramalu, K., 1984, Case histories of genetic variability in vitro. Potato. In: I.K. Vasil (Ed.). Cell Culture and Somatic Cell Genetics. Vol. 3. Academic Press, London. pp. 449–469.

Sumner, A.T., 1990, Chromosome banding, MRc Human genetics Unit, Edinburg, London. pp. 39–68.

Tashiro, Y., 1984, Genome analysis of *Allium wakegi*. J. Japan Soc. Hort. Sci. **52**: 399–407.

7

Regeneration and Characterization of Citrus Somatic Hybrids Derived from Asymmetric Somatic Hybridization

Jihong Liu* **and Xiuxin Deng**

National Key Laboratory of Crop Genetic Improvement, Huazhong Agricultural University, Wuhan 430070, People's Republic of China

ABSTRACT

Irradiation for 1 h or 1.5 h at 80 kV and 5 mA, especially the latter, inhibited the growth of callus of Dancy tangerine (*Citrus reticulata*). The protoplasts of Dancy, when irradiated for 1.5 h, did not divide but plasmolysed during the course of culture. UV, with the intensity of 56 μW/mm^2, decreased the viability of the protoplasts with a corresponding increase in irradiation time. Protoplasts of Page tangelo (*C. reticulata* × *C. paradisi*) were found to be more sensitive to UV than those of Murcott tangor (*C. reticulata* × *C. sinensis*). The protoplasts, when irradiated for 1 min and 5 min., did not divide. RAPD analysis revealed the difference in band pattern between the calli arising from the protoplasts of *Swinglea glutinosa* that were irradiated for 6 min. and the control.

X-ray irradiated embryogenic protoplasts of *Microcitrus papuana* Swing., Valacenia sweet orange (*C. sinensis* Osb.), Dancy tangerine and mesophyll protoplasts of citrange (*Poncirus trifoliata* × *C. sinensis*) were electrically fused with iodoacetic acid (IA)-treated embryogenic protoplasts of Newhall navel orange, Murcott tangor and Page tangelo, respectively. Callus was obtained for the first combination and embryoids for the last one, whereas plants were obtained for the other two combinations. A cytological examination of the regenerants showed that mainly diploid and aneuploid cells, together a with few tetraploid cells, were observed, indicating they were mixoploids. For example, chromosome counting of 4 cell lines derived from the fusion between *M. papuana* and Newhall showed that each cell line consisted of aneuploid (45.10%, 38.98%, 32.69% and 34.85%, respectively) and diploid cells (52.94%, 59.33%, 63.46% and 62.12%, respectively), whereas only a few tetraploid cells (1.96%, 1.69%, 3.85% and 3.03%, respectively) were observed. Analyses of RAPD (random amplified polymorphism DNA) were used to identify the hybridity of the aforementioned regenerants. Four, three, seven and four 10-mer arbitrary primers have been utilized for the above four combinations, respectively. Results of RAPD analysis confirmed that all of the calli, embryoids and plants derived from the donor-recipient fusion were somatic hybrids. The negative effects of ionizing irradiation on regeneration of embryoids and plantlets and possible agronomic interest of the mixoploid plants have been discussed in this chapter.

*Corresponding author: E-mail: Jhliu26@public.wh.hb.cn or liujihong@mail.hzau.edu.cn

INTRODUCTION

Citrus cultivars and their relatives are a vast but comparatively untapped reservoir, possessing resistance to both biotic and abiotic stresses. Genetic recombination between citrus and its relatives by sexual hybridization is impeded by sexual incompatibility, polyembryony, female and male sterility and high heterozygosity observed in many cultivars (Cameron and Frost 1968). Protoplast fusion can overcome these barriers and many somatic hybrids between citrus and related genera (Grosser et al., 2000), i.e. *Severinia* (Grosser et al., 1996), *Citropsis* (Ling and Iwamasa 1994), *Atalantia* (Louzada et al., 1993) and *Clausena* (Guo and Deng 1998), have been produced. However, most of these hybrids were allotetraploid plants that incorporated the complete genomes from the fusion partners. Integration of whole genomes from both fusion parents has negative impacts on the growth of the resulting hybrids owing to somatic incompatibility. A case in point is the somatic hybrid plants between *S. disticha* and citrus cultivars that were killed by a fungal disorder (Grosser et al., 1996). As a consequence, the practical use of these hybrids has been hampered, owing to somatic incompatibility at various levels and to linkage and correlation between the desirable and undesirable traits.

Donor-recipient fusion is an alternative to alleviate somatic incompatibility and transfer specific genes or chromosomes (Trick et al., 1994; Yamashita et al., 1989). Cybrids were regenerated between *Nicotiana tabacum* and *N. sylvestris* via donor-recipient fusion (Zelcer et al., 1978). Dudits et al., (1980) obtained the first asymmetric hybrids between tobacco (*N. tabacum*) and parsley (*Petroselium hortense*) by this fusion method. Recently, significance has been attached to asymmetric fusion in many crops such as tobacco (Vlahova et al., 1997; Atanassov et al., 1998), rice (Wang et al., 1998; Liu et al., 1999), potato (Oberwalder et al., 1997; Oberwalder et al., 1998), tomato (Ratushnyak et al., 1993), wheat (Xia et al., 1996; 2003), *Cichorium* (Varotto et al., 2001), *Medicago* (Tian and Rose 1999) and *Brassica* (Gerdemann-Knock et al., 1995; Forsberg et al., 1998a, b). To date, a lot of asymmetric somatic hybrids and cybrids (callus or plants) have been obtained (Gleba et al., 1988; Glimelius and Bonnett 1986; Liu and Deng, 2002; Oberwalder et al., 1998; Rasmussen et al., 2000; Tian and Rose 1999; Varotto et al., 2001; Xiang et al., 2003; Yamagishi et al., 2002; Zhou et al., 2001; Zubko et al., 2002). By this method, Gerdemann-Knorck et al., (1995) successfully transferred genes resistant to black leg and club root disease from *Brassica nigra* into *B. napus*. With citrus, only 5 cybrids have been produced via donor-recipient fusion (Vardi et al., 1987, 1989).

Microcitrus papuana Swing. confers resistance to drought, flooding, the burrowing nematode (O'Bannon and Ford 1977) and *Phytophthora*

(Carpenter and Furr 1969). Newhall navel orange [*Citrus sinensis* (L.) Osb.], a superior cultivar, is sensitive to *Phytophthora*. Moreover, it is lacking in functional pollen and viable ovules (Hodgson, 1967). *Swinglea glutinosa* is reported to be resistant to citrus tristeza virus (CTV), which has caused a great loss to the citrus industry in the world in the previous century. Dancy is reported to be tolerant to citrus excortis virus (CEV), which causes a splitting of the graft union. Citrange possesses resistance to *Phytophthora* caused foot rot and root rot, which have been found to be rampant in the citrus groves along Yangtze River.

In the present research, on the one hand, the donor-recipient fusion system was established between interspecific combinations. Donor-recipient fusion between *Citrus* and the above mentioned genera was carried out, with the intention of creating a novel germplasm for citrus cultivar improvement.

MATERIALS AND METHODS

Plant Materials

Embryogenic cultures of *M. papuana*, Valenica sweet orange (*C. sinensis* Osb.), Murcott tangor (*C. reticulata* × *C.sinensis*), Page tangelo (*C. reticulata* × *C. paradisi)*, Dancy tangerine (*C. reticulata* Blanco), *Swinglea glutinosa* and Newhall navel orange (*C. sinensis* Osb.) were maintained on semi-solid MT basal medium (Murashige and Tucker 1969), containing 50 g l^{-1} sucrose. Seeds of citrange (*Poncirus trifoliata* × *C. sinesis*), kindly provided by Mr. Huo at Fruit Trees Research Institute in Guangdong Province, were surface-sterilized and sown in a MT medium. Prior to isolation of the protoplasts, the embryogenic cultures were grown in liquid MT medium supplemented with 0.5 g l^{-1} malt extract, 1.5 g l^{-1} glutamine and 50 g l^{-1} sucrose and subcultured for three times at 2-week intervals on a rotary shaker (110 rpm).

Protoplast Isolation

Approximately 1 g of embryogenic culture or tender leaves that have been grown for almost 20 days was mixed with 1.5 ml enzyme solution containing 1% cellulase Onozuka R-1, 2% pectinase and equal volumes of 0.7 M EME medium as described by Grosser and Gmitter (1990). The isolated protoplasts were passed through two layers of stainless steel sieves and then centrifuged in 25% sucrose and 13% mannitol gradient (1000 rpm, 6 min.).

Pre-treatment of the Protoplasts

Calli and protoplasts of Dancy tangerine were irradiated with X-ray for 1 and 1.5 h at 80 kV and 5 mA to investigate the effects of X-ray irradiation

(Li and Deng 1997). UV, with the intensity of 56 $\mu W/mm^2$, was used to irradiate the protoplasts of Page and Murcott for 30 s, 1 min. and 5 min. Protoplasts of *S. glutinosa* were irradiated for 6 min. to investigate the liability and repeatability of UV. 0.25 mmol l^{-1}/I Iodoacetic acid (IA) with a concentration of 0.25 mmol/I was used to treat the protoplasts of Page and Murcott for 15 min.

For donor-recipient fusion, the purified embryogenic protoplasts of *M. papuana* were irradiated for 1.5 h; those of Dancy tangerine for 1 h, 1.5 h and 2 h while those of Valencia for 45 min. Protoplasts of the recipient, Newhall navel orange, Page and Murcott, were treated with 0.25 mM IA for 15 min. (Li and Deng 1997). Mesophyll protoplasts of citrange were irradiated for 45 min. Protoplasts were then washed once in electrofusion solution containing 0.7 M mannitol and 0.25 mM $CaCl_2$ and then adjusted to the density of 5×10^5 m l^{-1}.

FUSION AND CULTURE OF THE PROTOPLASTS

The fusion combinations carried out in the research were *M. papuana* and Newhall, citrange and Page, Valencia and Murcott, Dancy and Page. The protoplasts of the donor and the recipient were mixed at a 1:1 ratio. Somatic hybridizer SSH-2 (Shimadzu Corporation, Kyoto, Japan) was utilized to mediate the fusion of the protoplasts. About 1.6 ml of the mixed protoplasts were pipetted into the fusion chamber (FTC-04). The fusion procedure was modified from other reports (Liu et al., 1999; Liu and Deng 2000; Liu et al., 2000). The mixed protoplasts were aligned with an AC field of 95 V cm^{-1} at a frequency of 1 MHz for 60 s. DC, causing reversible breakdown of the aligned protoplasts with a field strength of 1250 V cm^{-1} and duration of 40 μs, was applied five times at 0.5 s intervals in order to induce protoplast fusion. Following fusion, the protoplasts were kept still for 20 min. so that the fusion products could regain normal shape, followed by centrifugation at 800 rpm for 5 min.

The supernatant was removed and the protoplast pellet was resuspended in BH3 medium with a final density of $1 \times 10^5/ml^{-1}$ (Grosser and Gmitter 1990). Liquid cultures were employed for the last three combinations, whereas low-melting agarose embedding culture was employed for the first one (Deng et al., 1988). The mini-culture derived from protoplasts was transferred to EME500 medium for further growth. MT basal medium added with 2% Glycerol, MT and MT + 0.45M sucrose + 0.15M sorbitol were employed for embryoid induction. Media for embryoids regeneration and shoot and root induction were the same as depicted elsewhere (Liu et al., 2000). The shoots that cannot root were cut and subjected to in vitro grafting (Deng et al., 1993).

CYTOLOGY

For chromosome counting, hematoxylin staining method was applied (Grosser and Gmitter 1990). Fresh calli or shoot tips were harvested and pretreated with 1,4-dichlorobenzene for 3 h, fixed in Carnoy (1 part glacial acetic acid:3 parts ethanol) for 16 to 20 h, then softened in 5 M HCl for 12 min., followed by treatment with 4% ammonium sulfate for 1 h (washing with distilled water was necessary after each step). The rinsed materials were stained with 0.5% hematoxylin for 3 h, and then smeared for chromosome examination.

RAPD ANALYSIS

Total DNA extraction was carried out according to Xiao et al., (1995) and Shi et al., (1998) with minor modification. The DNA pellet was dissolved in 200 to 400 µl TE buffer (10 mM Tris-HCl and 0.1 mM EDTA). DNA Thermal Cycler 480 was used for PCR-RAPD analysis. 10-mer primers were used for amplification of the template DNA. Amplification was performed with the following conditions: 1 cycle of 93°C, 2 min.; 41 cycles of 93°C, 1 min., 35°C, 1 min., 72°C, 2 min. and 1 cycle of 72°C for 10 min. Electrophoretic separation of the amplified DNA fragment was carried out in 1.5% agarose gel containing 0.5 µg ml^{-1} ethidium bromide. The gels were visualized and photographed with UV light.

RESULTS AND DISCUSSION

Effects of X-ray on the Growth of Calli and Protoplasts

It could be seen from Table 7.1 that with an increase in irradiation time, the growth of callus was retarded more seriously, as indicated by the decrease in dry weight. For instance, the callus that was irradiated for 1.5 h lost its weight by 47.7% 1 month later, while the control increased its

Table 7.1 Effects of X-ray irradiation on the growth of callus of Dancy tangerine

Time of irradiation (h)	*Dry weight of callus (g)*			*Increase or decrease of dry weight*	
	The 1st day	*The 15th day*	*The 30th day*	*Increase (%)*	*Decrease (%)*
0	0.36	0.56	0.79	119	
1	0.33	0.22	0.25		24.2
1.5	0.45	0.35	0.24		47.7

weight by 119%. In addition, the callus color could also indicate the inhibitory effects of irradiation on callus growth. During the culture, the normal callus was white, whereas those irradiated for 1 and 1.5 h became brown and some in the latter even turned black (data not shown).

The control protoplasts could recover the first division within 7 days and develop into cell clusters. The protoplasts irradiated for 1 h divided later, but no difference from the control was observed after division. However, very few protoplasts irradiated for 1.5 h could divide. Instead, they plasmolyzed and broke in some cases (Fig. 7.1).

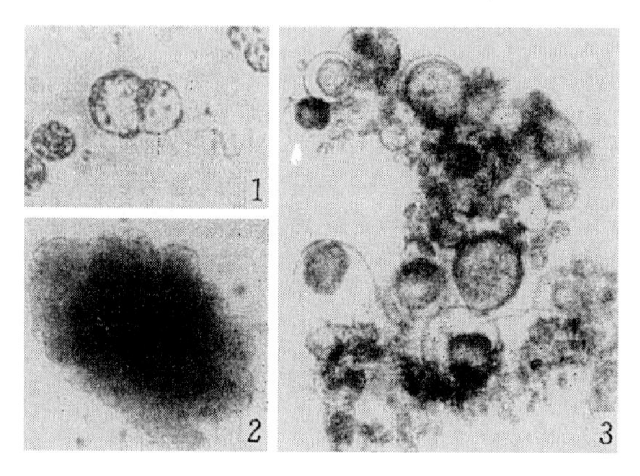

Fig. 7.1 Cultured protoplasts of Dancy tangerine 1. The 1st division of the control protoplast. 2. Formation of cell cluster from the cultured control protoplasts. 3. Plasmolyzed protoplasts irradiated for 1.5 h.

Effects of Iodoacetic Acid on Growth of Protoplasts

The viability of Page tangelo and Murcott tangor protoplasts was compared after treatment with 0.25 mmol/L Iodoacetic acid (IA) for 15 min. (Table 7.2). It showed that some difference in the viability was found between them, indicating varying sensitivity to IA. Although the protoplasts from both cultivars could not undergo division during the course of culture but broke (Fig. 7.2), the control protoplasts could develop into callus within 1 month.

Effects of UV Irradiation on Growth of Protoplasts

The result of UV irradiation demonstrated that the viability of the protoplasts decreased with an increase in the irradiation time as well. Protoplasts of Page tangelo were found to be more sensitive to UV than

Table 7.2 Viability and growth of Page and Murcott protoplasts treated with 0.25 mmol/L IA for 15 min

Species	*Time of treatment (min.)*	*Viability (%)*	*Regeneration*
Page tangelo	0	92.45	Normal
	15	88.68	Abnormal and broken
Murcott tangor	0	90.16	Normal
	15	46.55	Abnormal and broken

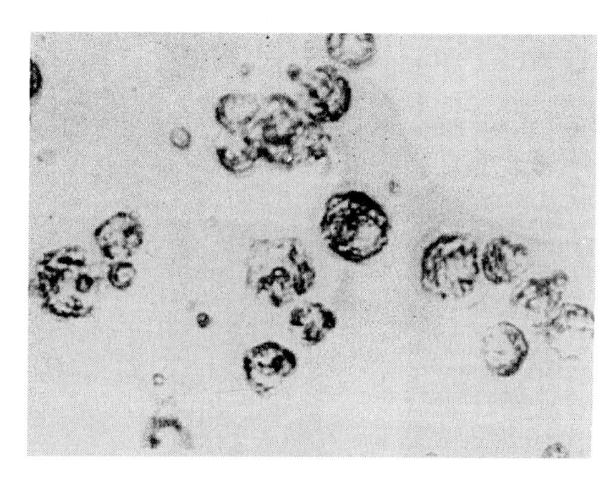

Fig. 7.2 IA-treated protoplasts of Page tangelo

those of Murcott tangor. The protoplasts, when irradiated for 1 min. and 5 min. did not divide. Instead, most of the protoplasts plasmolyzed and broke (Fig. 7.3). Nevertheless, those irradiated for 30 s and the control protoplasts divided and developed into minicallus.

However, as reported elsewhere, the UV irradiation was not repeatable in some cases. When the same irradiation dosage rate was employed to *Swinglea glutinosa*, division of the protoplasts could not be arrested even when they were irradiated for 6 min. Instead, some calli arose from the irradiation-treated protoplasts. One possible reason is that these calli are different from those of Page and Murcott in tolerance to irradiation. RAPD analysis of the regenerated callus revealed different band patterns arising from the UV-irradiated protoplasts and the control (Fig. 7.4). Compared with the control, the callus derived from the irradiated protoplasts lost some of the specific bands and meanwhile, some novel bands were found, indicating that change in the genome has possibly occurred in spite of the fact that division of the protoplasts was not arrested.

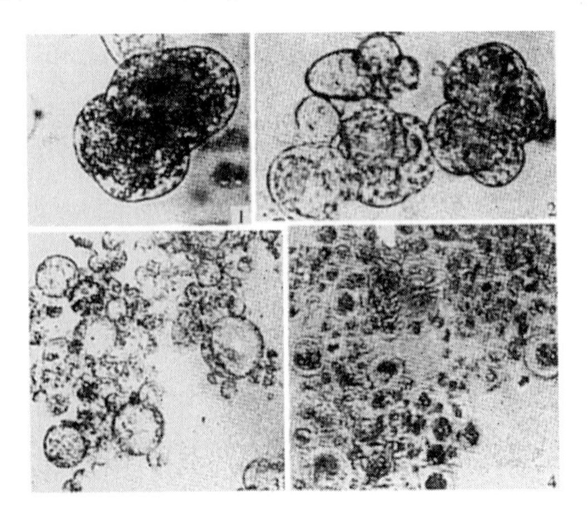

Fig. 7.3 Culture of protoplasts of the control (1 and 2) and the UV-irradiated protoplasts (3 and 4).

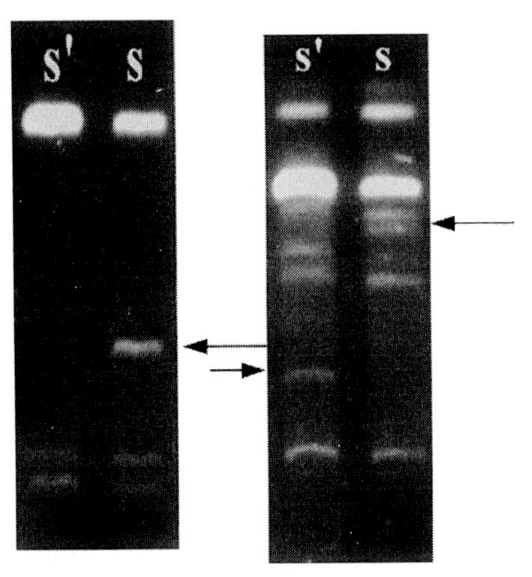

Fig. 7.4 Band patterns of callus of the control (S) and the callus (S') derived from UV-irradiated protoplasts of *Swinglea glutinosa* by OPA-04 and OPA-08, respectively. Arrowheads show specific bands.

Regeneration of Hybrid Embryoids between Leaf-derived Citrange Protoplasts and Embryogenic Page Tangelo Protoplasts

Leaf-derived protoplasts of citrus could not undergo division with the current culture system. The protoplasts of Page tangelo could not divide as well, owing to IA treatment. In theory, only the heterokaryons (Fig. 7.5) could divide and develop further. Calli and embryoids recovered from the fusion products, and some abnormal shoots could be found in the cultures (Fig. 7.6).

Cytological observation of the fusion-derived callus demonstrated that a lot of diploid (52.2%) and aneuploid cells (42.0%) together with few tetraploid cells (5.8%), were observed (Fig. 7.7). Four arbitrary primers, OPA-07 OPW-01 OPAN-07 and OPE-05, were selected to identify the hybridity of the shoots. In the band pattern of OPA-07, the embryoids possessed specific bands from both fusion parents, implying that they were somatic hybrids. Their hybridity was further confirmed by

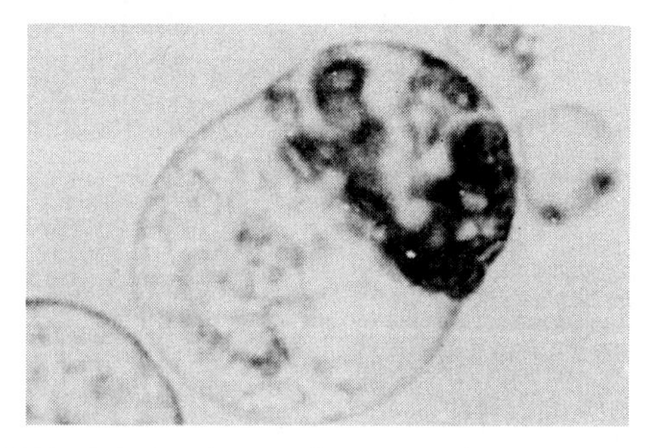

Fig. 7.5 A heterokaryon between citrange and Page.

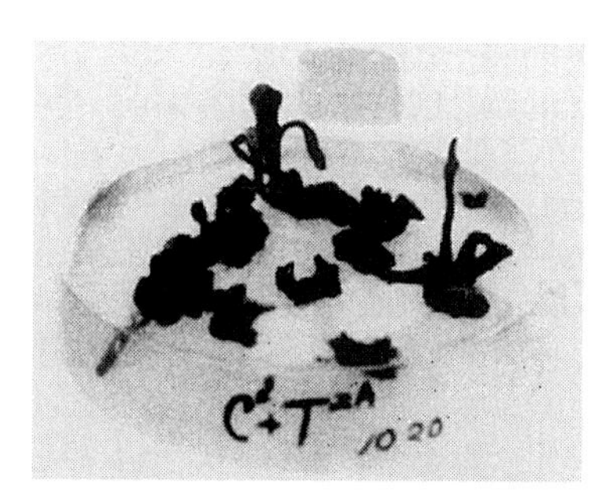

Fig. 7.6 Shoots derived from fusion between citrange and Page.

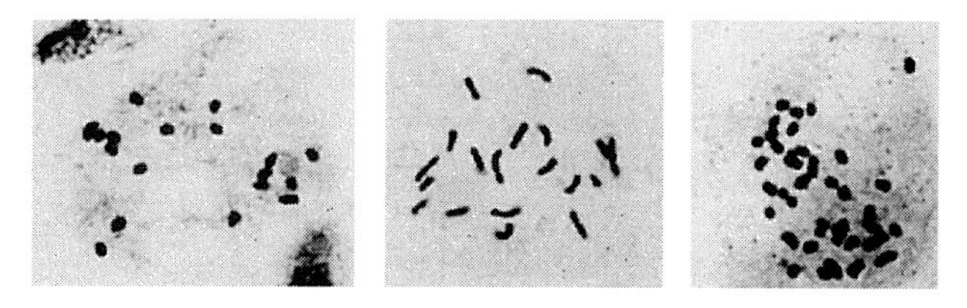

Fig. 7.7 Cytological observation of the callus derived from citrange and Page. From the left to the right are cells with 18, 19 and 36 chromosomes.

amplification results of the other 3 primers (Fig. 7.8). The above-mentioned results of RAPD analyses showed that the embryoids were somatic hybrids between citrange and Page tangelo.

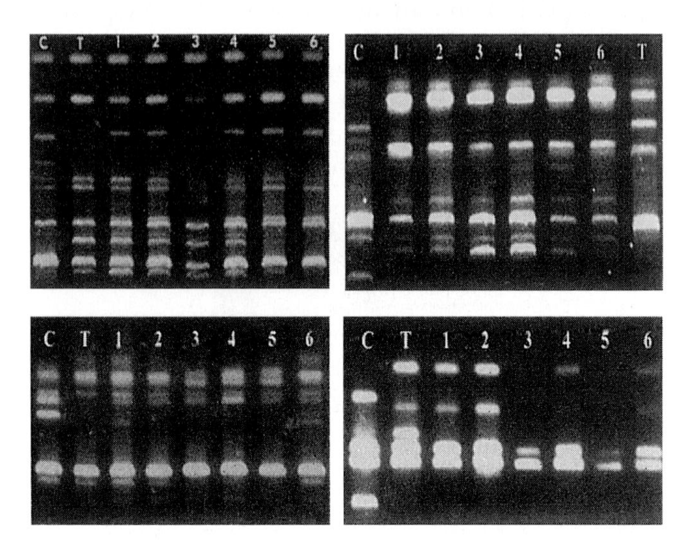

Fig. 7.8 RAPD analysis of the embryoids derived from donor-recipient fusion of citrange and Page. From top left to the bottom right are band patterns of OPAN 07, OPA 07, OPE 05 and OPW 01.

In the present research, mesophyll protoplasts of citrange were utilized as the donor, which lays groundwork for future work. On the one hand, when mesophyll protoplasts were used, it is not necessary to explore a division-arresting dosage since they could not divide with contemporary culture media. Unfortunately, the abnormal shoots could not develop into normal ones. Two possible reasons may explain this. First, irradiation of the protoplasts for 45 min. at the dosage rate may have caused great damage to the protoplasts, which poses negative impacts on the cells. As a consequence, the genomic and physiological complementation of the fusion parents could not be fulfilled, since the mesophyll and embryogenic protoplasts were isolated from different explants. Secondly, the irradiation has led to a high degree of asymmetry of the embryoids and the shoots, which in turn, affected the morphogenesis of the shoots.

For the Combination of *Microcitrus papuana* and Newhall Navel Orange

Four types of culture were utilized for this combination: control (CK), donor and recipient protoplasts without pretreatments; T1, X-ray irradiated protoplasts of *M. papuana* and IA-treated protoplasts of Newhall navel orange; T2, mixture of donor and recipient protoplasts without fusion treatment; T3, fusion- treated donor and recipient protoplasts.

Establishment of cell lines

The first division occurred fifteen days after the initial culture of the protoplasts in the control (CK), while no division was observed in T1, T2 and T3. On the twentieth day, a large number of cell clusters appeared in CK (Fig. 7.9A, cell clusters derived from Newhall navel orange not shown). Some dividing cells were observed in T3, indicating that the pretreatment delayed protoplasts division, which is very common in donor-recipient fusion (Vardi et al., 1987).

On the twenty-sixth day, many microcolonies appeared in CK. Most of the X-ray irradiated protoplasts were plasmolyzed and those that did not undergo plasmolysis were spherical (Fig. 7.9B), whereas almost all of Newhall navel orange protoplasts collapsed and then burst (data not shown). No division was observed in T2 and the protoplasts were either

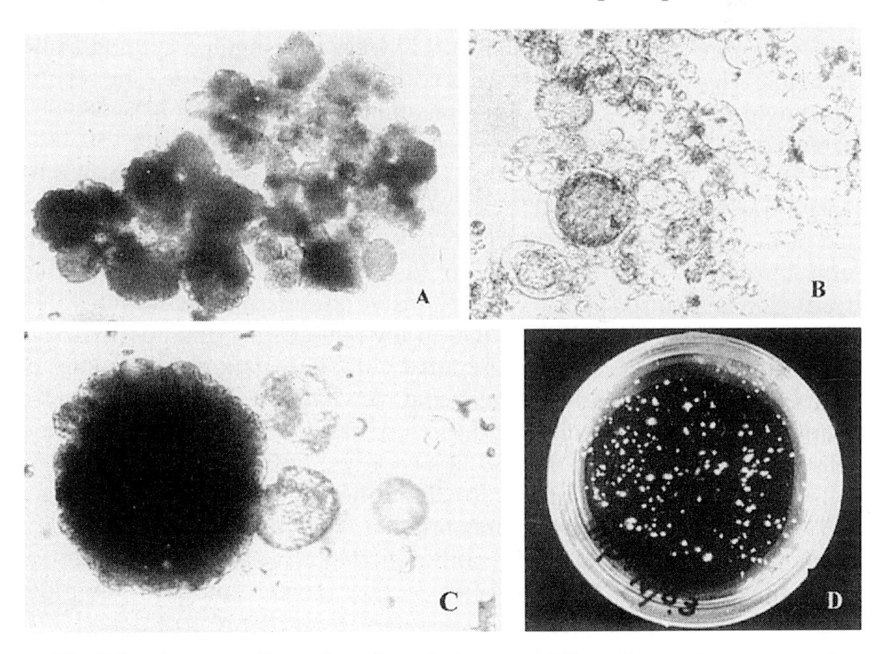

Fig. 7.9 Regeneration of callus between *Microcitrus papuana* and Newhall navel orange via asymmetric fusion: (A) cell clusters from protoplasts of *M. papuana* without irradiation; (B) X-ray irradiated protoplast of *M. papuana*; (C) cell clusters from the fusion-treated protoplasts; (D) Regeneration of mini-calli from fusion products.

spherical or plasmolyzed. Some microcolonies appeared in T3 (Fig. 7.9C), which were fewer here rather than in the control as well. Irradiation and IA treatment may have caused fatal damage to the donor and recipient protoplasts, respectively.

Colonies, 1~2 mm in diameter (Fig. 7.9D), were transferred to EME500 medium for further growth. Initially, they proliferated slowly for 15 days but later grew more rapidly. Altogether, 7 cell lines were established. But somatic embryogenesis was not observed. Belarmino et al., (1996) also observed cultures derived from donor-recipient fusion were non-morphogenic. Irradiation of the donor protoplasts can cause serious fragmentation and loss of chromosomes, which leads to an unbalanced genome in asymmetric hybrids (Xu et al., 1993) and poses a negative impact on embryogenesis. In other studies, 7 out of 25 combinations of donor-recipient fusion have produced somatic embryoids, and plants were obtained in only 2 interspecific combinations (Liu 1999; Liu and Deng, 2000, 2002).

Cytological determination of the cell lines

Four out of the 7 cell lines (1, 2, 3 and 5) were investigated cytologically. More than fifty cells in mitosis were counted in each cell line. The results are shown in Table 7.3. Each line has many aneuploid (45.10%, 38.98%, 32.69% and 34.85%, respectively) and diploid cells (52.94%, 59.33%, 63.46% and 62.12%, respectively), together with few tetraploid cells (1.96%, 1.69%, 3.85% and 3.03%, respectively) (Fig. 7.10). Aneuploid cells had primarily chromosome numbers close to 18, i.e. 19, 20 and 21. No cells with chromosome number close to 36 were found, which implies that irradiation has caused serious and extensive elimination of chromosomes.

Since the cytological determination involved callus, the possibility of somaclonal variation could not be ruled out. Thereafter, chromosomes of callus of fusion parents, *M. papuana* and Newhall navel orange, were also counted. There was a significant difference in the percentage of aneuploid cells in comparison with the cell lines. Only 2 out of 48 and 5 out of 66 cells counted were anueploids for the donor (4.17%) and the recipient (7.58%) respectively, which demonstrated that ploidy of the fusion parents is relatively very stable (diploid), although they have been subcultured on MT medium for more than five years. So, the high percentage of aneuploid cells in the cell lines must have resulted from X-ray irradiation and subsequent elimination of chromosomes in the fusion products.

RAPD analysis of the cell lines

Five cell lines (1, 2, 3, 4 and 5) were chosen for hybridity verification. All of the four primers selected for characterizing the cell lines could distinguish the donor from the recipient. The cell lines had band patterns

Table 7.3 Chromosome counting of the cell lines between *M. papuana* and Newhall navel orange

Cell lines	*No. of cells counted*	*No. of cells with different chromosome number*								
		Diploid cells	*Tetraploid cells*	*Aneuploid cells*						
		(2n = 18)	*(2n = 36)*	*(2n = 14)*	*(2n = 16)*	*(2n = 19)*	*(2n = 20)*	*(2n = 21)*	*(2n = 22)*	*(2n = 24)*
1	51	27	1		1	3	18	1		
2	59	35	1			4	16	3		
3	52	33	2	1		5	8	1	1	1
5	66	41	2	1	1	2	15	3	1	

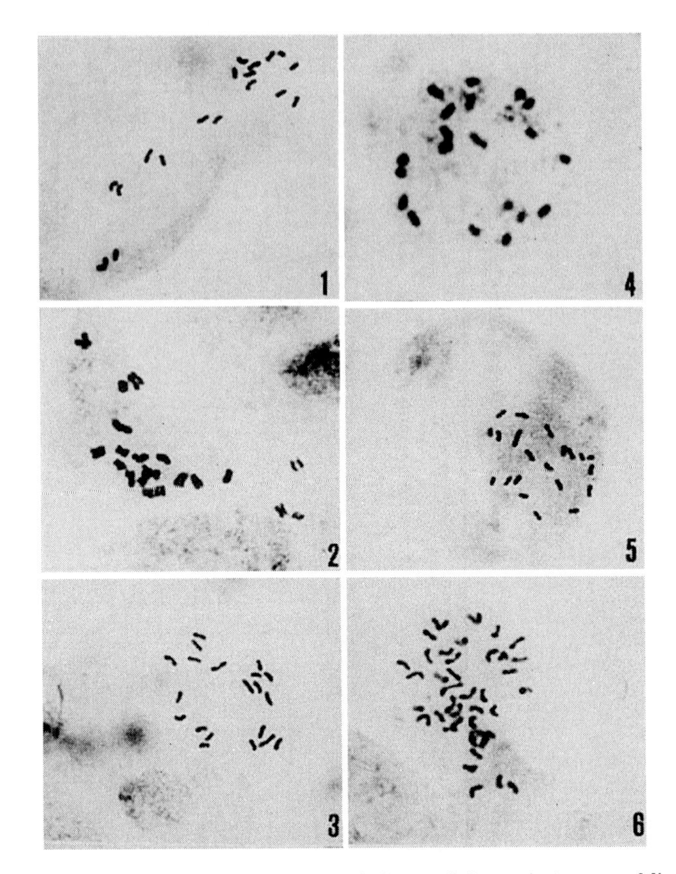

Fig. 7.10 Chromosome observation of the cell lines between *Microcitrus* papuana and Newhall navel orange 1, Diploid cell; 2–5, Aneuploid cells with 19, 20, 20 and 21 chromosomes. 6, Tetraploid cell.

similar to that of the recipient, although one band specific to the recipient was absent in the cell lines for primer OPAA-17(Fig. 7.11A). In case of primers OPA-07 (Fig. 7.11B) and OPS-13 (data not shown), the cell lines had the same band patterns as the recipient, which indicates that the genome of the recipient has been incorporated into the cell lines. The bands specific to the donor could not be detected in the cell lines for these three primers, whereas one band specific to the donor was present in the cell lines for OPV-07 (Fig. 7.11C). This implied that the donor, *M. papuana*, has contributed its genome to the cell lines as well. Therefore, the cell lines were confirmed as somatic hybrids. No variation in band profiles among the cell lines was observed, indicating stable integration of donor or recipient DNA in the hybrid cell lines (Rasmussen et al., 1997).

RAPD has been employed efficiently to characterize somatic hybrids derived from asymmetric somatic hybridization (Samoylov and Sink 1996;

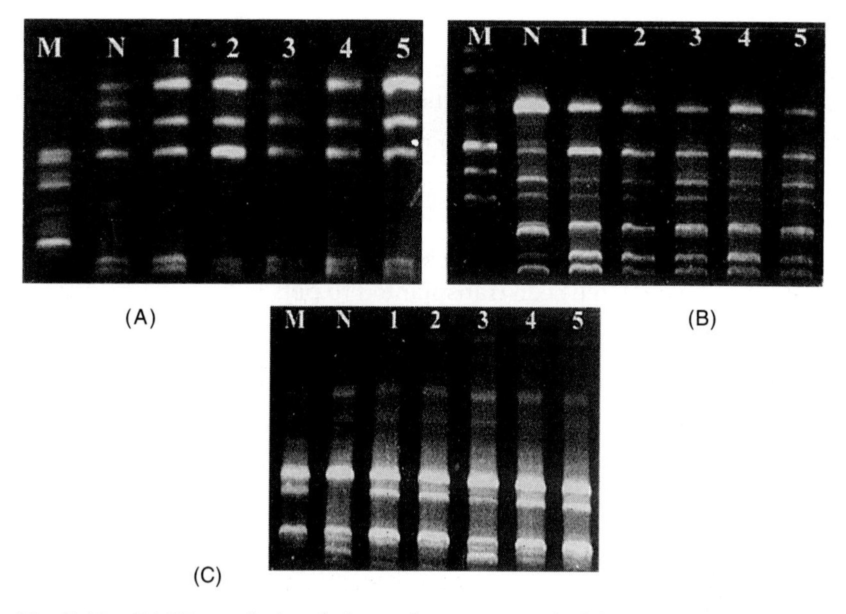

Fig. 7.11 RAPD analysis of the callus as revealed by 10-mer arbitrary primers OPAA-17(A), OPA-07(B) and OPV-07(C), respectively. In the band profiles M, N and 1–5 are *M. papuana*, Newhall navel orange and the cell lines, respectively.

Rasmussen et al., 1997; 2000; Xiang et al., 2003) by investigating the interrelationship between the number of specific RAPD bands and donor DNA present in the hybrids. In the band patterns of the 4 primers, there were 8 bands specific to the donor, of which only one was present in the cell lines. Rasmussen et al., (1997) concluded that the number of donor-specific RAPD bands present in the hybrids indicated the extent of donor DNA introduced into the hybrid genome. We can thus conclude that *M. papuana* has contributed little DNA to the hybrid cell lines. Results of RAPD, in combination with cytological determination, proved that the cell lines described herein were true asymmetric somatic hybrids.

For the Combination of Valencia Sweet Orange and Murcott Tangor

Regeneration of plants

The protoplasts were aligned with AC and fused when DC was exerted. Cell clusters appeared 25 days after fusion treatment. The clusters grew into visible mini-calli within 60 days, but could not develop into green embryoids on being transferred to the medium of EME500. Of the media tested, only MT added with 2% glycerol was effective to induce the calli to grow into embryoids for this combination. No embryoids were regenerated in the other medium. But the duration of regenerating embryoids was very long. It took nearly 15 months for them to change into embryoids. Shoots were regenerated when the cotyledonous embryoids were cultured in shoot-induction medium. However, it is of difficulty for them to root in the root-induction medium. In the end, only one shoot rooted in the medium. Therefore, in-vitro grafting was employed to get plants (Fig. 7.12). Altogether, 12 grafted plants have been obtained, which, along with the self-rooting plant, have been transplanted to pots in the greenhouse (Fig. 7.13).

Cytology of the Regenerated Plants

Chromosomes of 10 regenerated plants, including the self-rooting plant (No. 1) and 9 grafted plants, were investigated in this research. The results are listed in Table 7.4, where it can be seen that an extensive distribution

Fig. 7.12 A grafted plant

Fig. 7.13 Plants derived from protoplast fusion between Valencia sweet orange and fusion Murcott tangor.

Table 7.4 Chromosome determination of plants between Valencia sweet orange and Murcott tangor

Plants	*No. of cells observed*	*No. of cells with different ploidy*		
		Diploid (2n=2x=18)	*Tetraploid (2n=4x=36)*	*Anueploid*
1	57	42	2	13
2	10	9		1
4	15	11	2	2
5	7	7		
6	8	5		3
7	46	27	4	15
8	10	10		
9	10	10		
10	12	6		6
13	17	13		4
Total	192	140	8	44

of chromosomes was discovered in the plants. In 3 of the plants, No. 5, 8 and 9, only diploid cells were observed. In 4 plants (No. 2, 6, 10 and 13), both diploid and aneuploid cells were observed, whereas in the remaining 3 plants, No. 1, 4 and 7, diploid and aneuploid cells, together with few tetraploid cells were found (Fig. 7.14). The aneuploid cells usually had more than 18 chromosomes. No cells with less than 18 chromosomes or more than 36 chromosomes were observed. In 192 cells of the ten plants, 140 were diploid cells (72.9%) and, 44 (22.9%) were aneuploid cells, whereas only 8 tetraploid cells (4.2%) were observed. Aneuploid cells were still observed 40 days after the self-rooting plant was transplanted to a pot.

RAPD ANALYSIS OF THE PLANTS

Seven 10-mer arbitrary primers were screened, of which only three primers, OPA-04, OPAN-07 and OPE-05, showed polymorphism between Valencia and Murcott. The self-rooting plant (No. 1) and plant Nos. 2, 4, 5, 6, 7 and 13 were chosen for the verification of hybridity. In the amplified product of OPA-04 (Fig. 7.15A), bands specific to both fusion parents could be detected in the regenerated plants, indicating that genetic component of Valencia and Murcott has been incorporated into the plants. Therefore, these regenerated plants could be confirmed as putative hybrids. Their hybrid characteristic was further proved by the band patterns of OPAN-07 and OPE-05 (Fig. 7.15B and C). There was no significant difference between the plants regarding their band profiles although different chromosome distribution was found between them, as

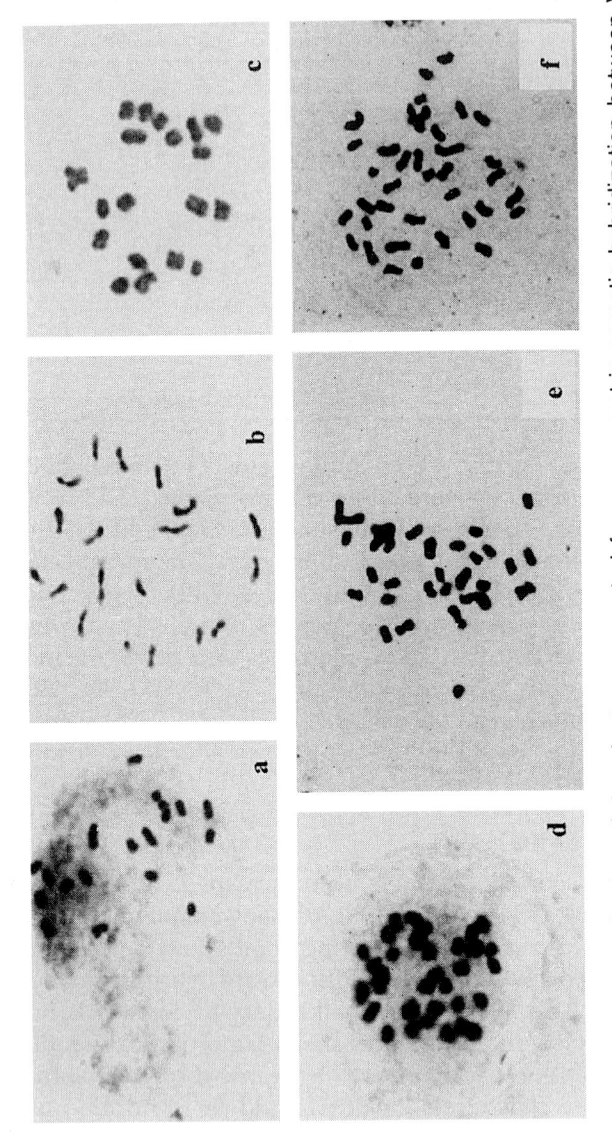

Fig. 7.14 Cytological investigation of the plants regenerated from asymmetric somatic hybridization between Valencia sweet orange and Murtcott tangor; (a) a cell with 18 chromosomes; (b and c) a cell with 19 chromosomes; (d) a cell with 20 chromosomes; (e) Two cells with 30 chromosomes; (f) a cell with 36 chromosomes.

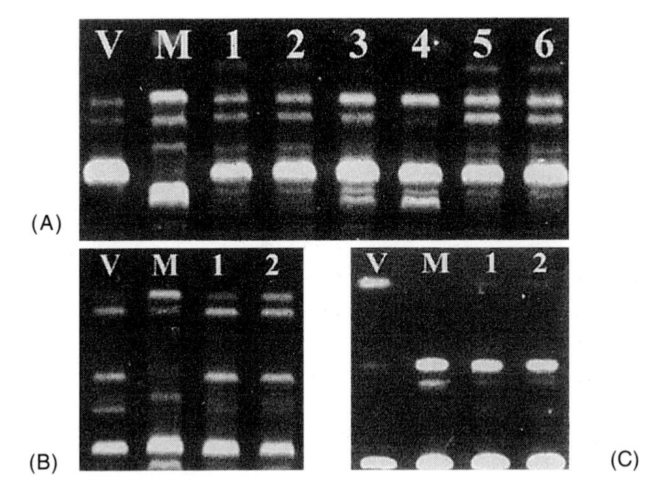

Fig. 7.15 RAPD analysis of the regenerated plants. From the left to the right are band patterns of OPA-04, OPAN-07 and OPE-05 respectively (V, M are Valencia, Murcott, respectively). 1 is the self-rooting plant; 2–6 are the regenerated plant Nos. 2, 4, 5, 6, 7 and 13 respectively.

shown in Table 7.4. RAPD analyses demonstrated the production of true interspecific hybrid plants.

For the Combination between Dancy Tangerine and Page Tangelo

Regeneration of embryoids and plants

Division of the donor protoplasts irradiated with 228 krad was not prevented but delayed. Protoplasts of the control resumed their first division 7 days after culture, while the irradiated protoplasts began mitotic division on the fifteenth day. However, protoplasts irradiated with 342 krad and 456 krad did not divide at all. Instead, they plasmolyzed or broke. The protoplasts of Page treated with IA shrank or broke during the culture (data not shown).

Three types of fusion were attempted for this combination. T1: Dancy (1 h) +Page (IA), T2: Dancy (1.5 h) + Page (IA), T3: Dancy (2 h) + Page (IA). The mixed protoplasts were aligned to form pearl chains under alternating current and then were fused quickly when treated with direct current (Fig. 7.16). Table 7.5 is the summary of regeneration of different combinations. Cell clusters were observed within one month in T1 and T2, whereas in T3 only two months later could some clusters be observed, which is similar to other reports. Xu et al., (1993) and Rasmussen et al., (1997) reported that an increase in the irradiation dose of the donor

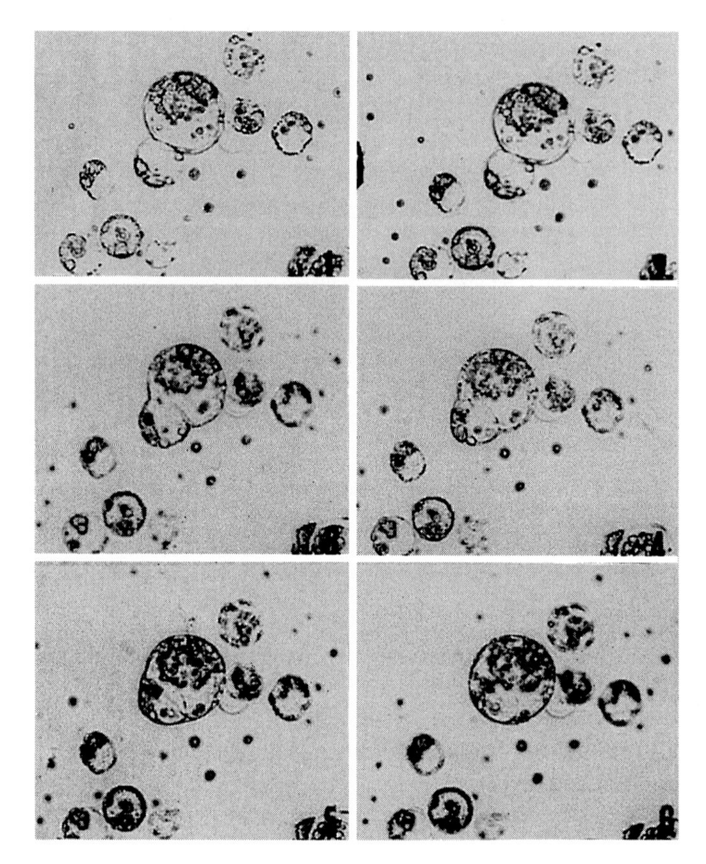

Fig. 7.16 Fusion process between Dancy tangerine and Page tangelo. The pictures were taken at intervals of 5 sec.

Table 7.5 Summary of regeneration of different types of fusion between Dancy and Page

Types of fuison	*Time for regeneration of cell clusters*	*Regeneration*		
		callus	*Embryoids and time needed*	*Shoots/ plants*
F_1	1 month	+	+2 months	+
F_2	1 month	+	+2 months	–
F_3	2 months	+	+24 months	–

Note: + indicates that callus or embryoids or shoots/plants were obtained while – indicates the opposite.

protoplasts could lead to a slow and reduced growth of the fusion products. The mini-calli in T1 and T2 could develop into embryoids within two months, whereas it took almost 24 months for those in T3. However, only embryoids in T1 could give rise to some shoots in the shoot-induction medium. Surprisingly, the shoots were recalcitrant to rooting when they were cultured in the root-induction medium. No self-rooting plants could be obtained. At last, the entire plants were obtained only by in vitro grafting. The viable grafted plants have been transplanted to the soil pots in the greenhouse (Fig. 7.17). The difficulty in regeneration of roots might possibly result from unbalanced genomes in the hybrids that had been reported elsewhere (Wijbrandi et al., 1990).

Fig. 7.17 Grafted plant (left) and the plants transferred to soil pots in the greenhouse (right).

Cytological Determination of the Regenerated Plants

Results of chromosome counting of nine grafted plants were listed in Table 7.6. The plants contained predominantly diploid and aneuploid cells with the exception of plant Nos. 2 and 3, in which a few tetraploid cells were observed. The aneuploid cells contained mainly 19, 20 and 21 chromosomes, 1 to 3 chromosomes more than the recipient. No cells with chromosome number close to 36 were found, which possibly meant that the irradiation has caused severe loss of chromosomes. Moreover, in plant No. 2, some cells with chromosomes less than 18 were observed (Fig. 7.18), indicating that extensive chromosome elimination has taken place. Plant Nos. 11 and 12 were derived from the same embryoid. Chromosome counting indicated that they possessed statistically identical chromosome distribution of diploid (38/56 and 43/66, respectively) and aneuploid (18/56 and 23/66, respectively) cells. In terms of the 10 plants in the 462 cells observed, 303 were diploids (65.58%), 154 were aneuploids (33.33%) and only 5 were tetraploids (1.09%). A chromosome examination revealed

Table 7.6 Chromosome distribution in the plants between Dancy tangerine and Page tangelo

Plant No.	*No. of cells examined*	*No. of cells with different ploidy*		
		Diploid cells (2n=2x=18)	*Tetraploid cells (2n=4x=36)*	*Aneuploid cells*
2	67	33	4	30
3	15	10	1	4
5	17	11	0	6
6	22	11	0	11
8	124	84	0	40
11	56	38	0	18
12	66	43	0	23
15	13	9	0	4
21	82	64	0	18
Total	462	303	5	154

that the plants were neither tetraploids nor diploids but mixoploids, which have been reported in asymmetric somatic hybridization of other crops (Rasmussen et al., 1997).

Regeneration of mixoploid plants from a single cell may result from a change in the nuclear DNA during callus growth, which then causes a non-homogenous DNA distribution in the callus from which the embryoids were obtained (Derks et al., 1992). Genetic instability and ongoing elimination of chromosomes may also lead to a regeneration of mixoploid plants (Bates 1990).

HYBRIDITY OF THE REGENERATED PLANTS

Seven 10-mer primers were screened, of which only three primers, namely OPAA-17, OPV-07 and OPE-03, could distinguish the fusion partners. Hybridity of four plants, Nos. 2, 6, 8 and 11 in Table 7.2, was analyzed using these three primers. The profile of the plants was identical to that of Dancy for the primer OPAA-17. These plants had the same band profile as Page for the primer OPV-07, whereas they had band pattern similar to Page for primer OPE-03, though one band specific to Page was not detected in the plants (Fig. 7.19). This confirms that four plants tested are true hybrids derived from fusion. The bands characteristic with Dancy could not be detected in the band patterns for the last two primers. In addition, no variation in the patterns was detected among the regenerated plants, the same as previous report (Liu and Deng 1999). Nevertheless, in the present research, the relationship was not quite clear. One possible reason is that not enough primers were used.

X-ray irradiation used in the present research had an inhibitory impact on the regeneration of shoots in a dose-dependent way. In many studies,

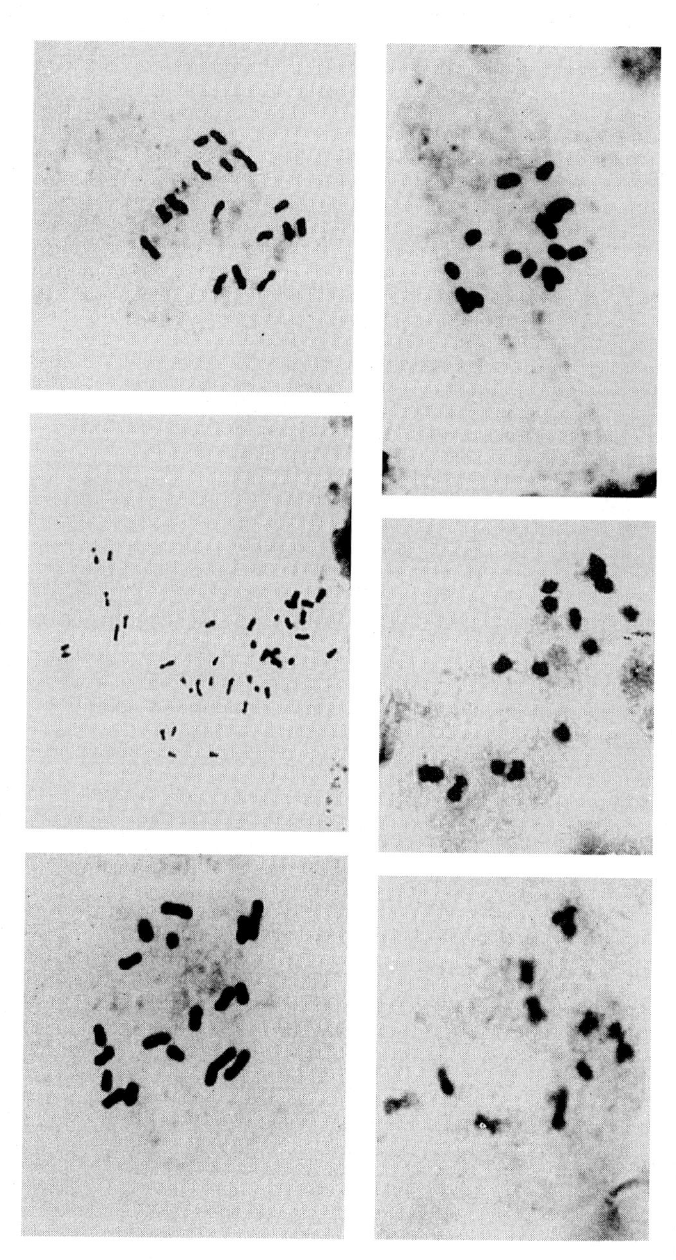

Fig. 7.18 Cytological observation of the regenerated plants. From up left to the bottom right are cells with 18, 36, 16, 10, 14 and 14 chromosomes, respectively.

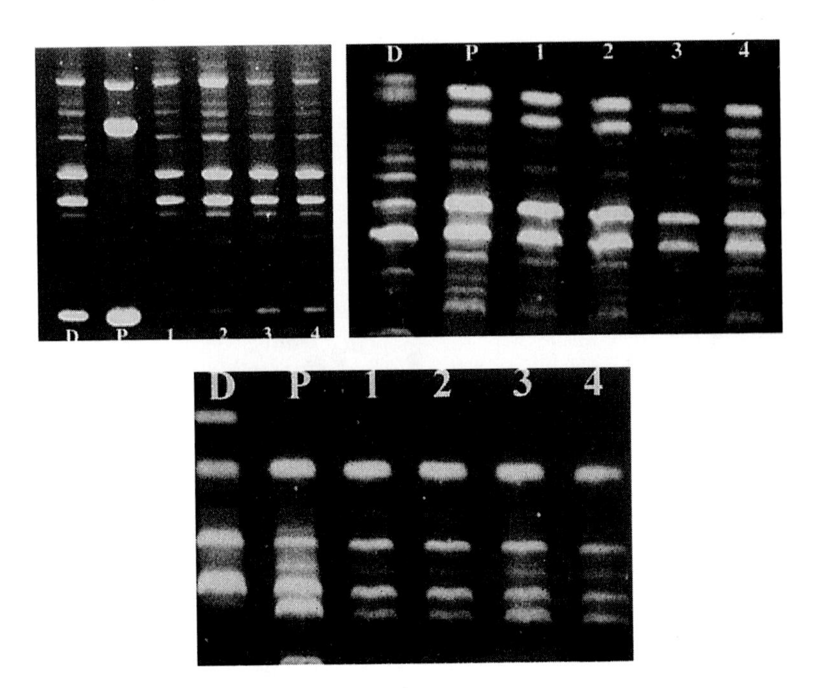

Fig. 7.19 RAPD analysis of the plants derived from donor recipient fusion between Dancy and Page. From the left to the right are band patterns of OPAA-17, OPV-07 and OPE-03, respectively. D, P, and 1–4 are Dancy, Page and the regenerated plants, respectively.

the regeneration of hybrids proceeded more slowly after a high irradiation dose than after a lower one. Oberwalder et al., (1997) conducted asymmetric somatic hybridizations between *Solanum tuberosum* and X-irradiated *S. bulbocastanum* and *S. circaeifolium*. Three irradiation doses, 70, 210 and 420 Gy, were tried. In the end, plants were recovered only from the former two doses, while no callus was recovered with the highest dose (420 Gy). Likewise, when protoplasts of *S. melongena* were fused with protoplasts of *Lycopersicum esculentum* × *L. pennellii* irradiated with 100, 250, 500, 750 and 1000 Gy, hybrid plants could arise only from the fusion combination with the lowest dose (Samoylov and Sink 1996).

Researchers always did their utmost to find a dosage that divisions of the protoplasts were arrested when carrying out asymmetric somatic hybridization (Spangbenberg et al., 1994). Nevertheless, this dose may have exceeded the dose leading to the fragmentation and subsequent elimination of most DNA. When the protoplasts were irradiated at the division-arresting dosage, they might have suffered the greatest harm. As a consequence, genomic and physiological complementation between the donor and the recipient could not be fulfilled to guarantee a normal growth of the fusion products. As is shown herein, though the protoplasts of Dancy irradiated for 1.5 h and 2 h did not divide at all, no shoots could be obtained when they were involved in the asymmetric fusion. Likewise, no embryoids and shoots have been obtained from the fusion between *M. papuana* and Newhall when protoplasts of the former fusion parent were irradiated for 1.5 h. It seems that only one conclusion could be drawn that the high irradiation dosage inhibited normal embryogenesis and morphogenesis (Forsberg et al., 1998a; Wang et al., 2003). It has been well documented that chromosome elimination in the hybrid is dependent not only on irradiation dosage but also on other factors such as donor species, genetic distance, culture condition, ploidy level of fusion partners and the regenerants (Oberwalder et al., 1997; Rasmussen et al., 1997; Samoylov and Sink 1996; Liu et al., 1999). Furthermore, an increase in the irradiation level did not always lead to increased chromosome elimination (Rasmussen et al., 1997). Therefore, it is not the prerequisite for the exploitation of a lethal or sublethal dosage to inhibit the division as far as asymmetric somatic hybridization is concerned. Low irradiation dose, on the one hand, causes less serious damage to the cells while it can still produce asymmetric or mixoploid hybrid, as is demonstrated herein and in other reports (Oberwadler et al., 1997; Samoylov and Sink 1996; Liu and Deng 2000).

To our knowledge, this is the first report on recovering mixoploid hybrid plants via asymmetric somatic hybridization in *Citrus*. It can be seen that in most cases, the cells in the mixoploid hybrids in the present research contained only one to three chromosomes more than either fusion partner. It possibly means the occurrence of chromosome elimination of the donor. Highly asymmetric somatic hybrids have been obtained elsewhere (Hinnisdaels et al., 1991; Vlahova et al., 1997). Adding a few or no amount of chromosomes from the donor has some advantages over adding the complete sets of chromosomes. On the one hand, it does not lead to an increase in the ploidy of the subsequent somatic hybrids (for cybrids). On the other hand, it decreases the negative impact from undesirable traits that are transferred to the somatic hybrids accompanying the desirable ones in standard symmetric fusions (for asymmetric hybrids). In addition, it can, to some degree, ease the somatic incompatibility encountered with some symmetric fusion, which is conducive to regeneration and development of the asymmetric hybrids. For example, the mixoploid plants derived from Dancy and Page may possess the tolerance to CEV derived from Dancy tangerine. Since the genetic background of citrus is not clear as yet, it is unknown in which chromosome(s) the tolerance to CEV is carried. As it is demonstrated that chromosome elimination is a random process in asymmetric somatic hybridization, it is also not clear that the chromosomes carrying the CEV tolerance have been maintained in the mixoploid plants. Therefore, further work should be carried out in order to investigate the tolerance of the mixoploid plants to CEV.

CONCLUSIONS

1. Mitotic division of protoplasts could be arrested when they were irradiated with X-ray for 1.5 h at 80 kV and 5 mA.
2. Mixoploid hybrids (calli, embyoids and plants) have been obtained from four donor-recipient fusion combinations. Cytological observation showed that mainly diploid and aneuploid cells were detected, together with few tetraploid cells, implying that irradiation has led to extensive chromosome eliminations.
3. It was found that embryogenesis and morphologenesis were more recalcitrant for the donor-recipient fusion compared with symmetric fusion, indicating that irradiation poses negative effects on the regeration of fusion products.

ACKNOWLEDGMENTS

This project was supported by NSFC (30200189) and International Foundation for Science (IFS) in Stockholm, Sweden as a grant to Dr J.H. Liu (D/3001-1 and D/3001-2). We would like to extend our gratitude to Dr J.W. Grosser at the Citrus Research and Education Center (University of Florida) for providing the embryogenic cultures of Dancy tangerine, *S. glutinosa, M. papuana,* Murcott tangor and Page tangelo. Thanks should also be given to Dr X.X. Deng, Mr Per Ekman and other staff of IFS to help finishing the project. Courtesy to Mr H.H. Huo at Fruit Trees Research Institute of Guangdong Province for providing the seeds of citrange.

REFERENCES

Atanassov II, Atanassov S.A., Dragoeva A., Atanassov A.I., 1998, A new CMS source in *Nicotiana* developed via somatic cybridization between *N. tabacum* and *N. alata*. Theor. Appl. Genet., **97**: 982–985.

Bates G.W., 1990, Asymmetric hybridization between *Nicotiana tabacum* and *N. repanda* by donor recipient protoplast fusion: Transfer of TMV resistance. Theor. Appl. Genet. **80**: 481–487.

Belarmino M.M., Toshnori A., Sasahara T., 1996, Asymmetric protoplast fusion between sweet potato and its relatives and plant regeneration. Plant Cell, Tiss. and Organ Cult., **46**: 195–202.

Cameron J.W., Frost H.B., 1968, Genetics, breeding and nucellar embryony. In: Reuther W., Webber H.J., Batchelor L.D. (Eds), The Citrus Industry, **2**: 325–370.

Carpenter J.B., Furr D.P.H., 1969, Evaluation of resistance to root rot caused by *Phytophthora parasitica* in seedlings of citrus and related genera. Phytopathology, **52**: 1277–1285.

Deng X.X., Zhang W.C., Wan S.Y., 1988, Studies on the isolation and plant regeneration of protoplasts in *Citrus* (in Chinese). Acta Hort. Sin., **15**: 99–102.

Deng X.X., Xiao S.Y., Deng Z.A., Zhang W.C., 1993, Interspecific somatic hybrid of Ichang papeda with Valencia sweet orange. Chinese J. Biotech, **9**: 103–107.

Derks F.H.M., Hakkert J.C., Verbeek W.H.J., Colijn-Hooymans C.M., 1992, Genome composition of asymmetric hybrids in relation to the phylogenetic distance between the parents, nucleus-chloroplast interaction. Theor. Appl. Genet., **84**: 930–940.

Dudits D., Fejer U., Hadlaczky G.Y., Konez C., Horwath G., 1980, Intergeneric gene transfer mediated by protoplast fusion. Mol. Gen. Genet., **179**: 283–288.

Forsberg J., Dixelius C., Lagercrantz U., Glimelius K., 1998a, UV dose-dependent DNA elimilation in asymmetric somatic hybrids between *Brassica napus* and *Arabidopsis thaliana*. Plant Sci. **131**: 65–76.

Forsberg J., Lagercrantz U., Glimelius K., 1998b, Comparison of UV light, X-ray and restriction enzyme treatment as tools in production of asymmetric somatic hybrids between *Brassica napus* and *Arabidopsis thaliana*. Theor. Appl. Genet., **96**: 1178–1185.

Gerdemann-Knorck M., Nielen S., Tzscheetzsch C., Iglisch J., Schieder O., 1995, Transfer of disease resistance within the genus *Brassica* through asymmetric somatic hybridization. Euphytica, **85**: 247–253.

Gleba Y.Y., Hinnisdaels S., Sidorov V.A., Cherp N.N., Parokonny A.S., 1988, Intergeneric asymmetric hybrids between *Nicotiana plumbaginifolia* and *Atropa belladonna* obtained by gamma fusion. Theor. Appl. Genet., **76**: 760–766.

Glimelius K., Bonnett H.T., 1986, *Nicotiana* cybrids with *Petunia* chloroplast. Theor. Appl. Genet., **72**: 794–798.

Grosser J.W., Gmitter F.G. Jr., 1990, Protoplast fusion and citrus improvement. Plant Breed Rev., **8**: 339–374.

Grosser J.W., Mourao-Fo F.A.A., Gmitter F.G. Jr, Louzada E.S., Jiang J., Baergen K., Quiros A., Cabasson C., Schell J.L., Chandler J.L., 1996, Allotetraploid hybrids between *Citrus* and seven related genera produced by somatic hybridization. Theor. Appl. Genet., **92**: 577–582.

Grosser J.W., Ollitrault P., Olivares-Fuster O., 2000, Somatic hybridization in citrus: an effective tool to facilitate variety improvement. In Vitro Cell Dev. Biol. Plant, **36**: 434–449.

Guo W.W., Deng X.X., 1998, Intertribal hexaploid somatic hybrid plants regeneration from electrofusion between diploids of *Citrus sinensis* and its sexually incompatible relative *Clausena lansium*. Theor. Appl. Genet., **98**: 581–585.

Hinnisdaels S., Bariller L., Mouras A., Sidorov V., Del-Favero J., Veuskens J., Negrutiu I., Jacobs M., 1991, Highly asymmetric intergeneric nuclear hybrids between *Nicotiana* and *Petunia*: evidence for recombination and translocation events in somatic hybrid plants after 'gamma-fusion'. Theor. Appl. Genet., **82**: 609–614.

Hodgson R.W., 1967, Horticultural varieties of Citrus. In: Reuther W., Webber H.J., Batchelor L.D. (Eds), The Citrus Industry, **1:** 431–589.

Li F., Deng X.X., 1997, A preliminary study of asymmetric fusion of citrus protoplasts. J. Huazhong Agri. Univ., **16**: 87–90 (in Chinese).

Ling J.T., Iwamasa M., 1994, Somatic hybridization between *Citrus reticulata* and *Citropsis gabunensis* through electrofusion. Plant Cell Rep,. **13**: 493–497.

Liu B., Liu Z.L., Li X.W., 1999, Production of a highly asymmetric somatic hybrid between rice and *Zizania latifolia* (Griseb) evidence for inter-genomic exchange. Theor. Appl. Genet., **98:** 1099–1103.

Liu J.H., 1999, Studies on protoplast symmetric and asymmetric fusion in Citrus. Ph.D dissertation of Huazhong Agricultural University, Wuhan, China.

Liu J.H., Deng X.X., 1999, Production of hybrid calluses via protoplast asymmetric fusion between *Microcitrus papuana* and Newhall navel orange. Plant Cell Tiss. and Organ Cult., **59**: 81–87.

Liu J.H., Deng X.X., 2000, Production of interspecific somatic hybrid plants of *Citrus* via protoplast asymmetric fusion. Acta Bot. Sin., **42**: 1144–1148.

Liu J.H., Deng X.X., 2000, Regeneration and characterization of plants derived from protoplast asymmetric fusion in *Citrus*. Acta Bot. Sin., **42:** 1144-1149.

Liu J.H., Deng X.X., 2002, Regeneration and analysis of citrus interspecific mixoploid hybrid plants from asymmetric somatic hybridization. Euphytica, **125:** 13-20.

Liu J.H., Hu C.G., Deng X.X., 1999, Regeneration of diploid intergeneric somatic hybrid plants between *Microcitrus* and *Citrus* via electrofusion. Acta Bot. Sin., **41:** 1177–1182.

Liu J.H., Hu C.G., Deng X.X., 2000, Production of intergeneric somatic hybrid plants via protoplast electrofusion in *Citrus* (in Chinese). Acta Bio. Exp. Sin., **33:** 325–332.

Louzada E.S., Grosser J.W., Gmitter F.G. Jr., 1993, Intergeneric somatic hybridization of sexually incompatible *Citrus sinensis* and *Atalantia ceylanica*. Plant Cell Rep., **12**: 687–690.

Murashige T., Tucker D.P.H., 1969, Growth factor requirements of citrus tissue culture. In: Chapman H.D. (Ed.) Proc. 1st Citrus Symp. Vol. 3. University of California. Riverside, pp. 1155–1161.

O'Bannon J.H., Ford H.W., 1977, Resistance in citrus rootstocks to *Rodopholus similis* and *Tylenchulus semipenetrans* (Nematoda). Proc. Int. Soc. Citricult, **2**: 544–549.

Oberwalder B., Schilde-Rentschler L., Ruoß B., Wittmann S., Ninnemann H., 1998, Asymmetric protoplast fusions between wild species and breeding lines in potato-effect of recipients and genome stability. Theor. Appl. Genet., **97**: 1347–1354.

Oberwalder B., Ruoß B., Schilde-Rentschler L., Hemleben V., Ninnemann H., 1997, Asymmetric fusion between wild and cultivated species of potato (*Solanum* spp.)—detection of asymmetric hybrid and genome elimination. Theor. Appl. Genet., **94**: 1104–1112.

Rasmussen J.O., Waara S., Rasmussen O.S., 1997, Regeneration and analysis of interspecific asymmetric potato—*Solanum* ssp. hybrid plants selected by micromanipulation or fluorescence-activated cell sorting (FACS). Theor. Appl. Genet., **95**: 41–47.

Rasmussen J.O., Lossl A., Rasmussen O.S., 2000, Analysis of the plastome and chondriome origin in plants regenerated after asymmetric *Solanum* ssp. protoplast fusions. Theor. Appl. Genet., **101**: 336–343.

Ratushnyak Y.I., Cherep N.N., Zavgorodnyaya A.V., Latypov A.S., 1993, Fertile asymmetric somatic hybrids between *Lycopersicon esculentum* Mill and *Lycopersicon peruvianum* var. *dentatum* Dun. Mol. Gen. Genet., **236**: 427–432.

Samoylov Y.M., Sink K., 1996, The role of irradiation dose and DNA content of somatic hybrid calli in producing asymmetric plants between an interspecific tomato hybrid and eggplant. Theor. Appl. Genet., **92**: 850–857.

Shi Y.Z., Guo W.W., Deng X.X., 1998, Establishment of RAPD system and identification of citrus somatic hybrid (in Chinese). Acta Hort. Sin., **25**: 105–110.

Spangenberg G., Valles M.P., Wang Z.Y., Montavon P., Nagel J., Potrykus I., 1994, Asymmetric somatic hybridization between tall fescue (*Festuca arundinaceae* Schreb.) and irradiated Italian ryegrass (*Lolium multiforum* Lam.) protoplasts. Theor. Appl. Genet., **88**: 509–519.

Tian D., Rose R.J., 1999, Asymmetric somatic hybridization between the annual legumes *Medicago truncatula* and *Medicago scutellata*. Plant Cell Rep., **18**: 989–996.

Trick H., Zelcer A., Bates G.W., 1994, Chromosome elimination in asymmetric somatic hybrids: effect of gamma dose and time in culture. Theor. Appl. Genet., **88**: 965–972.

Vardi A., Arzee-Gonen P., Frydman-Shani A., Bleichman S., Galun E., 1987, Protoplast-fusion-mediated transfer of organelles from *Microcitrus* into *Citrus* and regeneration of novel alloplasmic trees. Theor. Appl. Genet., **78**: 741–747.

Vardi A., Breiman A., Galun E., 1987, *Citrus* cybrids: Production by donor-recipient protoplast-fusion and verification by mitochondrial-DNA restriction profiles. Theor. Appl. Genet., **75**: 51–58.

Varotto S., Nenz E., Lucchin M., Parrini P., 2001, Production of asymmetric somatic hybrid plants between *Cichorium intybus* L. and *Helianthus annuus* L. Theor. Appl. Genet., **102**: 950–956.

Vlahova M., Hinnisdaels S., Frulleux F., Claeys M., Atanassov A., Jacobs M., 1997, UV irradiation as a tool for obtaining asymmetric somatic hybrids between *Nitcotiana plumbaginifolia* and *Lycopersicon esculentum*. Theor. Appl. Genet., **94**: 184–191.

Wang L.J., Ni D.A., Wan X.S., Xia Z.A., 1998, Regeneration of rice (*Oryza sativa*) fertile somatic hybrids after protoplast inactivation and fusion. Acta Biol. Exp. Sin., **31**: 413–421 (in Chinese).

Wang Y.P., Sonntag K., Rudloff E., 2003, Development of rapeseed with high erucic acid content by asymmetric somatic hybridization between *Brassica napus* and *Crambe abyssinica*. Theor. Appl. Genet., **106:** 1147-1155.

Wijbrandi J., Posthuma A., Kok J.M., Rijken R., Vos J.G.M., Koornneef M., 1990, Asymmetric somatic hybrids between *Lycopersicon esculentum* and irradiated

Lycopersicon peruvianum. 1: Cytogenetics and morphology. Theor. Appl. Genet., **80**: 305–312.

Xia G.M., Chen H.M., 1996, Plant regeneration from intergeneric somatic hybridization between *Triticum aestivum* L. and *Leymus chinensis* (Trin) Tzvel. Plant Sci., **120**: 197–203.

Xia G.M., Xiang F.N., Zhou A.F., Wang H., Chen H.M., 2003, Asymmetric somatic hybridization between wheat (*Triticum aestivum* L.) and *Agropyron elongatum* (Host) Nevishi. Thero. Appl. Genet., **107:** 299-305.

Xiang F.N., Xia G.M., Chen H.M., 2003, Effect of UV dosage on somatic hybridization between common wheat (*Triticum aestivum L.*) and *Avena sativa* L. Plant Sci., **164:** 697-707.

Xiao S.Y., Gmitter F.G. Jr., Grosser J.W., Huang S., 1995, RAPD analysis: a rapid method to identify citrus somatic hybrids (in Chinese). Hereditas, **17**: 40–42.

Xu Y.S., Murto M., Dunckley R., Jone M.G.K., Pehu E., 1993, Production of asymmetric hybrids between *Solanum tuberosum* and irradiated *S. brevidens*. Theor. Appl. Genet., **85**: 729–734.

Yamagishi H., Landgren M., Forsberg J., Glimelius K., 2002, Production of asymmetric hybrids between *Arabidopsis thaliana* and *Brassica napus* utilizing an efficient protoplast culture system. Theor. Appl. Genet., **104:** 959-964.

Yamashita Y., Terada R., Nishibayashi S., Shimamoto K., 1989, Asymmetric somatic hybrids of *Brassica*: partial transfer of *B. campestris* genome into *B. oleracea* by cell fusion. Theor. Appl. Genet., **77**: 189–194.

Zelcer A., Aviv D., Galun E., 1978, Interspecific transfer of cytoplasmic male sterility by fusion between protoplasts of normal *Nicotiana sylvestris* and X-irradiated protoplasts of male sterile *N. tabacum*. Z. Pflanzenphysiol., **190**: 397–407.

Zhou A., Xia G., Zhang X., Chen H., Hu H., 2001, Analysis of chromosomal and organellar DNA of somatic hybrids between *Triticum aestiuvm* and *Haynaldia villosa Schur*. Mol. Genet. Gen., **265:** 387-393.

Zubko M.K., Zubko E.I., Gleba Y.Y., 2002, Self fertile cybrids *Nicotiana tabacum* (+*Hyoscyamus aureus*) with a nucleo plastome incompatibility. Theor, Appl. Genet., **105:** 822-828.

8

Microspore and Anther Culture Improvements for Wheat Breeding

J. Pauk[1*], M.S. Hasan[1,2], M. Puolimatka[3], Cs. Lantos[1,2], R. Mihály[1], Á. Mesterházy[1], Z. Kertész[1] and J. Matuz[1]

[1]*Cereal Research Non-profit Company, Wheat Genetics and Breeding Department, 6701 Szeged, POB 391, Hungary.*

[2]*St. István University, Department of Genetics and Plant Breeding, 2103 Gödöllö, Pater K. u. l., Hungary.*

[3] *Plant Production Inspection Centre, Seed Testing Department, 32201 Loimaa, POB 111, Finland.*

ABSTRACT

The study of androgenesis of small grain cereals is one of the most important research fields of CR Np Co. in Szeged. Our company-derived anther culture wheat DH varieties are on seed market in Hungary. From the CR Np Co., the first haploid-derived wheat variety was released and patented in 1992. This chapter summarizes the present situation of the results of microspore and anther culture of wheat in our laboratory and breeding programme.

Culture of directly-isolated microspores of wheat (*Triticum aestivum* L.) were studied in various defined liquid culture media differing in their nutrient composition. Microspores were isolated from anther culture responsive wheat genotypes 'Mahti' and 'Hja 24201' using a microblendor. The isolated microspore cultures were then cultured with different explant types and their regeneration effect was also tested. The effect of different explants was found to be necessary for the development of embryo-like structures. The effect of ovary co-culture duration and regeneration effects were discussed. Microspore culture-derived green plants were either fertile or partially fertile and were dominantly sterile. In contrast, in cultures without different explants, only initial cell divisions were observed, but the induced multicellular structures did not develop beyond two weeks.

Complete anther culture breeding protocol is summarized in second part of this chapter. The breeding perspectives were discussed in case of using of doubled haploids in early (F1-F3) and late (F4-F6) segregated generations. Finally, the most important DH varieties from CR Np Co., Szeged, released and patented in Hungary have been presented.

INTRODUCTION

After a fairly long period of experimentations around the world on haploid production, different methods are now available to the geneticists

[*]HAS-SIU Breeding Group for Molecular Plant Breeding

and breeders. For us in Szeged, in vitro androgenesis is one of the most efficient systems of homozygote line production that is widely applicable in wheat, and could make a significant contribution to the breeding of new varieties by saving time and increasing selection efficiency via doubled haploid (DH) lines (Snape 1981; Baenziger et al., 1984).

During the development of large-scale DH production into a routine technique, many associated problems have been solved (Schaeffer et al., 1979; Lazar et al., 1984; Orshinsky et al., 1990; Kertész et al., 1991), and the protocol has been suggested for achieving new breeding goals (Barnabás et al., 2000). Some of them, which made the DH production more efficient and effective, are discussed here.

Induction medium is one of the most important factors for the induction of androgenesis and the subsequent development of embryoids and plantlets. In existing literature, about 20 different media for anther culture have already been published. On the basis of about 20 years of wheat anther culture research, the potato (P-4) medium published by Ouyang et al., (1983) appeared to be the most effective haploid induction medium following some minor modifications (P-4mf, see Materials and Methods).

Scientists showed the beneficial effects of amino acids on androgenesis (Xu 1986). It was also shown that glutamine and asparagine enhance androgenesis (Mordhorst 1993) which can partially replace the potato extract in potato medium. The positive effect of anther wall in *Nicotiana* anther culture was imputed particularly to the presence of serine (Nitsch 1974). Marsolais et al., (1986) described an efficient method of plant production by anther culture of barley (*Hordeum vulgare* L.) and wheat (*Triticum aestivum* L.) on Ficoll media. The use of liquid media with Ficoll led to an increase in embryo induction. Ficoll prevented anthers from sinking into the medium and avoided the presence of substances inhibitory to androgenesis from agar. Its applications later on opened up new vistas in wheat anther cultrure.

Researchers worldwide conducted a lot of experiments aimed at increasing the frequency of embryo formation; temperature shock, through a sudden rise in temperature regime, is one of them. Among the conditions and variables used in anther culture system, incubation temperature and stage of microspore development comprise important factors (Lazar et al., 1985). Cold pre-treatment breaks down normal microspore development and increases the frequency of symmetrical mitosis, which causes a further increase in the appearance of spontaneous diploid formation (Amssa et al., 1980).

The replacement of sucrose by other sugars, notably by maltose, enhanced the success of anther culture (Chu et al., 1990). Experiments on cereals generally showed an increase in embryo induction or an improvement in embryo development and plant regeneration by the use of maltose instead of sucrose (Orshinsky et al., 1990).

Programmes on shortening the breeding time and increasing the efficiency of the breeding are very important requirements in crop plant improvement. In the last decades of twentieth century, several new methods have been developed based on different plant tissue culture techniques. The excellent results of anther culture—and in recent years of microspore culture—have prompted new approaches in plant breeding (Mejza et al., 1993).

Valuable breeding lines and new varieties have amply demonstrated the practical aspects of DH breeding. The first wheat cultivar of anther culture origin, Jinghua No. 1, was released in China in 1986 (Hu et al., 1986). In Europe, 'Florin' was the first androgenic DH wheat variety released in France (De Buyser et al., 1987). In Hungary, 'GK Délibáb' was the first DH wheat variety registered and patented in 1992 (Pauk et al., 1995). Since then, in Hungary, new DH varieties (GK Szindbád, GK Tündér, Mv Szigma, and Mv Madrigál) have been released from the laboratory and breeding programme of the Cereal Research Non-Profit Co., Szeged and Agricultural Research Institute of the Hungarian Academy of Sciences at Matonvásár (Bedő et al., 1996; Barnabás et al., 2000).

However, besides these achievements, still many problems exist, such as genotype dependence (though today it is more moderate than was 20 years ago), plant regeneration, albinism, etc., which must be solved so as to make DH production accepted as a more routine technique and a 'technique of choice'.

MATERIALS AND METHODS

Plant Material

For the microspore and anther culture, the donor plants were grown both in the nursery and in a greenhouse. For the nursery, the seeds were sown in October and the donor tillers were collected for the experiments in first week of May. For the greenhouse conditions, seeds were germinated at 25°C and vernalized in the dark for 6 weeks at 4–6°C under constant dim light. The plantlets were transplanted to 17 cm diameter plastic pots. Each pot contained 2 plants. Plants were grown in natural light, supplemented with high-pressure sodium lights (400 W, Philips) in the greenhouse, where the temperature was controlled between 17°C and 25°C, depending on the phenophase.

MICROSPORE CULTURE

To synchronize the microspores into late uninucleated and early binucleated stage, the anthers were incubated for 3 days onto 0.3 M mannitol solution in the dark at 28°C. Following the incubation, the anthers were plated onto a 160-μm sterile nylon sieve and intensively scraped by a glass rod and repeatedly washed through the filter with 0.3 M manitol using a Pasteur pipette. The crude microspore suspension was filtered through 160 and 80-µm nylon sieve, respectively. The filtrate was then transferred to 10 mL tubes and centrifuged at 80 g for 5 min. After removing the supernatant, the microspores were resuspended in 2 mL of 0.3 M mannitol. The suspension was carefully layered on top of 21% autoclaved maltose solution, and centrifuged at 60 g for 10 min. The viable microspores in the mannitol/maltose interphase were taken up by a Pasteur pipette and washed again in 0.3 M mannitol; they were centrifuged at 60 g for 5 min.

The purified microspores were resuspended in 1 mL of culture medium. The number of microspores was estimated by a Burker haemocytometer. The microspore viability was tested by staining with fluorescein diacetate (Widholm 1972). The culture density was adjusted to 0.4–0.5 × 10^5 microspores per mL by adding the culture medium. Then, 2 mL of microspore suspension was poured into a Greiner 35 mm pre-sterilized plastic Petri dish. Parafilm-closed cultures were carried out in darkness at 80% relative humidity at 28°C. Embryo-like structures (ELS) of about 1.5–2.0 mm were transferred on Gelrite-solidified microspore culture medium under 16 h fluorescent light.

For the regeneration of ELS, 190-2 medium (Zhuang and Jia 1983) was used wherein $CuSO_4 \cdot 7\,H_2O$ doses was increased 20 times as compared to the prescribed concentration in the original MS medium.

Anther culture

The donor tillers containing microspores at the late uninucleate stage were cut between the second and third nodus and kept in an Erlenmayer flask with tap water. It was covered by a PVC bag to maintain high humidity and kept in the dark at 4°C for about 2 weeks for cold pre-treatment. Before the isolation, the microspore developmental stage was checked microscopically by squashing the anthers in a drop of water and examining without a coverglass. After the determination of microspore develpoment, spikes containing mid- and uninucleated microspores were surface-sterilized with 2% NaOCl containing 2 drops of Tween-80 for 20 min. and rinsed 3 times with sterile distilled water. Excised anthers were plated onto a modified P-4 induction medium. The P-4 basic culture medium (Ouyang et al., 1983) was modified by 2 important ingredients (P-4mf): (i) instead of sucrose, maltose was used; and (ii) agar agar (9%) was replaced with 10% Ficoll. Cultures were kept in a dark growth room at 32°C for 3 days, following which they were transferred to 28°C for embryoid induction. ELS of 1.5–2 mm size were put onto the regeneration media. A modified 190-2Cu was used as a regeneration media (see above).

When the embryoids developed into about 0.5 cm shoots, they were transferred to a tube for rooting. Well-rooted plantlets were potted in soil in 5 cm plastic net pots and transferred to a greenhouse. To maintain the ideal humidity, the regenerants were covered by PVC bags. The plantlets were grown up in 16 h daylight and at 18°C/15°C day/night temperature regime.

To determine the ploidy level of the plantlets, the length of the 10 mm distal leaf segment's stomatal guard cells was checked. Chlorophyll was extracted in 70% ethanol and the leaf segments were mounted in a drop of water on a glass slide with a coverlid. The length of the stomatal guard cells was measured by an ocular micrometer and found to be 40–50% shorter than that of control hexaploid plants. The chromosomes of the haploid plants were doubled by treatment with 0.2% colchicine in 2% DMSO solution for 5 h, followed by an overnight washing with tap water. Then the DH plants were put for 42 days into a cool chamber (4°C) for vernalization. Following this process the plants were transferred to the greenhouse under standard growing condition.

RESULTS

To generate in vitro haploid plants in wheat, there are 2 different cell and tissue culture-based methods: isolated microspore culture and anther culture. Historically, the method of wheat anther culture is about 30 years old. The first positive result of wheat microspore culture was published only ~10 years ago. The review of our results commences with microspore culture improvements, and followed by anther culture results, rather than application of haploid technique in wheat breeding is discussed. It is estimated that from the isolated divided microspore, a new in vitro haploid-derived variety can be achieved via about 1 year's laboratory work and some years of nursery selection.

Microspore Culture of Wheat

The aim of our experiment was to investigate the effect of different co-culture explants, their duration and initiation time on microspore culture of wheat (*Triticum aestivum* L). The experiment was carried out by using ovaries, glumes, florets and anthers. Genetically non-related 2 spring wheat genotypes, Mahti and Hja 24201, which are responsive to anther culture, were investigated.

In general, the responsive microspores had dense cytoplasms (Fig. 8.1) with co-culture and developed into ELS. Whereas the 'only initially responsive' microspores developed without a co-culture, the dividing cells possessed less dense cytoplasm and their cell division stopped within 2 weeks; there was no further development of microspores into ELS.

If we consider two types of ELS size produced—the large (>2 mm) and the small—the ovary and floret co-culture produced both, while the glume

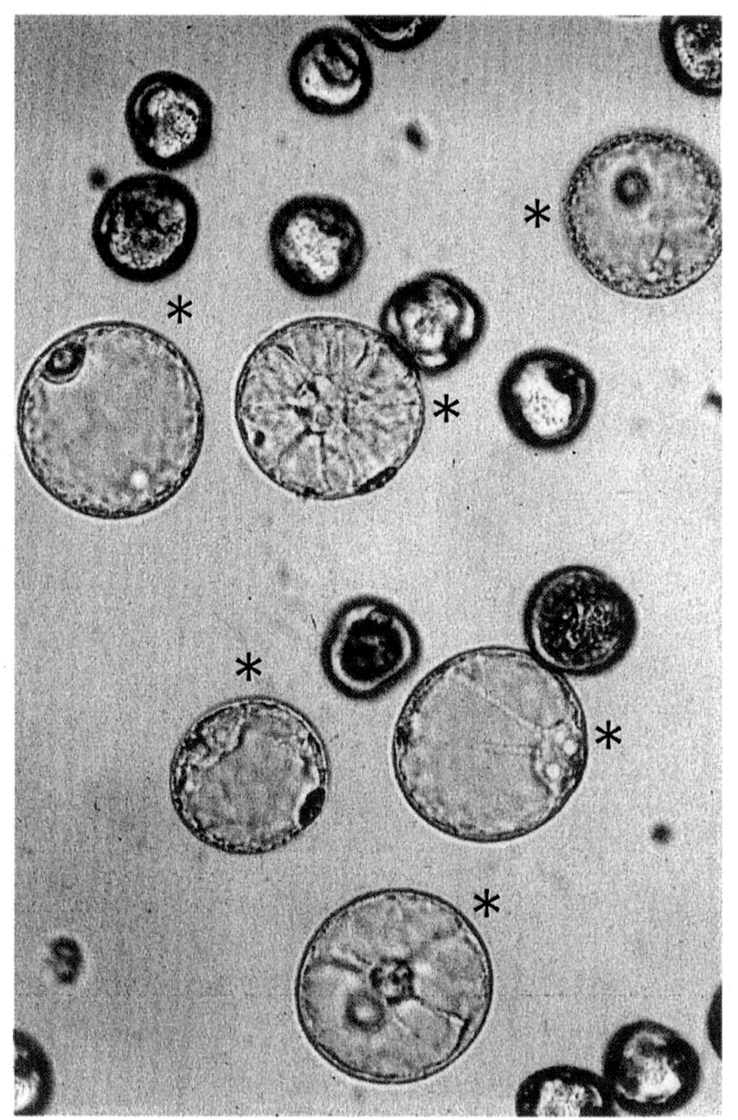

Fig. 8.1 One-day-old isolated wheat microspores in culture medium. Two different microspores are visible: cytoplasm dense living microspores (*) and dead ones which are shriveled.

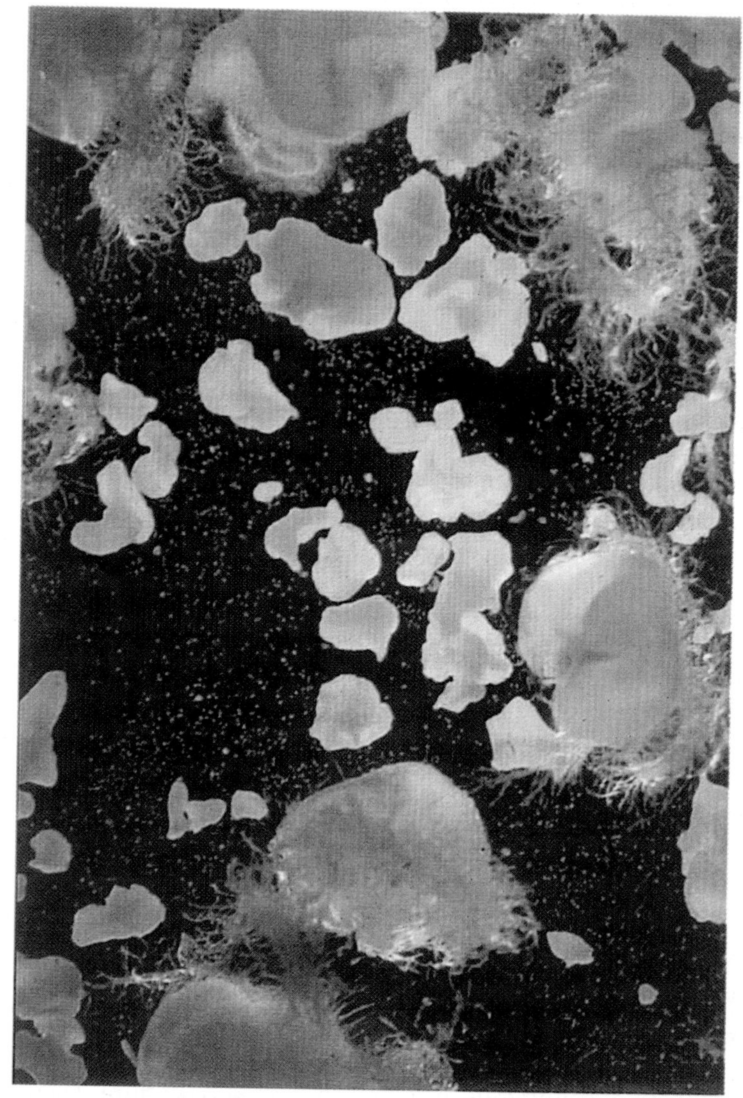

Fig. 8.2 Four-five weeks old wheat microspore ovary co-culture with ELS. The ovaries are well visible near the marginal part of the picture.

co-culture produced mainly large embryogenic structures (Fig. 8.2). Floret co-culture with large and small structures showed less embryogenic morphology than did those from the ovary or glume co-culture.

EFFECT OF CO-CULTURE EXPLANT TYPE ON THE NUMBER OF ELS

According to the statistical analysis, co-culture explants differed with regard to ELS production in Mahti, but not in Hja 24201. Significant differences were observed in Mahti when checking the ovary, floret, and anther co-culture. But, ovary, glume, floret and anther co-culture did not differ significantly from each other in Hja 24201. It was observed that the anther culture without isolated microspore produced more ELS than it was achieved through an isolated one.

EFFECT OF CO-CULTURE EXPLANT TYPE ON REGENERATION

No plant was gained in Mahti through the effect of microspore co-culture with florets. But, in all the other cases, regenerants were produced from each genotype/co-culture explant combination. According to the statistics, the explant types affected in the number of regenerants in Mahti, but not in Hja 24201. Ovary co-culture had significantly more regenerants than in anther co-culture. Glume co-culture showed no significant differences from either floret or anther co-culture. Anther culture without isolated microspores produced on average 4.8 regenerants per 100 anthers in Mahti, while Hja had 3.2. The average percentages of regenerable transferred ELS ranged from 0% in floret co-culture to 29.9% in ovary co-culture in Mahti and from 5.5% in floret co-culture to 7.3% in anther co-culture of Hja 24201. According to the P-values, the explant types did not affect the percentages of regenerable ELS in any genotype. The average percentage of green regenerants ranged from 15.1% in ovary co-culture to 100% in floret co-culture in Mahti and from 0% in floret co-culture to 33.3% in glume co-culture in Hja 24201. The 100% green plant production from anther co-culture in Mahti is based on a total of only 4 green regenerants. The floret co-culture in Hja 24201 produced 23 regenerants, all of which were albinos.

EFFECT OF OVARY CO-CULTURE DURATION ON THE NUMBER OF ELS AND REGENERATION

The results showed that ovary co-culture duration and initiation time affected the number of ELS. Five days of co-culture produced more ELS than did those of continuous or 10 and 15 days of co-culture. There were no significant differences among the continuous, 10 and 15 days of ovary co-culture.

According to a statistical test, the 2 genotypes responded differently to the addition or removal of ovaries at the fifth day in culture. The addition of ovaries on the 15th day after the isolation produced no ELS. The initiation of ovary co-culture on the fifth day produced more ELS than did finishing the co-culture at the same day in Mahti. The statistical test showed that the co-culture duration and initiation time affected the number of regenerants in Mahti, while it was not observed in Hja 24201.

Anther Culture of Wheat

In vitro androgenesis has been proved to be an efficient system of homozygote (DH) line productions (Fig. 8.3). Due to the significant contribution to the breeding of new varieties by saving time, anther culture is widely applied in a number of countries. In Szeged, from 1984 attempts were made to combine the conventional and in vitro haploid and DH methods; the results are summarized below.

The entire process is started with crossings, where the main breeding goal is to increase the genetic variability of basic material. As microspores of F_1 generation have already been segregated, the anther culture can be theoretically started from F_1 plant material. But after our breeding practice, we don't advice it, due to the lack of selection in F_1 population. It was more recommended to initiate the donor head collection from the segregated F_2 population. A good breeder can select good mother plants for donor head collection in F_2 population. The first traditional selection can be very valuable, because the absolutely negative variants (tall, disease sensitive, etc.) are neglected from the programme. The donor spike collection can be recommended from F_2 generation. The tillers should be put in an Erlenmeyer flask almost full with water and the heads should be covered by a PVC bag to maintain high humidity (Fig. 8.3). After our experiments, we can propose about 2-week cold pre-treatment of donor spikes at 4 °C, because this relatively long treatment has a positive effect on the spontaneous diploid production. After 2 weeks of cold pre-treatment, anther isolation can be started. Several methods are applied around the world. The Chinese-origin technique (tik-tak) appears to be the most effective. In case of induction media, there is a wide range of choice. A lot of media formulations were published in the last 20–25 years. After testing of different media, the modified P-4 medium (Ouyang et al., 1983) was found to be the most effective (Fig. 8.4). The modification was made by 2 components: 9% maltose is used instead of sucrose, and the other component was 10% Ficoll in place of agar. Depending on genotypes, after 5–6 weeks of isolation, embryo-like structures (ELS) can be visible in liquid medium as white beads. Usually, the ELS of 1.5–2 mm should be transferred to a regeneration media (Fig. 8.5A, B), as the bigger ones are less capable of regeneration or mostly regenerate into albino plants. A modified 190-2 medium (Zhuang and Jia 1983) proved to be more useful

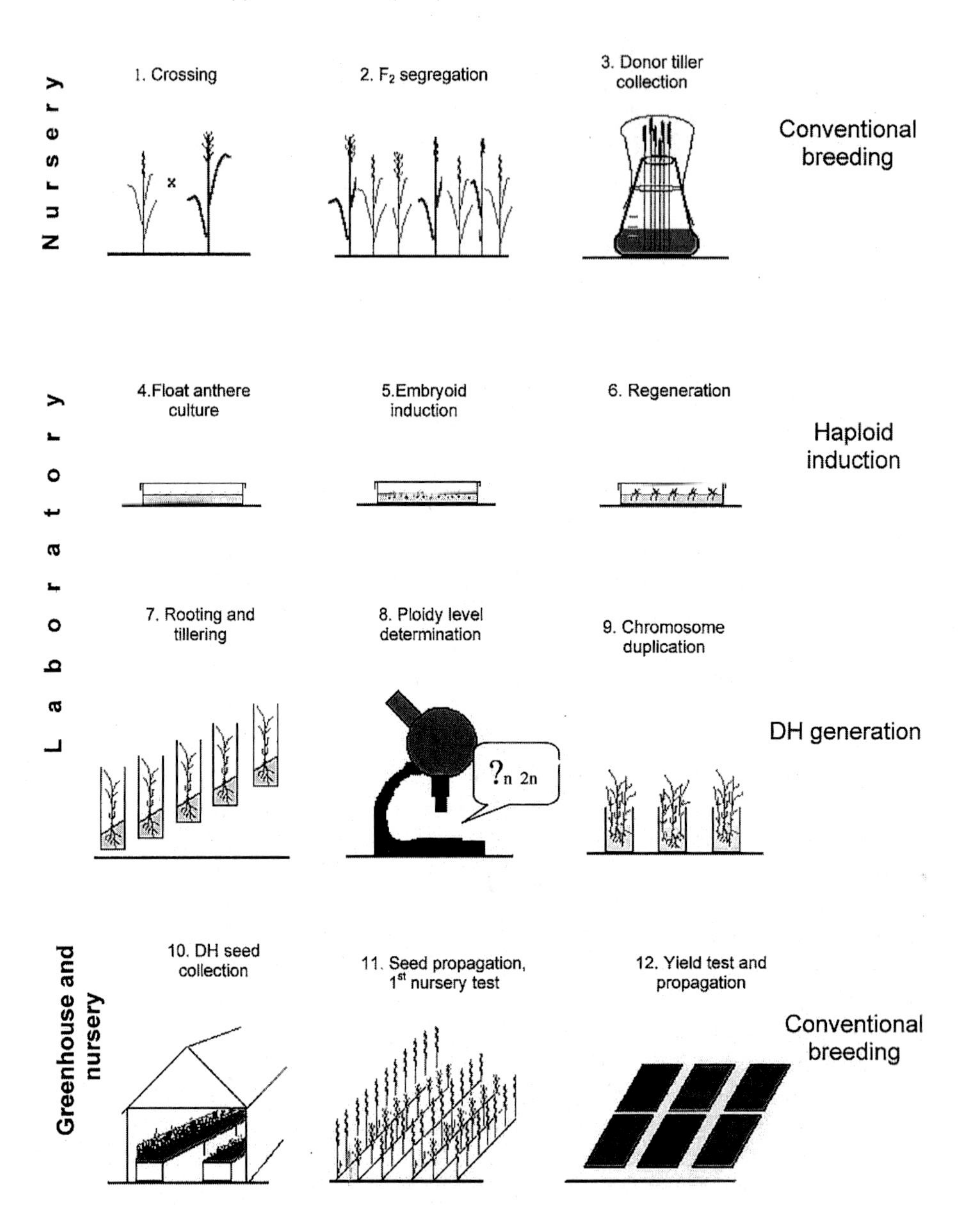

Fig. 8.3 Combining of in vitro androgenesis and conventional selection on wheat improvement.

than any other media. We found a significant effect of Cu^{++} on rooting in plantlets. Here, in the 190-2Cu medium, the doses of $CuSO_4{\cdot}7H_2O$ was increased 20 times as compared with the similar component of the MS

Fig. 8.4 Embryo-like structures (ELS, white globular structure) in float wheat anther culture using P-4mf medium.

medium. Germinated embryos (Fig. 8.5C) developed into about 0.5 cm shoots and were transferred to glass tubes for rooting and tillering. Followed by 4–5 weeks, the well-rooted plantlets (Fig. 8.5D) were potted in normal (non-sterile) soil in greenhouse and the ploidy level was determined. There are 2 common methods for the ploidy level determi-

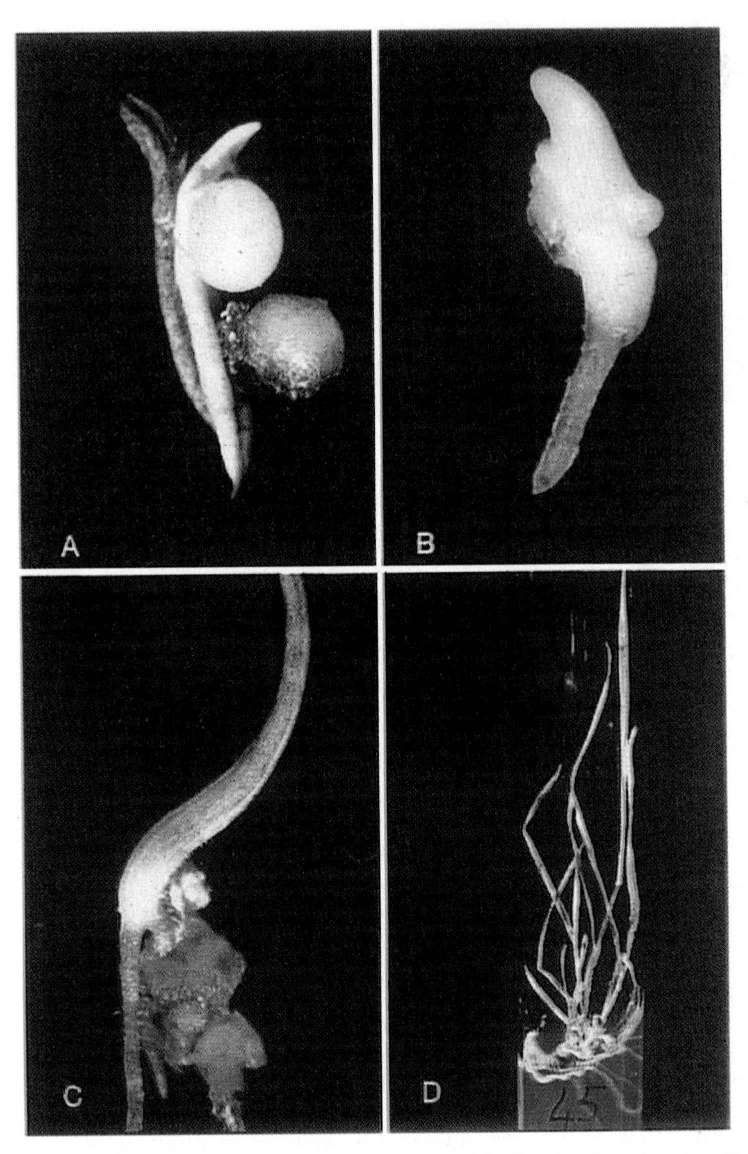

Fig. 8.5 From the microspore to the haploid plantlet in wheat anther culture: Appearing globular embryoids on the surface of anther (A). Well developed adventitious embryo, ready for germination (B). Germinated in vitro embryo onto regeneration medium (C). Well-rooted and tillered haploid plantlet in glass tube before greenhouse transplant (D).

nation. The classical root tip cytology under the microscope and the more practical measurement of stomatal guard cell length.

The chromosome number should be doubled to get homozygous fertile plants. In order to achieve this target, plantlets were treated with 0.2% colchicine in 2% DMSO solution for 5 h, followed by overnight washing under tap water. The plantlets were then put in a cool chamber for vernalization at 4°C, under dim light. After 42 days of vernalization, the plantlets are transferred to a greenhouse under standard conditions. The seeds of the DH plants were harvested and sown into ear-to-row system in nursery (Fig. 8.6).

With the previous step, the in vitro history of plants was stopped and the conventional selection gets started again. After all these stages, the breeders have to decide through objective experiments the value of DH lines. Positive DH lines become valuable breeding materials, and the superior ones can be variety candidate and after official agronomic tests they can be released, registered, and patented to be a new variety.

By combining in vitro androgenesis with the conventional breeding steps (Fig. 8.3) since 1985, in Szeged, 3 registered and patented new winter wheat varieties could be released in Hungary.

Use of Doubled Haploids in Wheat Breeding

Design of Breeding Programmes and Application of Haploid Induction

The first question which was raised simultaneously with the publication of first promising paper on in vitro haploid methods, was how could we produce large-scale pure lines in a laboratory by applying tissue culture methods. Can we develop improved breeding lines via haploid and DH induction?

Our first experiments confirmed the differential response of genotypes in anther culture and the superior induction result of segregating breeding materials (F_2) over the parental lines or varieties. These results prompted the breeders to opt for the anther culture method.

Each breeding programme is based on a large germplasm collection. A collection of breeding material is tested for response to induction and regeneration of the varieties. In the design of breeding programme, the best responsive varieties are integrated into crossings. The in vitro breeding method provides another selection criteria to the breeders in their breeding strategy.

To confirm the applicability of the in vitro haploids, about 100–150 DH analogues of different elite varieties were induced through anther culture. From the 5 varieties that we could regenerate, DH lines with the same agronomic characters were selected as the donor genotypes. We could not

Fig. 8.6 Nursery field test of DH lines in Szeged, Hungary.

measure essential gametoclonal variation, showing that the tissue culture method is unaffected by stress factors.

Breeding from early generation (F_1–F_3) via DH lines: 'GK Délibáb'

F_2 segregated populations were screened for agronomically-ideal donor elite individuals (mother plants). These were used for haploid induction (Fig. 8.7, 1st arrow). DH lines obtained via colchicine treatment were multiplied and tested further in nursery. After 3 years of official tests at the National Institute for Agricultural Quality Control for registration, the DH line 773 named GK Délibáb (Fig. 8.8A and B) was registered and released as the first winter wheat DH cultivar in Hungary.

Benefits of this breeding approach comprise perfect homogeneity, shorter breeding time, easy and simple variety maintenance, whereas the disadvantages are tedious laboratory processes, relatively high number of unusable DH lines because of lack of selection for agronomic characters.

Breeding from late generation (F_4–F_6) via DH lines: 'GK Szindbád' 'GK Tündér', use of DH lines in disease-resistant programmes

The main dream of the breeders is to produce and release an ideal variety (high and stable yield, horizontal and/or vertical disease resistance, good baking quality, etc.), which bears a long cultivation time. To reach this ideal situation, we need a lot of testing (e.g. marker-assisted selection) under the selection. When we know from the nursery and laboratory tests about what are the best populations or individuals for breeding, the haploid and DH induction can be an effective support in the process of breeding.

In our resistance breeding programme, after the most important resistance and preliminary yield tests from F_4 and F_6 generations, 2 varieties were produced (Fig. 8.7, 2nd arrow). The GK Szindbád has an excellent fungus resistance to the most important Hungarian diseases (powdery mildew, red and black rust). The high level of resistance, combined with good yield and bread quality, are found together in GK Szindbád; the variety was released in 1996.

Another new variety, GK Tündér, was evolved from a special disease resistance programme, which was similar as in the case of GK Szindád, but the tests were supplemented with head blight (*Fusarium* spp.) in an early generation test. This new variety had a broad resistance to most of the important wheat diseases in Hungary. The GK Tündér is a promising modern variety, which bears important agronomic characters in itself.

Benefits of the above breeding approach are perfect homogeneity, more useable regenerants in breeding compared with the early generation (F_1–F_3) regenerants, simple variety maintenance; the disadvantage is that there is no significant time saving in the breeding process.

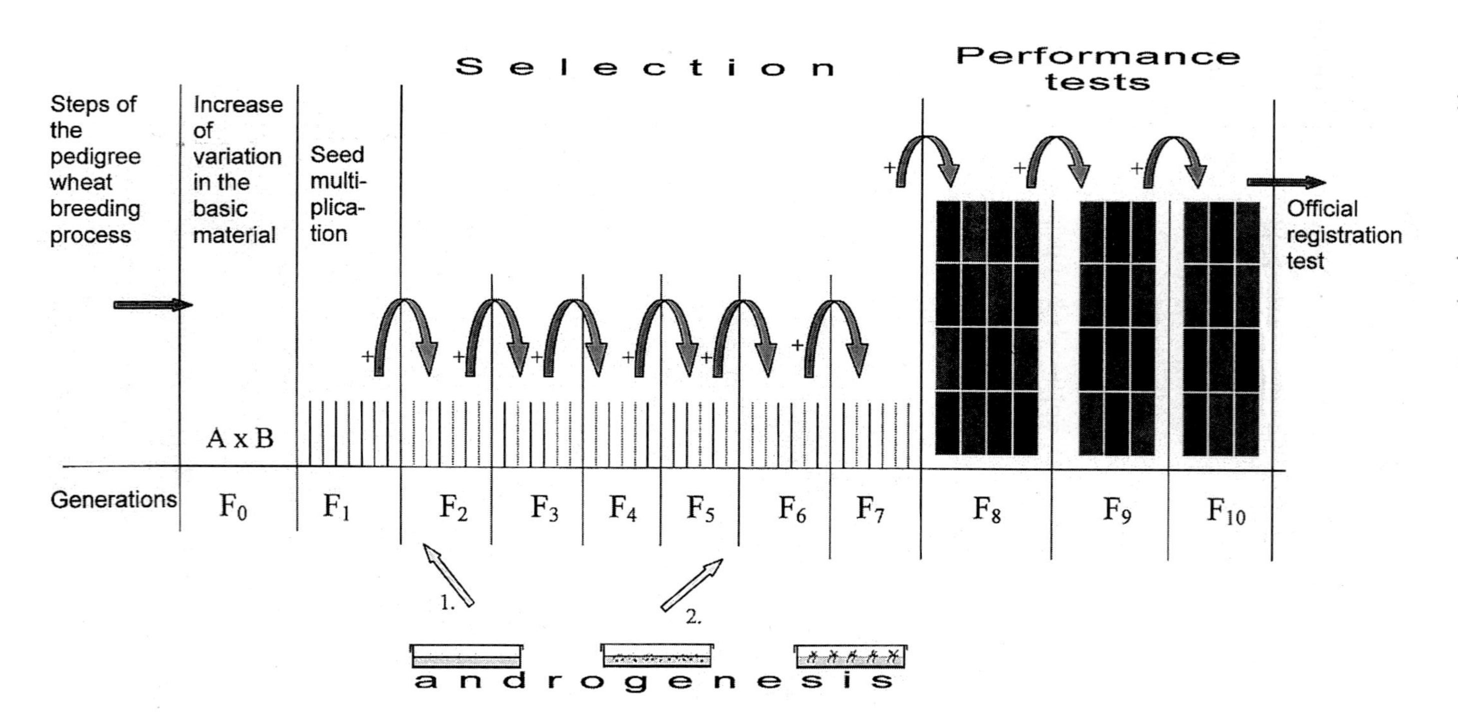

Fig. 8.7 Connections between conventional pedigree breeding method and in vitro androgenesis.

DH Wheat Varieties

In CR Co., Szeged from 1984, the combination of conventional breeding and in vitro androgenesis is used in winter wheat breeding programmes, focussing on the variety improvement. Our first androgen-derived variety: GK Délibáb was released and patented in 1992. It was followed by GK Szindbád and GK Tündér in 1996 and 2001, respectively ('GK' before the name of our variety is the Hungarian abridge of our institution name). In CR Co., Szeged, the Wheat Cell and Tissue Culture Laboratory is integrated into the Wheat Genetics and Breeding Department. It serves an excellent chance for cooperation with different breeding programmes. Here, there is no 'scale up' problem between the basic and applied research projects. Our first DH variety was bred from quality and the 2 others from resistance breeding programmes. The most important agronomic characters of DH varieties from our company are summarized below.

'GK Délibáb' (1992)

GK Délibáb is a winter wheat variety, released in Hungary in 1992, and later on, patented in the same year. This extremely early variety has a high yield, premium baking quality, excellent winter hardiness, and good adaptability to dry conditions under non-irrigated Eastern European conditions.

GK Délibáb was improved and released in 9 years (6 + 3 = breeding + registration), combining conventional plant breeding and in vitro methods. This variety is derived from a multiple cross (MM/3/Jbj/SadovoS/2/MM/Mv12) made in 1984. After F_2 conventional plant selection, haploid plants were induced from greenhouse origin F_3 lines by anther culture. Fertile DH plants, obtained via colchicine treatment, were multiplied and the obtained lines were further tested in screening nurseries, yield trials, and finally in trial network over 10 locations before registration.

The main agronomic characteristics of our new variety are the following: winter wheat, with a prostrate early growing habit over the winter. The leaves are narrow, almost completely waxy; the stems are semi-dwarf, 70–80 cm tall, elastic, but strong enough to carry the short, awnless heads, with an average sink capacity. It has a medium level of resistance to powdery mildew and stem rust, and low-level resistance to leaf rust, but it can escape the severe epidemic of fungal diseases by virtue of its earliness. GK Délibáb is the earliest cultivar now in Hungary.

This variety has a grain-yielding capacity of about 7–9 tons per hectare without irrigation, under Hungarian conditions. The grain is red, relatively small, with a 1000 kernel mass of 40 g, test weight is high, 80–

Fig. 8.8 Green head population (A), seed and bread (B) of the first Hungarian DH winter wheat variety: GK Délibáb, released in 1992.

85 kg/hl. The grain protein content is >14%, with a gluten content of 30–33%. Its baking quality is also excellent, one of the best in Hungary, with a high loaf volume of 1400 cm^3, and a falling number over 350s.

Our first DH winter wheat variety, after release, had reached a position among the 12 most popular grown wheat varieties in Hungary for 1996, and is still on the market as a DH winter wheat variety.

'GK Szindbád' (1996)

'GK Szindbád' is a high-yielding winter wheat (*Triticum aestivum* L.) variety, released and patented in Hungary, in 1996 and 1999 respectively, as a really valuable bread wheat cultivar. The origin of this variety goes back to a cross of GK Kincsö//Kremena/Avrora. Starting from the F_2, strict pedigree selection has been carried out over 3 successive generations. A peculiar F_5 line having high resistance to leaf rust, stem rust and powdery mildew was fixed in homozygous condition by anther culture, which resulted in a DH bread wheat candidate line, ready to be released.

The main characteristics of the variety are the following: winter wheat, with a semi-erect, or prostrate early growing habit over the winter. Leaves are narrow, almost non-waxy; stems are semi-dwarf, 80–90 cm tall, strong enough to carry very fertile but short, awnless heads, with very high sink capacity. It has a high level of resistance to powdery mildew, stem and leaf rust, as well as winter frost and drought during the grain filling period, in early summer. Its ripening time is intermediate.

GK Szindbád has a grain yielding capacity of about 10 tonnes per hectare without irrigation, under Hungarian conditions. The grain is red with a 1000 kernel mass of 41 g; test weight is 78–80 kg/hl. The grain protein content is >13%, with a gluten content of 30–50%. Its baking quality is good to excellent, with a high loaf volume of 1000 cm^3, and a falling number over 350s.

Our second DH winter wheat variety has now an introductory share from the Hungarian market and is still on the market, as a DH winter wheat cultivar.

GK Tündér (2001)

GK Tündér is the latest released DH winter wheat variety, coming from the Szeged wheat breeding programme. Our variety originated from a cross combination of GK Zugoly/85.50. Starting from a peculiar conventional F_5 line, it was fixed in homozygous condition by anther culture.

The main characteristics of our new variety are the following: Winter wheat, with a semi-erect growing habit over the winter; leaves are narrow, almost devoid of any wax; stems are semi-dwarf, 80–90 cm tall, strong, carrying fertile but short, white, awnless heads. It has a very high level of resistance to powdery mildew, stem and leaf rusts. Its ripening time is intermediate.

GK Tündér has a grain yielding capacity of about 10–11 tonnes per hectare without irrigation, under Hungarian conditions. The grain is red, relatively small, with a 1000 kernel mass of 38 g; test weight is 78–80 kg/hl. The grain protein content is 14%, with a gluten content of 33–35%. Its baking quality is good, with a high loaf volume of 1100 cm^3, and a falling number over 280s.

Our third DH winter wheat variety has now entered the Hungarian market as a new, DH winter wheat cultivar, with a prospective future.

CONCLUSIONS

The in vitro haploid induction and colchicine-induced DH production (Fig. 8.3) is routinely used for winter wheat breeding (Fig. 8.7) in Cereal Research Non-profit Company, Szeged in Hungary. In the past 15–16 years, 3 patented DH varieties were released (GK Délibáb, GK Szindbád, and GK Tündér). Above-mentioned results confirmed the applicability of the in vitro induced haploid and DHs in breeding.

REFERENCES

Amssa M., De Buyser J., Henry Y., 1980, Origine des plantes diploides obtenues par culture in vitro d'anthères de Blé tendre (*Triticum aestivum* L.): influence du prétraitement au froid et de la culture in vitro sur le doublement. C. R. Acad. Sc. Paris, t. 290: 1085-1097.

Baenziger P.S., Kudirka D.T., Schaeffer G.W., Lazar M.D., 1984, The significance of doubled haploid variation. In: Gustafson JP (Ed.), Gene Manipulation in Plant Improvement. pp. 385–414. (Plenum Press: New York).

Barnabás B., Szakács É., Karsai I., Bedő Z., 2000, In vitro androgenesis of wheat from fundamentals to practical application. In: Bedő and Láng (Eds.), Wheat in a Global Environment. pp. 517–525. (Kluwer Academic Publishers: Dordrecht/Bostonb/London).

Bedő Z., Karsai I., Láng L., Vida G., 1996, In vitro haploid production in higher plants. In: Mohan Jain S, Sopory SK, Veilleux RE (Eds), Current Plant Science and Biotechnology in Agriculture. Vol. 2, pp. 93–109. (Kluwer Academic Publishers: Dordrecht/Boston/London)

Chu C.C., Hill R.D., Brule-Babel A.L. 1990, High frequency of pollen embryoid formation and plant regeneration in *Triticum aestivum* L. On monosaccharide containing media. Plant Science **66**: 255–262.

De Buyser J., Henry Y., Lonnet P., Hertzog R., Hespel A., 1987, 'Florin': A doubled haploid wheat variety developed by the anther culture method. Plant Breeding **98**: 53–56.

Hu D., Yuan Y., Liu J. 1986, Jinghua No. 1—A winter wheat variety derived from pollen sporophyte. Scientia Sinica Series B **XXIX**: 733–745.

Kertész Z., Pauk J., Barabás Z. 1991, Production and utilization of doubled haploid wheat mutants in hybrid and conventional breeding. Proceedings of the FAO/IAEA Meeting. Katowice (M. Maluszynski, Z. Barabas, Eds). Cereal Research Communications **19**: 109–117.

Lazar M.D., Baenziger P.S., Schaeffer G.W., 1984, Cultivar and cultivar × environment effects on the development of callus and polyhaploid plants from anther cultures of wheat. Theoretical and Applied Genetics **67**: 273–277.

Lazar M.D., Schaeffer G.W., Baenziger P.S., 1985, The physical environment in relation to high frequency callus and plantlet development in anther cultures of wheat (*Triticum aestivum* L.) cv. Chris. J. Plant Physiology **121**: 103–109.

Marsolais A.A., Wheatley W.G., Kasha K.J., 1986, Progress in wheat and barley haploid induction using anther culture. N.Z. Agronomy Special Publication No. **5:** 240–234.

Mejza S.J., Morgant V., DiBona D.E., Wong J.R., 1993, Plant Regeneration from isolated microspores of *Triticum aestivum*. Plant Cell Reports **12**: 149–153.

Mordhorst A.P., Lörz H., 1993, Embryogenesis and development of isolated barley (*Hordeum vulgare* L.) microspores are influenced by the amount and composition of nitrogen sources in culture media. Journal of Plant Physiology **14**: 485–492.

Nitsch C. 1974, Pollen culture: a new technique for mass production of haploids and homozygous plants. In: Kasha KJ (Ed.) Haploids in Higher Plants: Advances and Potential. Proceedings of the first Symposium. pp. 123–135. University of Guelph, Ontario, Canada.

Orshinsky B.R., McGregor L.J., Johnson G.I.E., Hucl P., Kartha K.K., 1990, Improved embryoid induction and green shoot regeneration from wheat anthers cultured in medium with maltose. Plant Cell Reports **9**: 365–369.

Ouyang J.W., Zhou S.M., Jia S.E., 1983, The response of anther culture to culture temperature in *Triticum aestivum*. Theoretical and Applied Genetics **66**: 101–109.

Pauk J., Kertész Z., Beke B., Bóna L., Csõsz M., Matuz J., 1995, New winter wheat variety: "GK Délibáb" developed via combining conventional breeding and in vitro androgenesis. Cereal Research Communications **23**: 252–256.

Pauk J., Manninen O., Mattila I., Salo Y., Pulli S., 1991, Androgenesis in hexaploid spring wheat F2 populations and their parents using a multiple-step regenarion system. Plant Breeding **107**: 18–27.

Schaeffer G.W., Baenziger P.S., Worley J., 1979, Haploid plant development from anthers and in vitro embryo culture of wheat. Crop Science **19**: 697–702.

Snape J.W., 1981, The use of doubled haploids in plant breeding. In: Induced Variability in Plant Breeding. Proceedings of the International Symposium. Eucarpia, Wageningen. (Pudoc: Wageningen).

Widholm J.M., 1972, The use of fluorescein diacetate and phenosafranine for determining viability of cultured plant cells. Stain Technology **47**: 189–194.

Xu Huijun, 1986, Plantlets in anther culture of wheat. Acta Botanica Sinica **28**: 1–15.

Zhuang J.J., Jia X., 1983, Increasing differentiation frequencies in wheat pollen callus. In: Cell and Tissue Culture Techniques for Cereal Crop Improvement. Proceedings of a workshop cosponsored by the Institute of Genetics, Academia Sinica and the IRRI, pp. 431–432. (Science Press: Beijing).

9

In vitro Plant Regeneration and Preliminary Studies on *Agrobacterium*-mediated Genetic Transformation of Mungbean

R.H. Sarker* and Murshida K. Siddiqua

Department of Botany, University of Dhaka, Dhaka-1000, Bangladesh

ABSTRACT

The best response regarding callus induction was obtained by culturing the immature embryos on MS medium containing 0.5 mg/l picloram in three varieties of mungbean [(*Vigna radiata* (L.) Wilczek)]. However, these calluses failed to regenerate the shoots in a medium containing a wide range of hormonal supplements, including 6-benzylaminopurine (BAP), kinetin (Kn) and naphthaleneacetic acid (NAA). The highest frequency of adventitious-shoot morphogenesis without the intervention of callus was achieved from the cotyledonary node. The best response in multiple shoot induction from this explant was found on MS medium containing 2.0 mg/l BAP + 10% coconut water in all the varieties. The highest response regarding direct multiple shoot regeneration from the cotyledonary node was obtained in BARI mug-3 (a mungbean variety). Root induction at the base of the isolated regenerated shoots was found to be the best when shoots were cultured on full strength MS medium containing 1.0 mg/l indole-3-butyric acid (IBA) in all varieties. These in vitro regenerated plantlets were successfully established in the soil. Histological sections prepared by using freezing microtome revealed the formation of several shoot buds from the cotyledonary node explant as well as the organization of vascular tissue within the developing shoots. *Agrobacterium tumefaciens* strain containing binary plasmid pBI-121 with β-glucuronidase (GUS) and neomycin phosphotransferase (*npt*II) genes was used for co-cultivation of explants. A number of GUS positive regions were obtained within co-cultured cotyledonary node, cotyledon and leaf segment following GUS histochemical localization, indicating the transformation ability of these explants of mungbean.

INTRODUCTION

Several different grain legumes are cultivated throughout the tropics, subtropics and temperate zones of the world. Grain legumes are commonly known as pulses and have been considered to be one of the most important sources of vegetable dietary protein in many developing

*Corresponding author: E-mail: rhsarker@bd.drik.net

countries where malnutrition is associated with the acute shortage of animal protein production (Bressani 1973).

Moreover, grain legumes have the ability to fix atmospheric nitrogen and thus improve the soil fertility (Bajaj and Gosal 1981). Unlike cereal, pulses deposit significant amounts of organic matter into the soil, enriching it with much needed carbon. Mungbean is one of the most important pulse crops grown mainly for its protein-rich edible seeds (Poehlman 1991). Due to low yield potentials, most of the farmers of developing countries are giving less priority to mungbean cultivation and are allocating marginal land for this purpose. Considering its importance, it is an essential task to improve the agronomic performance of this crop. Its improvement in terms of higher seed yield, disease resistance and wider adaptability is very important in the context of our agricultural need.

Efficient plant regeneration system is required for the successful crop improvement programmes using modern techniques of biotechnology. In case of grain legumes, improvement through in vitro techniques is mostly hampered due to the recalcitrant nature of leguminous tissue. Little attention has been paid to tissue culture research in mungbean. Transfer of desired traits is an important step toward crop improvement. Among the techniques used for gene transfer, A*grobacterium*-mediated transformation has been considered as an efficient way to introduce the desirable traits in crop plants (Gardner 1993; Islam et al., 1993). This transformation method has been demonstrated in a number of leguminous crop, including soybean (Hinchee et al., 1988), pea (Schroeder et al., 1993), subterranean clover (Khan et al. 1994), white clover (White and Greenwood 1987), chickpea (Kar et al., 1996; Krishnamurthy et al., 2000) and lentil (Sarker et al., 2003).

The present investigation has been carried out as a preliminary step to perform *Agrobacterium*-mediated genetic transformation in mungbean varieties. Since the genetic transformation technology depends on the efficient in vitro regeneration system of that particular crop, the present study was also attempted to develop a protocol for in vitro regeneration of mungbean varieties prior performing transformation experiments. Moreover, efforts were also made to understand the anatomical features involved during the in vitro regeneration from suitable explants of mungbean.

MATERIALS AND METHODS

PLANT MATERIAL

Seeds of the three varieties of mungbean (*Vigna radiata* (L.) Wilczek) namely, BARI mug-2, BARI mug-3 and MK-72 used in the present study were collected from the Bangladesh Agricultural Research Institute (BARI), Joydebpur, Gazipur. These materials were maintained in the Department of Botany, University of Dhaka, Bangladesh.

Explants of shoot tip, nodal segment, epicotyl, hypocotyl, mature embryo, leaf segment and cotyledonary node of the afforsaid materials were collected from aseptically grown three- to seven-day-old seedlings as well as from germinated seeds. Medium containing 3% sucrose and 0.8% agar was used to germinate the seeds and to grow seedlings aseptically. Immature seeds were collected from the plants grown in the net house as well as in the field of the Department of Botany, University of Dhaka. Immature embryos obtained from 15- to 18-day-old pods were used during this investigation. These pods were washed in running tap water and surface sterilized with 0.1% $HgCl_2$ for ten min. The pods were then washed four to five times with sterile distilled water. Following washing, the immature seeds were taken out of the pods and the seed coats were removed carefully. The cotyledons were then separated with a pair of forceps to gather the embryos.

Callus induction and plantlet regeneration

Experiments were conducted using MS (Murashige and Skoog 1962) as well as MSB (macro- and micro elements of MS with vitamins of B_5 medium, Gamborg et al., 1968) medium, supplemented with various concentrations and combinations of 2,4-dichlorophenoxy acetic acid (2,4-D), picloram, kinetin (Kn), 6-benzylamino purine (BAP) and coconut water (CW). Regeneration of plantlets was attempted through callus culture as well as without the intervention of callus, using various explants. For all media, pH was adjusted to 5.8 before autoclaving. The cultures were maintained under fluorescent light on a 16-h photoperiod at 25 ± 2°C.

To obtain a complete plantlet, in vitro raised shoots were subjected to root induction. Regenerated shoots of 2–3 cm long were dissected and placed vertically on the medium for root induction. Both half and full strength MS medium, supplemented with different concentrations of indole-3-butyric acid (IBA), were tried for root induction.

Histological studies

During the formation of adventitious shoots, anatomical studies were carried out using a freezing microtome (Coldtome, Sakura, Japan). Small segments (4 × 4 mm) of regenerating tissue were placed on a specimen holder with a small drop of distilled water at –20°C to change the tissue segment into a frozen block. The sample thus prepared was sliced into thin sections of 15 – 20 μm also under –20°C. The sections were stained with either Coomassie brilliant blue R-250 or decolorized aniline blue solution. The stained sections were also observed under Nikon photomicroscope, using an epi-fluorescence illumination with a combination of UV-2A and V-2A filter cassette.

AGROBACTERIUM-MEDIATED TRANSFORMATION

Agrobacterium tumefaciens strain LBA-4404 used in this study was obtained by the courtesy of Professor Zeba Islam Seraj of the Department of Biochemistry, University of Dhaka. The *Agrobacterium* strain with the binary plasmid pBl-121 containing both β-glucuronidase (GUS) and neomycin phosphotransferase (*npt*II) genes was used (Fig. 9.1). Stocks of this *Agrobacterium* harboring the binary plasmid was grown and maintained in YMB (yeast extract mannitol broth) agar medium, having 50 mg/l kanamycin with pH adjusted at 7.0 – 7.2. For co-cultivation, *Agrobacterium* suspension was made using the same components of YMB medium except agar. This suspension culture was prepared by inoculating a single colony from the stock of *Agrobacterium* in the YMB medium and maintained for 19 hours in a shaking incubator.

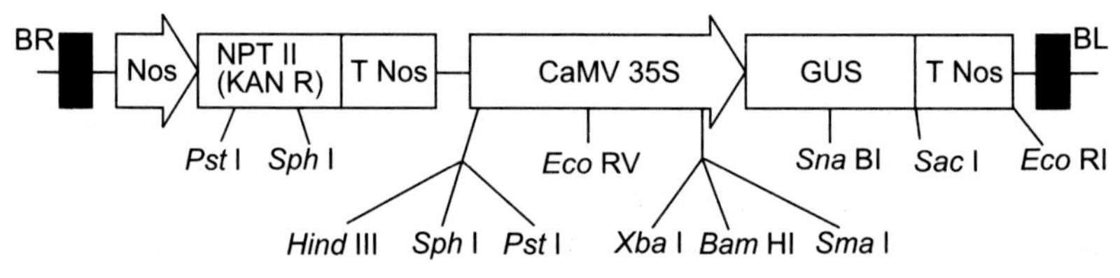

Fig. 9.1 Region between left (BL) and right (BR) boarder of pBI121 showing *npt*II and GUS genes.

The transformation ability of the various explants, such as cotyledonary node, cotyledon, hypocotyl, epicotyl, leaf segment and shoot tip was examined from BARI mug-3. The cut explants were dipped in *Agrobacterium* suspension for different periods and transferred to Petri dishes containing co-cultivation medium. Prior to transfer to the co-cultivation medium, all the explants were soaked with a sterile filter paper for a short period to remove excess bacterial suspension. The co-cultivation medium is composed of regeneration medium containing MS with 2.0 mg/l BAP and 10% CW. For co-cultivation, the Petri dishes containing the cultures were incubated under fluorescent illumination having 16-h photoperiod. Following co-cultivation for 72 h, the explants were subjected to histochemical GUS assay.

RESULTS AND DISCUSSION

Various explants, such as immature embryo, shoot tip, hypocotyl, epicotyl, cotyledonary node, nodal segment, mature embryo and leaf segment were used for regeneration of plantlets through callus culture. For this purpose, explants were cultured on MS and MSB media containing hormonal supplements of 2,4-D (0.2–10.0 mg/l), picloram (0.2–10.0 mg/l), Kn (0.1–2.0 mg/l) and BAP (0.5 mg/l). Several concentrations of these hormones either alone or in combination with others were used to induce callus using explants from three varieties of mungbean.

Among all the explants, immature embryo produced the highest percentage of callus which was followed by cotyledonary node explant. The best response in callusing from immature embryo was observed when cultured on MS medium with the supplement of 0.5 mg/l picloram (Table 9.1). The calluses were mostly compact in nature and were whitish, creamy and light green in color. Very little or no response regarding callus induction was recorded in case of the explants namely, hypocotyl, epicotyl, nodal segment, shoot tip, leaf segments and mature embryo. The overall response of callus induction by all three varieties used in this study was almost equal.

Calluses derived from immature embryo and cotyledonary node were used to obtain the regeneration on MS medium containing variable concentrations and combinations of BAP (0.5 – 1.0 mg/l), Kn (0.5 – 1.0 mg/l) and NAA (0.2–5.0 mg/l). Such attempts towards the induction of organogenesis through callus culture were not successful. Although many scientists reported a good amount of callus formation out of legume tissue culture, regeneration of complete plants from callus still presented as a big problem (Gosal and Bajaj 1979; Mroginski and Kartha 1984; Singh et al. 1986; Bhadra et al., 1989). Earlier, Bose et al., (1992) reported that all strains of *V. radiata* negatively responded to shoot differentiation via callus.

Apart from these studies, culture conditions were developed for shoot regeneration directly from the explant without the intervention of callus in three varieties of mungbean. MS and MSB media containing different concentrations of BAP (0.2 – 1.0 mg/l) alone or in combination with Kn (0.1 – 0.5 mg/l) and NAA (0.2 – 0.5 mg/l) were employed for this purpose. Moreover, coconut water (0 – 25%) was also included as an ingredient of the medium.

Results obtained in this study demonstrated that the combination of BAP and coconut water in MS medium was effective in regenerating multiple shoots from all the explants, including cotyledonary node (Table 9.2). The best hormonal combination from multiple shoot regeneration found to be 2.0 mg/l BAP + 10% CW for all the varieties of mungbean (Figs. 9.2 and 9.3). Lawrence and Arditi (1964) observed that a lower amount of coconut milk enhanced the seed germination and seedling growth in orchids. Present investigation clearly demonstrated that coconut water certainly possessed the capacity of inducing multiple shoots in mungbean.

Table 9.1 Effects of different concentrations of picloram, Kn and BAP in MS medium on callus induction from immature embryo of *Vigna radiata* (L.) Wilczek var BARI mug-3.

Hormonal supplements (mg/l)			*No. of explants* inoculated	*Percentage of explants* responded to *callus induction*	*Days to callus initiation*	*Nature of callus*
Picloram	*Kn*	*BAP*				
0.2	–	–	40	95.0	3 – 4	Creamy
0.5	–	–	40	97.5	3 – 4	Whitish to creamy and greenish
1.0	–	–	40	75.0	3 – 4	Creamy and semi-friable
2.0	–	–	40	90.0	3 – 4	Whitish, creamy and semi-friable
5.0	–	–	40	80.0	3 – 4	Creamy
1.0	0.2	–	50	96.0	3 – 8	Whitish creamy
1.0	0.5	–	50	90.0	3 – 8	Creamy and whitish friable
0.5	0.1	–	50	80.0	3 – 8	Brownish
5.0	2.0	–	50	90.0	3 – 8	Creamy
0.5	–	0.5	40	87.5	3 – 4	Creamy and greenish

Table 9.2 Effect of different concentrations of BAP and coconut water (CW) in MS medium on multiple shoots induction from cotyledonary nodal explant of *Vigna radiata* var. BARI mug-3.

Supplements		*No. of explants inoculated*	*Days to shoot initiation*	*Percentage of multiple shoots forming explants*	*No. of shoots/ explant*	*Mean no. of shoots/explant*
BAP (mg/l)	*CW (%)*					
1.0	10	36	8 – 12	50.0	1 – 2	1.5
1.5	10	36	8 – 12	75.0	3 – 4	3.5
2.0	10	36	8 – 12	80.0	6 – 12	9.0
2.5	10	36	8 – 12	40.0	2 – 3	2.5
3.0	10	36	8 – 12	62.5	3 – 5	4.0
4.0	10	36	8 – 12	14.3	1 – 2	1.5
5.0	10	36	8 – 12	50.0	3 – 6	4.5

Among the various explants studied, the best response in multiple shoot formation has been achieved from cotyledonary node. Cotyledonary nodes also reported to be effective in producing multiple shoots without a callus phase (Mathews 1987). Such cotyledonary node also produced multiple shoots in case of chickpea (Kar et al., 1996; Sarker and Awal 1999) and in *Vigna mungo* (Gill et al., 1987; Mathews and Rao 1984). Although the frequency of shoot formation was low, the explants like nodal segment and shoot tip were also found to produce multiple shoots. During the present investigation, MS medium was found to be effective for both callus induction and plant regeneration. This is in agreement with that the findings of Reddy and Reddy (1993) and Kartha et al., (1981).

Anatomical and histological studies were performed for better understanding of the nature of morphogenesis during in vitro regeneration (Yeung 1999). Only a few reports are Known regarding the histological studies in grain legumes (Gill et al., 1987; Mendoza et al., 1993). However, no histological evidence is available to illustrate the regeneration of shoots in mungbean without the intervention of callus. Histochemical studies regarding the regeneration of shoots without a callus phase is helpful in selecting the suitable explants as well as the required culture conditions for optimum regeneration. Therefore, efforts have been made to carry out histological investigations during the present study. The observation obtained during this study clearly demonstrated the nature of in vitro morphogenesis from the mungbean tissue. Histological studies clearly demonstrated the formation of a number of shoot buds (Fig. 9.4), directly following the culture of cotyledonary nodes. Moreover, the technique of freeze sectioning is very simple and does not require the complicated process of conventional microtomy. Another advantage of this method is that the sections obtained through freezing microtome can be suitably used for histochemical analysis, including the fluorescent microscopy.

To get a whole plantlet, the development of root from the regenerated shoots is no doubt very important. For root induction, regenerated shoots were cultured on MS and half strength of MS media, supplemented with different concentrations of IBA. Full-strength MS medium was found to be more effective than half-strength MS medium. Inclusion of 1.0 mg/l IBA in MS medium was found to be the best in producing roots, in all the three varieties of mungbean. Root developed from the base of the regenerated shoot is shown in Fig. 9.5. Following sufficient development of roots, the regenerated plantlets were successfully transferred into the soil. Gradually, the plantlets were found to adapt to the soil (Fig. 9.6).

A number of *Agrobacterium*-mediated genetic transformation experiments were conducted to find out the transformation ability of various explants of BARI mug-3 using the GUS gene as a screenable marker. Following

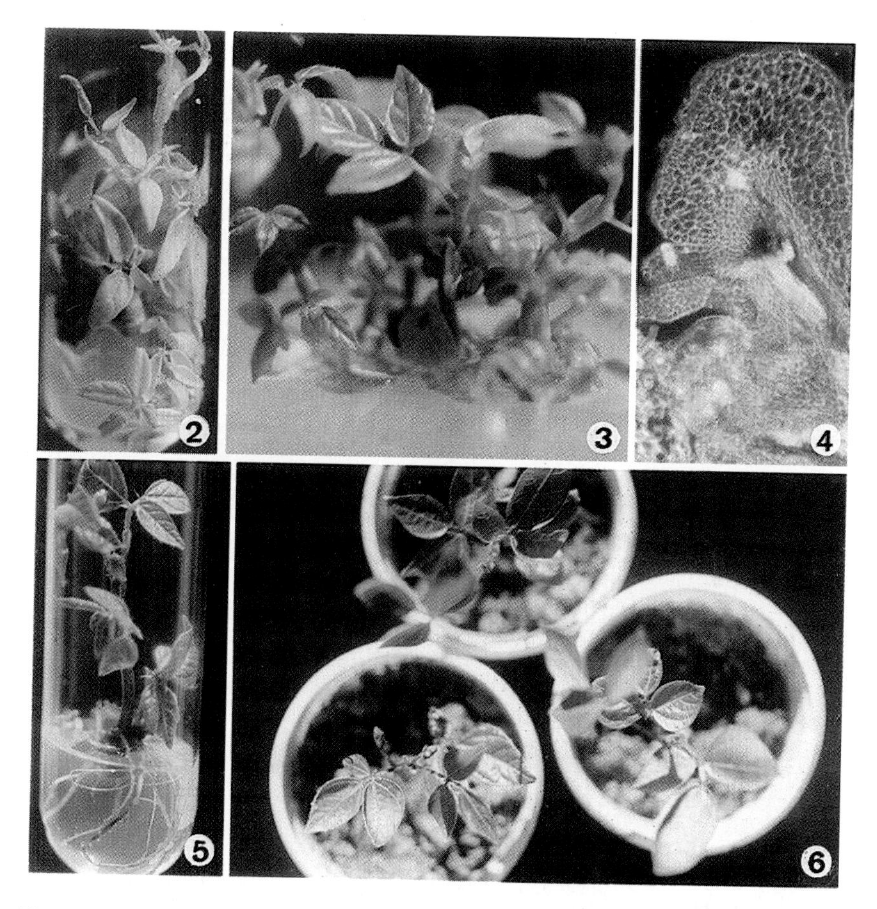

Fig. 9.2 – 9.6 In vitro regeneration of plantlets in mungbean. 9.2. Regeneration of multiple shoots from cotyledonary node of BARI mug-3 on MS medium containing 2.0 mg/l BAP and 10% CW. 9.3. Same as Fig. 9.2 but in case of MK-72. 9.4. Longitudinal section through the developing shoot buds showing the compact nature of cells. ×58. 9.5. Roots developed from the base of the regenerated shoots of BARI mug-3. 9.6. Plantlets growing in small pots containing soil.

co-culture, the explants were subjected to histochemical assay in order to verify the transient expression of GUS. A good number of cotyledonary node, cotyledon, leaf segment and hypocotyl explants were found positive through transient GUS assay (Table 9.3). Conspicuous GUS positive blue color regions were detected along the cut ends as well as within the internal tissue of the explant (Figs. 9.7 – 9.10). GUS expression was found in cotyledonary node, cotyledon, leaf segment and hypocotyl segment. But

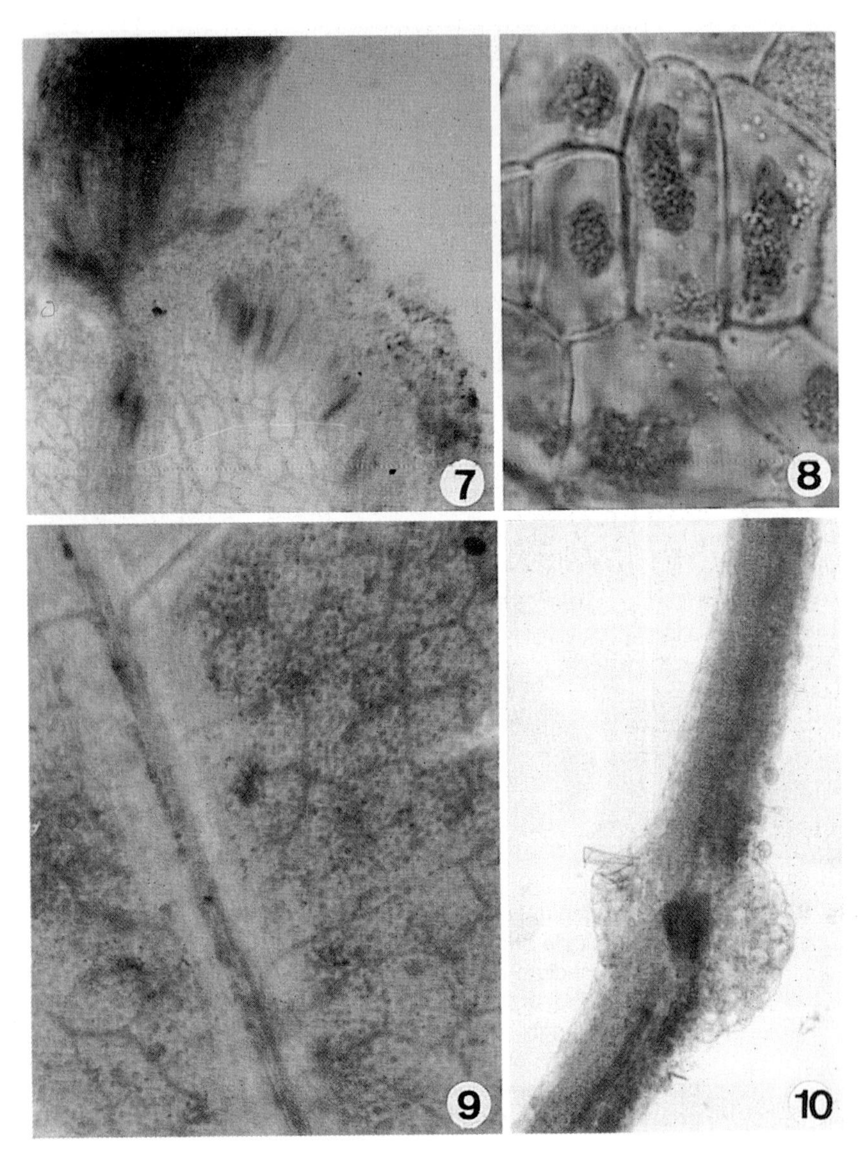

Fig. 9.7 – 9.10 Expression of GUS gene in various explants of mungbean (BARI mug-3). 9.7 Micrograph showing GUS positive blue color cells from the cotyledonary node. × 29. 9.8. Magnified view of GUS activity in cotyledon. ×480. 9.9. Leaf segment tissues showing GUS expression. ×58. 9.10. Transverse section of leaf showing GUS positive (blue color) zone. × 120.

Table 9.3 Responses of various explants of BARI mug-3 towards histochemical GUS assay following co-cultivation with *Agrobacterium* strain containing binary plasmid pBI121.

Name of explants	*No. of explants inoculated*	*No. of explants assayed for GUS*	*No. of GUS positive explants*	*Percentage of GUS positive explants*
Cotyledonary node	110	20	10	50.0
Cotyledon	80	16	9	56.25
Shoot tip	50	5	–	–
Hypocotyl	60	14	5	35.71
Epicotyl	40	10	–	–
Leaf segment	80	22	10	45.45

none of the epicotyl and shoot tip was found positive towards GUS activity.

The preliminary transformation experiments revealed that a few mungbean explants were susceptible to *Agrobacterium* infection. Such studies might be helpful in developing a transformation protocol for mungbean. The regeneration system developed in this study without the intervention of callus may also be exploited in developing transgenic plants of mungbean. Among the explant studied, cotyledonary nodes were susceptible to *Agrobacterium* and simultaneously, regeneration was achieved from this explant. Therefore, this explant may be used in transformation experiments. Further transformation protocol development for mungbean largely depends on achieving the regeneration of the plantlet from suitably transformed explant tissue.

REFERENCES

Bajaj, Y.P.S. and Gosal S.S., 1981, Introduction of genetic variability in grain legumes through tissue culture. In: Proc. COSTED Symposium on tissue culture of economically important plants. A.N. Rao (Ed.) Singapore. pp. 24 – 40.

Bhadra, S.K., Hammatt N. and Davey M.R., 1989, Prospects for the use of in vitro techniques in the improvement of *Vigna* pulses. SABRAO J. **21(2):** 75 – 91.

Bose, M., Sarker R.H., Hoque M.I. and Haque M.M., 1992, Investigation into the possible causes for failure of in vitro regeneration in mungbean. Plant Tissue Cult. **2(2)**: 81 – 88.

Bressani, R., 1973, Legumes in human diet and how they might be improved. PAG of United Nations System. pp. 15 – 42.

Gamborg, O.L., Millers R.A. and Ojima K., 1968, Nutrient requirements of suspension cultures of soybean root cells. Exp. Cell. Res. **50**: 151 – 158.

Gardner, R. C., 1993, Gene transfer into tropical and subtropical crops. Scientia Hort. **55**: 65 – 82.

Gill, R., Eapen S. and Rao P.S., 1987, Morphogenic studies of cultured cotyledons of Urd bean (*Vigna mungo* (L.) Hepper.). J. Plant Physiol. **130**: 1 – 5.

Gosal, S. S. and Bajaj Y.P.S., 1979, Establishment of callus tissue cultured and the induction of organogenesis in some grain legumes. Crop Improv. **6**: 154 – 160.

Hinchee, M.A.W., Cannor-Ward D.V., Newell C.A., Mc Donell R.E., Sato S.J., Gasser C.S., Fischhoff D.A., Re D.R., Fraley R.T. and Horsch R.B., 1988, Production of transgenic soybean plants using *Agrobacterium*-mediated DNA transfer. Biotechnology **6**: 915 – 922.

Islam, R., Farooqui H. and Riazuddin S., 1993, In vitro organogeneisis of chickpea and its transformation by *Agrobacterium tumefaciens.* Plant Tissue Cult. **3(1)**: 29 – 34.

Kar, S., Johnson T.M., Nayak P. and Sen S.K., 1996, Efficient transgenic plant regeneration through *Agrobacterium*-mediated transformation of chickpea (*Cicer arietinum* L.). Plant Cell Rep. **16**: 32 – 37.

Kartha, K.K., Pahl K., Leung N.L. and Mroginski L.A., 1981, Plant regeneration from meristems of grain legumes: soybean, cowpea, peanut, chickpea and bean. Can. J. Bot. **59**: 1671.

Khan, M.R.I., Tabe L.M., Heath L.C., Spencer D. and Higgins T.J.V., 1994, *Agrobacterium*-mediated transformation of subterranean clover (*Trifolium subterraneum* L.). Plant Physiol. **105**: 81 – 88.

Krishnamurthy K.V., Suhasini K., Sagari A.P., Meixner M., de Kathen A., Pickardt T., and Schieder O., 2000, *Agrobacterium* - mediated transformation of chickpea (*Cicer arietinum* L.) embryo axes. Plant Cell Reports **19:** 235 - 240.

Lawrence, D. and Arditti, J., 1964, New medium for the germination of orchid seed. Amer. Orchid Soc. Bull. **33**: 766 – 768.

Mathews, V.H. and Rao P.S., 1984, In vitro production of multiple seedlings from single seeds of mungbean (*V. radiata* L. Wilczek). Z. Planzenphysiol. **113**: 325 – 239.

Mathews, V.H., 1987, Morphogenic responses from in vitro cultured seedling explant of mungbean (*Vigna radiata* (L.) Wilczek). Plant Cell. Tissue Org. Cult. **11**: 233 – 240.

Mendoza, A.B., Hattori K., Nishimura T. and Futsuhara Y., 1993, Histological scanning electron microscopic observations on plant regeneration in mungbean cotyledon (*Vigna radiata* (L.) Wilczek) cultured in vitro. Plant Cell. Tissue Org. Cult. **32**: 137 – 143.

Mroginski, L.A. and Kartha K.K., 1984, Tissue culture of legumes for crop improvement. In: Plant Breeding Rev. Inc. Westport, Cann. **2**: 215 – 264.

Murashige, T. and Skoog F., 1962, A revised medium for rapid growth and bio-assays with tobacco tissue culture. Plant Physiol. **15**: 473 – 497.

Poehlman, J.M., 1991, The Mungbean. Oxford & IBH Publishing Co. Pvt. Ltd., New Delhi, India.

Reddy, I.R., Reddy G.M., 1993, Factors affecting direct somatic embryogenesis and plant regeneration in groundnut, *Arachis hypogaea* L. Indian J. Expt. Biol. **31**: 57 – 60.

Sarker, R.H. and Awal S.T., 1999, In vitro morphogenesis in chickpea (*Cicer arietinum* L.). Plant Tissue Cult. **9(2)**: 141 – 150.

Sarker R.H., Biswas A., Mustafa B.M. Mahbub S., Hoque M.I., 2003, Agrobacterium- mediated transformation of in lentil (*Lens culinaris* Medik) Plant Tissue Cult. **13(1):** 1 – 12.

Schroeder, H.E., Schotz A.H., Wardly-Richardson T., Spencer D. and Higgins T.J.N., 1993, Transformation and regeneration of two cultivars of pea (*Pisum sativum* L.). Plant Physiol. **101**: 751 – 757.

Singh, R.P., Singh B.D. and Singh R.M., 1986, Organ regeneration in mungbean callus cultures. In: P.K. Gupta and J.R. Bahl (Eds), Genetics and Crop Improvement, Rastogi and Co., Meerut, pp. 387 – 393.

White, D.W.R. and Greenwood D., 1987, Transformation of forage legume *Trifolium repens* L. using binary *Agrobacterium* vectors. Plant Mol. Biol. **8**: 461 – 469.

Yeung, E.C., 1999, The use of histology in the study of plant tissue culture systems – some practical comments. In vitro Cell. Dev. Plant. **35**: 137 – 143.

10

Induced Mutations and Biotechnology in Improving Crops

S. Mohan Jain* **and M. Maluszynski**
FAO/IAEA Joint Division, Plant Breeding and Genetics Section, International Atomic Energy Agency, Box 100, A-1400, Wagramerstrasse 5, Vienna, Austria

ABSTRACT

Conventionally, induced mutations are generated by physical and chemical mutagens in producing a large number of new cultivars in crop species. Alternatively, tissue culture and insertional mutagenesis are able to generate a wide range of genetic variability in plant species, which could be incorporated in plant breeding programs. By in vitro selection, desirable mutants with useful agronomical traits, e.g. salt or drought tolerance or disease resistance can be isolated in a short period of time. Genetically, stable selected somaclones or induced mutants can directly be incorporated in breeding new cultivars. Another approach for plant improvement could be somatic cell fusion. Even though somatic hybridization has a great potential for gene transfer from one plant to other plant species, it is yet to be exploited in plant improvement.

INTRODUCTION

Rapidly growing human population, decreasing arable land, combined with environmental degradation are causes of concern to agriculturists in sustaining agriculture production and loss of genetic biodiversity. Plant breeders are under continuous pressure for finding new ways for sustainable agriculture by using natural and induced genetic diversity, which are essential to crop improvement. (Jain 2002a) Plant breeders have successfully recombined the desired genes from cultivated crop germplasm and related species by sexual hybridization, and have also been able to develop new cultivars with desirable traits such as high yield, disease, insect and pests, and drought resistance. No further scope of expanding arable land is foreseen. Moreover, rapid industrialization is mounting a heavy toll on the environment, including atmospheric ozone layer depletion, acid rain, erratic weather conditions, insect and pests, diseases, global warming and increase of ultra-violet (UV-B) radiation

*Corresponding author: E-mail: s.m.jain@iaea.org

level. The adverse environmental impact will gradually lead to water shortage and reduction in crop production. Thereby, plant breeders need to adopt both conventional and new technologies such as induced mutations, plant tissue culture, and recombinant DNA for sustainable food production in order to feed the world (Jain 2002a).

In recent years, plant cell and tissue culture technology has made tremendous progress in plant regeneration from a wide range of plants such as cereals, ornamentals, fruits and forest trees. Consequently, it has paved the way to apply in genetic engineering and plant improvement. Since genetic variation is an essential component for plant breeding, it can be induced by tissue culture, induced mutations with physical, chemical and biological mutagens. Both somaclonal variation and induced mutations can generate a wide range of genetically-stable useful variants or mutants (Skirvin et al., 1993; Jain et al., 1998; Jain 2001; Maluszynski et al., 2001). Induced mutations with gamma radiation and chemical mutagens are random changes in the nuclear DNA or cytoplasmic organelles, resulting in chromosomal or genomic mutations and that enable plant breeders to select useful mutants such as high yield, flower color, disease resistance, early maturity, etc., in crop, fruit and ornamental plants (Crino et al., 1994; Micke et al., 1990). A specific advantage of mutation induction is the possibility of obtaining unselected genetic variation, improvement of vegetatively-propagated plants when one or few characters of an outstanding cultivar are to be modified (Broertjes and van Hartan 1978). In this chapter, we will focus on somaclonal variation, mutagenesis and somatic cell hybridization and their future potential in plant breeding towards feeding the world.

INDUCED MUTAGENESIS

Physical and chemical mutagen treatments induce mutations for useful genetic changes such as high yield, flower color, disease resistance, and early maturation in crop, vegetables, medicinal herbs, fruit and ornamental plants (Micke et al., 1990; Crino et al., 1994). Mutagens cause random changes in the nuclear DNA or cytoplasmic organelles, resulting in gene, chromosomal or genomic mutations. To induce mutations, the most common mutagens used are: (a) Chemical (alkylating agents)—ethyl-nitroso-urea (ENH), methyl-nitroso-urea (MNH), sodium azide (SA), ethyl-methane-sulphonate (EMS), di-ethylsulphonate (DES), ethyleneimine (EI), ethidium bromide (EB); and (b) physical agents—X-rays, gamma rays, fast neutrons, ultra violet rays, electron beam, laser beam. (Jain 2002b) Physical mutagens are good dosimetry and show reasonable reproducibility, and also have a high and uniform penetration of multicellular system, particularly by gamma rays. By using chemical mutagens, the mutation rate is high, predominantly point mutations. The problems of using physical mutagens are high cost radiation source requirement, and simultaneous induction of chromosomal and gene mutations. The uses of chemical mutagens are carcinogenic in nature and penetrate non-uniformly in multi-cellular system (Jain, 2002b).

The International Atomic Energy Agency (IAEA) maintains a database of officially released mutant varieties in different parts of the world. So far, more than 2250 improved varieties of major crops such as rice, wheat, barley, sorghum, pulses, cotton and edible oilseeds—derived through induced mutations—have been officially released worldwide (Maluszynski et al., 2000; 2001). Of these, 605 were released in China, 259 in India, 210 in the former Soviet Union, 176 in The Netherlands, 128 in the USA, and 120 in Japan. The top six countries are China, India, the former USSR, The Netherlands, Japan, and the USA. The most populous countries, i.e. China and India, who need a continuous supply of food for their ever-increasing human population, have used induced mutations to the greatest extent. Rice stands first by having the largest number: 318 induced mutant varieties, followed by barley, wheat, maize, durum wheat, oat, millet, sorghum, rye and dura (Maluszynski et al., 2000; Donini and Sonnino, 1998). A high-yielding barley "naked grain" has been released in Altiplano, Peru by selecting mutants from gamma-irradiated seeds. This area is more than 3600 m above the sea level, and is stress-prone, characterized by short vegetation periods. This variety has improved adaptability in stress-prone areas and is well accepted by the consumers. Maluszynski et al., (1995) have pointed out that even though most of the induced mutations are recessive and deleterious from the breeding point of view, a significant improvement in plant improvement worldwide has been achieved by induced mutations. For example, salt-tolerant rice in China, high-yielding cotton in Pakistan, heat-tolerant wheat in India—all these varieties have been released. IAEA developed Yuandong No. 3 wheat variety in 1976 by gamma irradiation, which shows complete resistance to rust, powdery mildew and aphids and is also tolerant to saline, alkaline and other environmental stresses. This variety is cultivated in an area of 200,000 hectares, an increase from 1000 hectares in 1986. In India, Behl et al., (1993) reported having developed heat-tolerant mutants of wheat and their yield performance was much better than the heat-sensitive types under the field conditions.

Induced Mutations in Date Palm

Date palm is one of the most important fruit trees in the Saharan and Sub-Saharan regions of Africa and the Middle East. It plays an important role in food security and maintaining the ecosystems of North African countries. In some areas, it is the only tree that provides food, shelter and fuel to the population. In North Africa, the total number of date palm trees is estimated 16 million, with 7.8 million in Algeria, 4.5 million in Morocco and 3.5 million in Tunisia. The total date palm production is estimated at 350,000 tons per year with an average yield of 30–40 kg per tree. The outbreak of Bayoud disease epidemic in the Saharan region has resulted in heavy losses of yield (Sedra et al., 1998). Since 1995, FAO/IAEA initiated Technical Cooperation on date palm with North African countries, including Algeria, Morocco, and Tunisia with the objective to induce mutations for the isolation of mutants with improved fruit yield, dwarfness, and resistance to Bayoud disease. Progress has been made in somatic embryogenesis and organogenesis; Bayoud toxin isolation from *Fusarium oxysporum* f. sp. *albedini;* radiosensitivity of date palm embryogenic cultures; and long-term maintenance of embryogenic cultures by low dosage of gamma radiation treatment. Irradiated plants are being evaluated for Bayoud disease resistance.

Induced Mutations in Banana

Banana is a poor man's fruit crop in tropical and subtropical countries. Bananas and plantains are grown in more than 100 countries throughout the world with an annual production around 88 metric tons (INIBAP Annual Report, 1998). They are cheap to produce, and can grow in a range of environments and produce fruit year-round. The world's total banana production ranks 4th after cereals, which has reached around 95 million tons per year. Banana production is severely affected by several diseases and pests such as banana bunchy top virus, burrowing nematodes (*Radopholus similes*), Moko disease (*Ralstonia solanacaearum*), black sigatoka or black leaf streak (*Mycosphaerella fijiensis*), Fusarium wilt (*Fusarium oxysporum* f. sp. *cubense)*, and nematodes. Novak et al., (1990) used in vitro shoot tips of banana and plantain for mutation induction with gamma radiation, and an early flowering mutant was isolated. FAO/IAEA started the banana Coordinated Research Program (CRP) in 1994 with the general aim to integrate radiation-induced mutations in vitro culture and molecular genetics methods into the conventional breeding of banana so as to induce desirable variations such as disease resistance, dwarfism and earliness, and also to promote the development of methods for large-scale and rapid multiplication of the mutants/segregants through somatic embryogenesis and micropropagation. The Belgium Government decided to fund this CRP in 1996. A total of sixteen countries are involved in this CRP, namely, Austria, Belgium, Colombia, the Czech Republic, Cuba, Guyana, France, Germany, Israel, Malaysia, Mexico, Philippines, Sri Lanka and the USA. Since the start of FAO/IAEA banana CRP in 1994, several achievements have been accomplished, including development of biotechnological tools, germplasm characterization, cryopreservation, radiation-induced mutations, somaclonal variation and genetic engineering. Amongst CRP participating countries, Cuba, Malaysia, Philippines and Sri Lanka generated a series of improved clones by induced mutations that were screened for different traits such as early flowering, reduced height, large fruit size, and tolerance to *Fusarium*. 'Novaria', a banana mutant variety derived from mutation induction of 'Grande Naine', was released in Malaysia with improved traits such as early flowering, short stature, high yield potential and good flavor. Collaborations among CRP participating laboratories were established, including exchange of staff, training and technology transfer. This CRP benefited several young researchers in receiving advance degrees in Israel, the Czech Republic and Belgium. A total of 51 research papers and abstracts have been published in conference proceedings and international refereed journals. Further breakthroughs were dissociation of chimerism, detection of polyploidy, and analysis of the karyological stability of embryogenic cell suspension with DNA flow cytometer; development of somatic embryogenic cell suspension of several cultivars, including plantains (AAB); early screening of Fusarium wilt in double-tray system; selection system against Black sigatoka disease by using *Mycosphaerella fijiensis* purified fraction (juglone); and screening technique for nematode resistance. Also, significant progress was made in detection of DNA methylation polymorphism in banana-micropropagated plants with amplified fragment length polymorphism (AFLP); transposon mutagenesis for gene tagging and 28 allele-specific simple sequence repeat (SSR) markers were used to detect polymorphism between the A and B genomes of *Musa*, to identify hybrids, and trace back B genome in hybrids. Fluoresence in situ hybridization (FISH) protocol was developed for *Musa* for karyological analysis, providing distinct chromosome landmarks, gene localization, analysis of long-range chromosome structure, and linking to physical and genetic maps. Some selections of gamma-irradiated banana genotypes resulted in obtaining putative mutants, showing a tolerance to Fusarium wilt disease and Black sigatoka. These are being confirmed under field conditions.

Induced Mutations in Ornamental Plants

The worldwide floriculture industry is estimated over Euros 50 billions and increasing annually by 8–10 percent, especially in the developing countries. The governments of countries like Indonesia, Malaysia and Thailand, etc., have targeted the increase in land area for floriculture products and the expected floriculture industry will contribute a substantial income of the national exports. Small farmers will get economic benefits and, consequently, floriculture will become a 'cash crop'. Under MAL05024, one new radiation induced mutant variety of orchid, namely Dendrobium 'Sonia KeenaAhmadSobri' was released in Malaysia. This mutant has diamond-shaped petals with flowers comprising parental types and shelf life is 15 days. Dendrobium is one of the largest orchid genera, consisting of more than 1500 species with many hybrids. Malaysians released another radiation-induced mutant variety *Tradescantia spathcea* var. Sobril was also released, which is variegated with creamy stripes and best used as a potted specimen. *T. spathcea* is a fast-growing tropical succulent, normally used as a ground cover to edge boarders/pathways, or as a potted specimen. In Thailand, under THA05045 project, floriculture improvement using radiation has been successful, e.g. Portulaca, Canna. In *Portulaca* sp., three (KU1, KU2, KU3) and four in *Canna* sp. (Pink Peeranuch, Yellow Arunee, Cream Prapanpongse, and Orange Siranut), radiation-induced mutant varieties have been released for flower color trait.

In vitro Mutagenesis

The plant breeder has available several in vitro techniques such as micropropagation, protoplasts, embryo rescue, somatic embryogenesis, etc., which increase the efficiency in obtaining variation, selection and multiplication of the desired genotypes (Ahloowalia 1998). The frequency of somaclonal variation is low in cereals, and that has hindered the frequent use of somaclonal variation in cereal improvement (Maddock 1986). It is desirable to increase the genetic variability by combining mutagenesis and tissue culture for the breeders to exploit in crop improvement (Jain 1998). A few more benefits of in vitro mutagenesis are: (a) mutagen treatment can be given to large cell or protoplast or somatic embryos density; (b) fast multiplication of mutant plant material; (c) in vitro selection of mutation; and (d) less space required for shoot multiplication under the controlled conditions. Tissue culture of periclinal chimeras often results in segregation of the component genotypes, depending on the pattern of differentiation and proliferation of the shoots formed (Lineberger et al., 1993). They obtained new ornamental cultivars of *Saintpaulia* and *Rhododendron.* Gavazzi et al., (1987) compared somaclonal variation and chemically-induced mutagenesis and found differences in their effects, changing the spectrum and frequency of mutants and even in some cases, the pattern of segregation of mutant character in *Lycopersicon esculentum,* and in *Brassica napus* (Jain and Newton 1988; 1989). Jain (1997b) irradiated the axillary buds with gamma rays excised from in vitro-grown strawberry plants; 5% plants survived the selection pressure of *Phytophthora cactorum* crude extract and these plants were also able to withstand water holding for 5–6 days. It seemed that pathogen-related (PR) proteins may be responsible for both drought and disease tolerance. Banerjee and Kallo (1989) observed a high total phenol content in disease-

and pest-resistant wild type tomato (*Lycopersicon hirsutum* f. *glabratum*) lines as compared to susceptible cultivated tomato (*Lycopersicon esculentum*), and suggested the use of this parameter for the selection of disease and pest resistant lines in cultivated tomato. While studying somaclonal and in vitro mutagen-induced variability in the grapevine, Kuksova et al., (1997) found an increase in tetraploid plants among somatic seedlings after gamma irradiation, and also variability was seen among regenerated plants after field tests.

T-DNA INSERTIONAL MUTAGENESIS

The alternative approach to the use of chemical and physical mutagens is through the use of *Agrobacterium*-mediated T-DNA mutagenesis (Jain 2002a). T-DNA is inserted through *Agrobacterium* infection in the plant cell genome, causing mutations. It can also be termed as insertional mutagenesis (Feldmann 1991; Feldmann et al., 1989; Tomilov et al., 1999). The integration of T-DNA within the active endogenous plant gene is random and will inactivate that gene, which may in turn, cause phenotypic mutations or even result in silencing transgenes (Heinze and Schmidt 1995; Jain et al., 1992). The site of insertion in a phenotype can be genetically mapped (using markers contained within the T-DNA). The sequences for the inactivated endogenous gene isolated by virtue of their proximity to the inserted T-DNA, can be determined. Usually, to isolate the gene of interest, a population of transgenic plants is screened. The mutated gene can be identified from a genomic library of the mutated plants using T-DNA as a probe. Yanofsky et al., (1990) obtained a homoetic gene *Ag*, regulating the flower morphogenesis in *Arabidopsis thaliana* by T-DNA tagging. Feldmann et al., (1989) produced a dwarf mutant of *Arabidopsis thaliana* by T-DNA insertional mutagenesis. Kocabek et al., (1999) isolated a flower mutation with increased number of stamens and carpels (*scaf1*) in *Arabidopsis thaliana* by *Agrobacterium*-mediated transformation. The major problem with this technology is inefficiency of producing mutants in higher plants because of the abundance of noncoding DNA, which decreases the probability of insertion into a functional gene, and also would require a lot of space for growing thousands of transformants (Coury and Feldmann 1998). Despite of inefficiency of insertional mutagenesis, Tomilov et al., (1999) increased the rate of insertional mutagenesis by presonication of germinating seeds of *Arabidopsis thaliana*, followed by *Agrobacterium*-mediated transforma-tion. Yamazaki et al., (2001) observed that *Tos17* retrotransposon specifically integrates into low-copy-number regions of the rice genome. Based on their findings, they suggested that *Tos17* could be used as an efficient tool for insertional mutagenesis. Recently, Rus et al., (2001) identified two

Arabidopsis thaliana extragenic mutations that suppress NaCl hypersensitivity of the *sos3-1* mutant. They were identified while screening T-DNA insertional population in the genetic background of Col-og/1 *sos3-1*, and genome sequencing in the left flank of T-DNA, which indicated that *sos3-1 hkt1-1* and *sos3-1 hkt1-2* plants have allelic mutations in *AtHKT1*. *AtHKT1* mRNA is more abundant in roots rather than the shoots of wild type of plants but is not detected in plants of either mutant, indicating that this gene is inactivated by the mutations. Furthermore, their results indicated that AtHKT1 is a salt tolerance determinant that controls Na^+ entry and high affinity K^+ uptake into plant roots.

STIMULATORY AFFECT OF PHYSICAL MUTAGENS

A low dosage of radiation significantly stimulates growth of in vitro cultures. Under FAO/IAEA INS/5/027 project in Indonesia, low dosage 10 Gy stimulated the growth of orchid shoot cultures. Similar stimulatory response of low dosage of gamma radiation was noticed in isolated hypocotyls segments of pigeon pea (Zapata and Aldemita 1989). In another FAO/IAEA project RAF/5/035 in Sfax, Tunisia, low dosage (15–20 Gy) of gamma radiation had a positive effect on the maintenance of date palm somatic embryogenic potential for several years without losing plant regeneration potential; reduced subculture intervals; and also had some stimulatory effect of somatic embryogenesis. Al-Juboory et al., (2000) used successfully excimer laser radiation of 308 nm wavelength in reducing fungal and bacterial contamination in vitro cultures of date palm 'Medjol' and 'deglet-Nour' varieties. Variety 'Medjol' responded very well to 50 pulse/sec laser treatment, and enhanced the frequency of somatic embryo production by 25%. Physical mutagens could stimulate organogenic or somatic embryogenic potential—depending upon genotype—in recalcitrant plant species.

Haploids have proven to be an efficient way to produce homozygous inbred plants, detection and recovery of mutants and also genetic analysis and selection within small population (Jain et al., 1996). Anther culture is one of the methods for obtaining haploid plants, which is most widely used for haploid production. One of the main major problems in practical applications of this method in crop improvement is the low response of anthers in producing haploids. Zapata and Aldemita (1989) demonstrated the improvement of rice anther response by combining radiation and in vitro culture techniques. They showed that the minimal stress on anthers by irradiation may not induce irreversible changes but can also stimulate callus induction and plant regeneration. Also, albinism is a serious problem in plants regenerated from callus produced by rice anther culture. There are not too many regenerated green plants, which make the task difficult in crop improvement, especially in cereals such as rice, wheat, etc. Zapata and Aldemita (1989) observed that gamma irradiation of rice Basmati 370 variety triggered green plant regeneration.

INACTIVATION OF ALKALOID BIOSYNTHESIS

Another aspect of mutagenesis is to isolate the mutants with blockage of certain enzyme activity for inactivating or activating the pathway for biosynthesis of secondary metabolites in plants. In opium poppy plants, opium and opium-alkaloids, particularly morphine (source of heroin) are an addictive narcotic, which lead to menacing global drug abuse, and also have medicinal values. Researchers at the Central Institute of Medicinal and Aromatic Plants (CIMAP), India have been engaged in converting narcotic "opium poppy" into non-narcotic "seed poppy" through induced mutations resulting into an opiumless and alkaloid-free mutant LL-34. This mutant is opiumless where latex biosynthesis peters out as the plant reaches to the lancing stage. Its straw (capsule hull) is also free from the five major opium alkaloids—morphine, codeine, thebaine, papaverine, and narcotine. The seed yield is very high and the plants contain 52.6% vegetable oil (better than most of oil-seed crops), which is largely unsaturated, i.e. good for diet control of coronary disease and even diabetes arising from lipid abnormalities. This mutant LL-34 has been registered as var. *Sujata*, which is not only has a potential of oil crop, but also as a cheap solution to combat opium-linked social abuses across all over the world (Sharma et al., 1999).

SOMACLONAL VARIATION

The first report on incidence of tissue culture-derived variation in sugarcane came from Heinze and Mee (1971), who found individual sugarcane regenerants with heavier tillering, slower growth rate and increased erectness. Skirvin and Janick (1976) were among the first to emphasize the potential of clonal variation for improvement of horticultural cultivars. Shepard et al., (1980) were the first one to gain widespread recognition for the extensive variation observed among protoplast-derived potato plants. Larkin and Scowcroft (1981) coined the term "somaclonal variation" and defined it as genetic variation in tissue culture-derived plants. There are different ways to create somaclonal variation, which are: (a) callus and cell suspension growth for several cycles; (b) regeneration of large number of plants from long-term cultures; (c) screening of desirable traits in the regenerated plants and their progenies, e.g. in vitro selection to select agronomically important traits such as tolerance against abiotic and biotic stresses; (d) testing of selected somaclones in the subsequent generations for genetic stability; and (e) multiplication of genetically-stable somaclones for developing new cultivars (Brar and Jain 1998). The changes associated with somaclonal variation are point mutations, DNA methylation, altered sequence copy number, transposable elements, genotype, explant type, culture medium, age of the donor plants, single gene mutations, and chromosomal rearrangements (Jain 1997a). Somaclonal variations are unpredictable in nature and can be both heritable (genetic) and non-heritable (epigenetic) in regenerated plants. DNA methylation causes genetic instability in somaclones, which probably comes from epigenetic changes (Jain et al., 1998; Jain 2001). Usually, gene mutations occur frequently in tissue culture-derived plants, and therefore, tissue culture system can be regarded as a mutation process. Since somaclonal variation can broaden the genetic variation range of crop plants, a broader range of plant characteristic can be altered, including plant height, yield, number of flowers per plant, early flowering, grain quality, resistance to diseases, insect and pests, cold and drought, and salt (Jain et al., 1998). The reduction, and even the total loss of regeneration ability, is a general phenomenon observed during undifferentiated cell culture. Nehra et al., (1990) observed in strawberry callus cultures that after 24 weeks, calli formed from in vitro cultured leaves lose completely their regeneration capacity as a result of formation of cells with abnormal DNA content during unorganized cell growth. Infante et al., (1996) obtained for the first time stable cell suspension of a diploid *Fragaria vesca monophylla* and plant regeneration after two years of unorganized cell culture. There was variation in shoot regeneration capability. In tuber disc-derived plant populations of potato, Rietvald et al., (1991) observed selected somaclones exhibited desirable alterations for yield, tuber number and shape and vigor were stable over more than two consecutive asexual generations, and that should be useful for breeders. Stephens et al., (1991) did not find large differences in tissue culture-derived homozygous progeny regenerated by organogenesis of soybean. When compared to the unregenerated parent, a statistically significant variation ($P<0.05$) was found for maturity, lodging, height, or seed protein and oil, but not for seed quality, seed weight, or seed yield. Cytokinin habituation and chilling resistance are examples of epigenetic variation (Hammerschlag 1992). Gonzalez et al., (1996) suggested that the source of explant was more important than the genotype (cultivar) and the type of callus (morphogenic vs non-morphogenic) in the chromosomal stability of cultures as time increases in barley. In Japan, disease-resistant lines of rice, tomato and tobacco were isolated from somaclones (Nakajima, 1991). Ramos Leal et al., (1996) obtained sugarcane somaclones resistant to eyespot disease. Carrot somaclones were found to be resistant to leaf blight (Dugdale et al., 1993). However, Van den Bulk and Dons (1993) suggested that the potential of somaclonal variation as a source of resistance to bacterial canker is limited in tomato. Somaclonal variation can become a part of plant breeding, provided the somaclones are heritable and genetically stable. Bebeli et al., (1993) noticed a heritable variation in rye in vitro culture of immature embryos, which was more pronounced in lines lacking telomeric heterochromatin. Moon et al., (1997) identified and characterized an aluminum-sensitive maize plant, obtained by somaclonal variation after plant regeneration from type I callus that was established from immature embryos of the Al-tolerant inbred maize line Cat-100-6. Histological sections of aluminum-treated roots from tolerant and sensitive plants stained with hematoxylin, an aluminum marker, showed a progressive destruction of the root tips of the aluminum-sensitive genotype over a period of time and indicated that tolerance in Cat-100-6 could be due to an aluminum exclusion mechanism.

Somaclonal Variation at the Chromosomal Level

Numerous plant species have shown chromosomal variation in tissue culture-derived plants, and their progenies (Duncan 1997; Roth et al., 1997; Kaeppler et al., 1998; Gupta 1998), and higher ploidy and high-chromosome explant source show high variability than among lower ploidy and low-chromosome number species (Creissen and Karp 1985). Polyploidy in tissue culture-derived plants is generally the product of either endopolyploidization or nuclear fusion (Sunderland 1977; Bayliss 1980). The type of altered karyotypes in somaclones include chromosomal rearrangements as well as aneuploidy and euploidy. Aneuploidy could be the result of non-disjunction, aberrant spindles, lagging chromosomes, chromosome breakage that produce dicentric and acentric chromosomes (Sunderland 1977). Normal cell cycle controls, which prevent cell division before the completion of DNA replication, are supposedly disrupted by tissue culture, resulting in chromosomal breakage (Phillips et al., 1994). Chromosome breakage and its consequences (deletions, duplications, inversions, and translocations) cause common aberration events (Duncan 1997). The breakage events are non-random, but involve late replicating chromosome regions characterized by heterochromatin (Lee and Phillips 1987; Benzion and Phillips 1988). Chromosome breakage may create mutations directly through "Position effect" or an alteration in gene expression due to chromosomal rearrangement and placement either in close proximity or distanced from specific heterochromatic region (Psechke and Phillips 1992). Furthermore, altered levels of DNA methylation could trigger chromosomal breakage. Moreover, the degree of chromosomal instability in tissue culture also varies from one species to another. For example, rye exhibits more chromosomal instability than either barley or pearl millet, which is due to the repetitive sequences located in the heterochromatin in rye genome (Gupta 1998). The age of the callus also affects the frequency of chromosomal aberrations. In general, as the callus gets older, the frequency of chromosomal instability enhances, except in maize callus that did not show any aging influence on chromosomal changes. Roth et al., (1997) found in embryogenic cell cultures of *Abies alba* a change in malformation of suspensor cells and loss of maturation capacity, and chromosome counts of cells showed trisomic. Chromosomal variation can also be induced by genetic transformation. However, its extent is dependent on the crop plant. Choi et al., (2000) demonstrated an increase in chromosomal variation in barley transgenic plants compared with non-transgenic plants. One of the factors for this increase is increased stress factor during the transformation process. This aspect might as well be studied in other crop plants.

Molecular Basis of Somaclonal Variation

There has not been a conclusive molecular basis of somaclonal variation in unfolding the causes of tissue culture as a mutagenic system (Jain 2001). Tissue culture system itself acts as a mutagenic system because the cells experience traumatic experiences from isolation of a cell or cells of plant or callus (Phillips et al., 1994). McClintock (1984; cross reference Phillips et al., 1994) suggested that when cells face traumatic experiences that resetting the genome and may not follow the same orderly sequence that occurs under the natural conditions. Instead, the genome is abnormally reprogramed, or decidedly restructured. This restructuring can give rise to a wide range of altered phenotypic expressions in newly-regenerated plants. It may be safe to state that no two of the callus-derived plants are exactly alike. Transposable element activation would then probably be another type of variation induced by tissue culture. A transposable element is a DNA sequence with the capability for movement throughout the genome by a process of excision and reintegration (Chandlee 1990). Transposable element activity has been detected by outcrossing to a transposable element tester stock; the regenerant-derived lines showed no qualitative or unstable mutations. Few reports of unstable mutations or sectoring of mutated sectors in regenerant-derived progenies have been reported. Groose and Bingham (1986) identified an unstable flower color mutation which acts like a transposable element-induced mutation, but this hypothesis is as yet unproved. Kaeppler et al., (1998) suggested that transposable elements probably account for a relatively small proportion of tissue culture-induced variation. Jain (2001) suggested 'hot spots' in the nuclear genome of plants, and whenever transposable elements jumps around them, genetic changes occur in the regenerated plants. Perhaps these 'hot spots' are prone to somatic recombination resulting in genetic variation. Retrotransposons comprise another class of mobile genetic elements that propagate via reverse transcription of RNA intermediates before integration within the host genome (Kumar and Hirochika 2001) They are ubiquitous in eukaryotes and constitute a major portion of the genome. They are dispersed as interspersed repetitive sequences throughout most of the length of the host chromosomes. Long terminal repeats (LTR) retrotransposons have been found in more than 100 species of plants, and most of them are presumed to be inactive, because of their defective structures and the absence of transcription. Hirochiko et al., (1996) showed that three rice retrotransposons (*Tos*10, *Tos*17 and *Tos*19) were activated under tissue culture conditions. Yamazaki et al., (2001) observed that *Tos17* retrotransposon specifically integrates into low-copy-number regions of the genome, and 30% flanking sequences examined showed a significant homology to the known genes.

Phenotypic changes found in the regenerated plants and their progeny are most strikingly observed as qualitative mutants, which are single gene mutants and inherited as single Mendelian factors and most frequently recessive alleles (Phillips et al., 1994). Some of the single-gene visible mutants such as chlorophyll deficiency, dwarfs, seed traits, reproductive structures, necrotic leaves have been observed in maize (Phillips et al., 1994). Molecular variations were characterized among tissue culture-derived plants by DNA and protein levels. Variation at the DNA level has been most extensively studied using restriction enzyme analysis. In most cases, changes in restriction pattern appeared to be due to altered fragment size, rather than additions or losses of restriction fragments (Kaeppler et al., 1998). Studies using restriction enzymes sensitive to 5-methylcytosine modifications have found extensive and frequent change in RFLPs (Kaeppler and Phillips 1993). Variation at the protein level has been characterized with grain storage proteins and isozymes being most frequent. Characterized variation can be summarized into three categories: (a) altered electrophoretic mobility (Karp et al., 1987; Sabir et al., 1992); (b) loss or gain of protein band (Davies et al., 1986); and (c) altered levels of specific proteins (Sabir et al., 1992).

DNA methylaytion has been associated with gene expression in numerous plant and animal species (Kaeppler et al., 1998; Jaligot et al., 2000). The direct role of DNA methylation in gene expression is still a subject of debate, even though the correlation of cytosine methylation and gene expression in plants and animals is quite good. The cytosine methylation detection at a large-scale can be done with the methylation-sensitive amplification polymorphism (MSAP) technique, which is a modification of the amplified fragment length polymorphism (AFLP) method that makes use of differential sensitivity of a pair of isoschizomers to cytosine methylation. Xiong et al., (1999) used MSAP in rice genome, showing that a small proportion of sites was differentially methylated in seedlings and flag leaves; DNA from young leaves seedlings was methylated to a greater extent than that from flag leaves. Gene expression may be altered during somatic embryogenesis (Reinbothe et al., 1992), and micropropagation, e.g. in banana (Peraza-Echeverria et al., 2001). Methylation can enhance quantitative trait variation because several genes can be affected simultaneously (Phillips et al., 1990). Increases in methylation in vitro potentially escalate gene activity and regulation. Methylation of a gene inactivates its transcription and thereby controls gene expression during somatic embryogenesis (Duncan 1997). Lambe et al., (1997) proposed that the progressive loss of totipotency in long-term plant tissue culture is a common event, which is a typical trait of plant neoplastic progression. Methylation of genes relevant to cell differentiation and progressive elimination of cells capable of differentiation is proposed as being responsible for the progressive loss of organogenic potential.

Biochemical and Molecular Markers for Discrimination of Somaclonal Variants

Isozymes have proven to be useful markers for somaclonal variation among regenerants from apple root stocks (Martelli et al., 1993), as well as providing an additional marker in monitoring the introgression of foreign germplasm in interspecific onion breeding (Hou et al., 2001). Isozyme polymorphism was observed among regenerants and based on banding patterns, rootstocks and regenerants could be distinguished. Genetic variability among *Zea mays* was analyzed with isozyme analysis (Ilarslan et al., 2001).

There are a wide variety of molecular methods available to characterize plant genomes. These methods include: (a) Restriction fragment length polymorphism (RFLP); (b) Random amplified polymorphic DNA (RAPD); (c) DNA amplification fingerprinting (DAF); (d) Amplified fragment length polymorphism (AFLP); (e) Short sequence repeats (microsatellites) (SSR); (f) Temperature gradient gel electrophoresis (TGGE); and (g) Denaturing gradient gel electrophoresis (DGGE) (Powell et al., 1997; Russell et al., 1997; Henry 1998; Gupta et al., 1999). Most of them have the potential for in analyzing somaclonal variation, both RFLP and RAPD markers have been widely applied (Chowdhury and Vasil 1993; Levall et al., 1994; Piccioni et al., 1997; Henry 1998; Osipova et al., 2001). Veilleux et al., (1995) applied both RAPD and SSR techniques to characterize the genetic composition of anther-derived potato plants. Wolff et al., (1995) used RAPD, SSR, and RFLP markers to evaluate somaclonal variation in vegetatively-propagated chrysanthemum cultivars. Taylor et al., (1995) observed very few RAPD polymorphisms in sugarcane plants regenerated from embryogenic callus, indicating infrequent gross genetic changes during tissue culture. However, Heinze and Schmidt (1995) found no gross somaclonal variation in somatic embryos and somatic seedlings of Norway spruce (*Picea abies*) by RAPD analysis. Fourre et al., (1997) who also obtained similar results, however, suggested to use cytological and morphological approaches to identify chromosomal variation which otherwise can't be identified with RAPD markers. Rani et al., (1995) identified somaclones of *Populus deltoides* using RAPDs. Six of the 23 plants propagated from a single clone could be distinguished from the others and field-grown plants by 13 polymorphisms using five primers. These results showed that a single somatic mutation had been detected that preceded the cell division that resulted in cells, contributing to the six identical variants. Hashmi et al., (1997) demonstrated a variation among plants regenerated from embryo callus cultures of peach by RAPD analysis. In *Quercus serrata* somatic seedlings, 40 primers for RAPD analysis did not show any variation in somatic seedlings (Ishii et al., 1999; Thakur et al., 1999). Similar results were obtained in maize somaclones (Osipova et al., 2001). Afza et al., (2001) detected androclonal variation in plants regenerated from anther culture in rice with RAPDs. Rahman and Rajora (2001) used microsatellite DNA markers of ten simple sequence repeat (SSR) for examining somaclonal variation among micropropagated *Populus tremuloides* plants, and found microsatellite DNA variation among the micropropagated plantlets. Rani et al., (2000) demonstrated for the first time in coffee (*Coffea arabica* L.) the presence of subtle genetic variability and novel organizations in well-established commercially-somatic embryo-derived plants by using RFLPs, RAPDs and ISSRs. These results indicated that some species are inherently unstable than others during propagation, and thereby, molecular diagnostic kit should be made an integral component of any micropropagated system aiming at producing true-to type plants. AFLP method generates more markers and may allow the identification of relatively low levels of variation (Thomas et al., 1995; Guthridge et al., 2001; Arroyo-Garcia et al., 2001). Turner et al., (2001) generated a total of 95 fragments with AFLP technique by using three primer pairs, and showed no differences in *Anigozanthos viridis* plants following three storage methods (tissue culture, cold storage and cryostorage).

In vitro Selection

The occurrence of genetic variation is very common in tissue culture-derived plants as a result of in vitro mutations. These mutations are not always expressed and only when a few of the changes are able to express, it is evident as phenotypic and cytological modifications in the regenerated plants from the callus tissue (Remotti 1998). However, the frequency of the somaclonal variation rate should be high enough for the selection of agronomically-desirable variants, and the selected lines should perform well under multiple environments (Duncan 1997), e.g. selection of genetic variants for removing heavy metals from contaminated soils or phytoremediation (Davies et al., 2001). While making selections under the natural conditions, plant breeders don't have a precise tool to make the right selections for desirable traits and takes years and generations to accomplish their goals. In vitro selection shortens the time considerably for the selection of a desirable trait under the selection pressure without having any environmental influence, and should complement the field conditions. Duncan (1997) reported that subjecting somaclonal variants to performance evaluations in the field includes additional genotype × environmental interactions. Heath-Pagliuso et al., (1988) selected *Fusarium oxysporum* f. sp. *apii* resistant celery, and some of the selected somaclones were superior to the parental type. Table 10.1 indicates in vitro selection for tolerant lines against several diseases. Aluminum toxicity is well known for growth limiting factor for plants in many acid soils, particularly at pH 5.0 or below (Rout et al., 2001). There are several reports on the in vitro selection of aluminum tolerant lines (Table 10.1). Similarly, in vitro selection against salt tolerance is commonly achieved as a result of a temporary adaptation; the cells are able to compartment the excessive salts into the vacuoles, and survive by adjusting the osmotic pressure, but this adaptation causes reduction in cell division and expansion (Bressan et al., 1985). The mechanism of this relationship is not yet known. However, some plants adapt better to high salts to become more halophytic than others, and are unable to grow in the absence of salt selection pressure (Remotti 1998). Bressan et al., (1985) found reversible tobacco salt tolerant cell lines at 10 g/l NaCl concentration. As the salt concentration was raised to 25g/l, stable salt tolerance lines were obtained, indicating that the high salt concentration acted as a selective as well as mutagenic agents, and caused specific modifications of the plant morphology, including unbalanced polyploidization, sterility, longer reproduction phase, and dwarfism coupled with heritable salt tolerance. Moreover, when salt stress is given osmoprotectants (glycinebetaine; Sabry et al., 1995), proline (Stewart and Lee 1974), and 24 kDa protein, called osmotin-1 (Bressan et al., 1987) accumulate. Sabry et al., (1995) found accumulation of glycinebetaine and sucrose in salt-stressed wheat plants. Glycinebetaine is known as osmoprotectant, which is accumulated in certain plant species under salt and drought conditions. Remotti (1998) has listed in vitro selection against salt in several crop species, and greater long-term salt adaptation

Table 10.1 In vitro selection of somaclones with desirable traits (updated from Jain 2001)

Plants	*Agronomic traits*	*References*
Arabidopsis thaliana	Valine resistance	Gaj et al., 1999
Herbicide resistance		
Wheat	Atrazine, difenzoquat	Bozorgipour and Snape, 1997
Tobacco	picloram	Chaleff, 1980
	atrazine	Sato et al., 1988
	amitrole	Singer and MacDaniel, 1984
	paraquat	Furosawa, 1988
	chlorsulfuron	Caretto et al., 1993
Brassica	atrazine, phenmedipham	Jain and Newton, 1988
Maize, barley	glyphosate	Racchi et al., 1995; Escorial et al., 1996
Sugarbeet	chlorsulfuron	Saunders et al., 1992
Disease resistance		
Tomato	*Fusarium lycopersici, Phytophthora infestans*	Shahin and Spivey, 1986; Remotti, 1998;
Gladiolus × *grandiflorus*	Fusaric acid	Remotti et al., 1997
Tobacco	*Alternaria alternata*	Thanutong et al., 1983
Potato	*Fusarium solani, F.oxysporum, Phytophthora infestans, Alternaria solani,*	Behnke, 1979, 1980 a,b; Shepard et al., 1980
	potato leafroll luteovirus	Kawchuk et al., 1997
Strawberry	*Verticillium dahliae, Phytophthora cactorum, Rhizoctonia fragariae, Botrytis cinera*	Sowik et al., 2001; Battistini and Rosati, 1991; Jain, 1997, Remotti, 1998, Orlando et al., 1997

Contd.

Table 10.1 (Contd.)

Plants	Agronomic traits	References
Barley, asparagus	*Fusarium* spp.; *Fusarium oxysporum* f. sp. *asparagi*	Chawla and Wenzel, 1987a Pontaroli and Camadro, 2001
Apple	*Phytophthora cactorum*	Rosati et al., 1990
Triticale	Fusarium head blight	Goral and Arseniukk, 1997
Wheat	*Helminthosporium sativum*	Chawla and Wenzel, 1987b, 1989
Alfalfa	*Fusarium oxysporum* f. sp. *medicaginis*	Arcioni et al., 1987
Sugarcane	Eyespot disease	Ramos Leal et al., 1996
Peach	*Xanthomonas campestris* pv. *pruni*	Hammerschlag, 1988
Grape	*Elsinoe ampelina*	Jayasankar et al., 2000
Mango	*Colletotrichum gloeosporioides*	Jayasankar et al., 1998, 1999; Jayasankar and Litz, 1998
Banana	*Mycosphaerella fijiensis*	Busogoro et al., 2001
Banana	*Radopholus similes* (nematodes)	Elsen et al., 2001
Date palm	*Fusarium oxysporum* sp. *albedini*	Sedra (personnel communication)
Abiotic stresses		
Setaria italica	Zinc tolerance	Samantaray et al., 1999
Rice, maize, wheat, *Cynodon dactylon*	Aluminum tolerance	Jan et al., 1997; Moon et al., 1997; Dornelles et al., 1997; Ramgareeb et al., 1999

Contd.

Table 10.1 (Contd.)

Plants	*Agronomic traits*	*References*
Alfalfa, barley, rice, wheat	Salt tolerance	Boscherini et al., 1999; Winicov, 1991, 1996
Brassica, tomato, tobacco		Barakat and Abdel-Latif, 1996; Locy et al., 1996; Kirti et al., 1991; Jain et al., 1990, Boscherini et al., 1999; Mandal et al., 1999; Remotti, 1998; Forster, 2001
Proso millet, barley,	Drought tolerance	Toker et al., 1999; Remotti, 1998
Tagetes minuta		Mohamed et al., 2000
Wheat, Arabidopsis	Frost tolerance	Dorffling et al., 1993; Xing and Rajashekar, 2001
Maize, rice, strawberry	Cold tolerance	Duncan and Widholm, 1987; Bertin et al., 1997; Rugienius and Stanys, 2001
Sugarbeet	UV-B tolerance	Levall and Bornman, 1993

requires further studies so that salt tolerant selected plants grow more like halophytic plants. Perhaps all plant species may not response to long-term salt adaptation, therefore, we may have to screen plant species for these studies.

SOMATIC CELL HYBRIDIZATION

During the 1970s and the beginning of 1980s, protoplast technology was considered a novel approach for gene transfer from one species to another by making symmetric somatic hybrids, somatic cybrids and asymmetric somatic hybrids. Gamma or X-ray irradiation were used for producing somatic cybrids and asymmetric somatic hybrids, especially when two species are not known to interbreed either in nature or in the laboratory. This technology could not lift the ground in crop improvement due to greater attention and heavy investment in genetic engineering, both from the private and state investors. As a result, engineering took rapid strides in producing transgenic plants and few transgenic varieties were released to consumers, e.g. tomato, *Bt* cotton. Unfortunately, consumers have been reluctant to consume genetically-modified (GM) food due to fear of undesirable health-related consequences, generated by news media and other interest groups. Investments to the tune of several billion US dollars have gone in plant genetic engineering and so far there are no economic benefit returns on investments. There is a gradual shift to return to somatic cell fusion technology for desirable gene transfer, especially those controlled by polygenes or uncloned genes, from wild species to crop plants, circumventing sexual-crossing barriers and to generate novel gene combinations. Grosser et al., (2000) and Guo and Deng (2001) extensively reviewed on citrus somatic hybridization by protoplast fusion, and it has become an integral part of the citrus improvement program worldwide. Over 200 parental combinations of citrus have been used for producing somatic hybrids and cybrids in order to improve rootstocks, scions and use of tetraploid somatic hybrids for producing triploids. Laferriere et al., (1999) obtained fertile somatic hybrids between *Solanum tuberosum* and *S. commersonii*, showing resistance to bacterial wilt caused by *Ralstonia solanacearum*. This trait was transferred from bacterial wilt resistant *S. commersonii* plants. Keskitalo et al., (1999) produced somatic hybrid calli between tansy (*Tanacetum vulgare* L.) and pyrethrum [*Tanacetum cinerariifolium* (Trevir) Schltz-Bip.] for improving the oil content and composition of tansy for enhanced biological activities. Pyrethrum is known for producing environmentally-benign pesticides, the pyrithrins, whereas tansy produces lower terpenes of variable biological effectiveness. The calli of protoplast fusion product grew more vigorously than the calli derived from intra-specific tansy fusion, which could be due to heterosis. The small number of volatile compounds detected from both the calli coming from fusion and donor species indicated that the unorganized callus is unable to produce tissue-specific volatile compounds. However, this approach can still be used for producing useful chemical compounds.

Somatic cybrids

Somatic cybrids have cytoplasm from the donor and the nuclear genome' from the recipient parent, which morphologically look like the recipient parent. Cybrids are produced to transfer cytoplasmic genome inherited trait, e.g. atrazine herbicide resistance, cytoplasmic male sterility, antibiotic resistance (streptomycin). The donor parent protoplasts are irradiated with low dosage, 5000–7000 rads, of X-rays or gamma radiation to functionally "enucleate" the donor cells. Low dosages of X-rays do not seem to harm the plastid genome whilst preserving some nuclear functions, yet preventing further nuclear divisions. These irradiated protoplasts are fused with recipient donor protoplasts, and regenerate plants from the fusion protoplast products. Prof. Esra Galun group, Israel in 1978, was the first to transfer cytoplasmic male sterility by fusion between normal *Nicotiana sylvestris* and X-ray irradiated protoplasts of male sterile *N. tabacum.* Later on, Jain et al., (1981, see cf Gressel et al., 1982) adopted their technique and attempted for the first time to produce somatic cybrids between atrazine herbicide-resistant *Solanum nigrum* (developed resistance after repeated atrazine herbicide treatments in the maize fields in France) and highly atrazine susceptible tobacco *Nicotiana sylvestris.* The plants that were regenerated from the resulting calli were typically tobacco. They were grown up, self-pollinated and the resulting F1 seed were bio-assayed for traizine tolerance by germinating on an atrazine-containing medium. Seed from some of the fusion products showed much greater tolerance to the herbicide than the parent tobacco, but less tolerant than the resistant *S. nigrum.* Binding et al., (1982) regenerated plants from protoplast fusion between haploid *Solanum tuberosum* L. and an atrazine resistant biotype of *S. nigrum.* Quite a diversity was noticed among regenerated plants in relation to atrazine response, shapes and pigmentation of leaves, flower morphology, and chromosome numbers. Jain et al., (1988) succeeded in transferring atrazine herbicide resistance from donor atrazine resistant *S. nigrum* to tomato by irradiating the donor parent with UV light. Most of the selected plants were somatic hybrids and not somatic cybrids, and were sterile and showed resistance to a high level of atrazine level. The hybrid nature of plants was characterized with glutamate aspartic aminotransferase (GAT) isozymes. By backfusion, they attempted to completely eliminate the *S. nigrum* genome from somatic hybrids. However, they succeeded in partial genome elimination, as indicated by GAT isozyme pattern. Yemets et al., (2000) transferred amiprophosmethyl (APM) herbicide resistance from *Nicotiana plumbaginifolia* in *Nicotiana sylvestris* and *Atropa belladonna* by somatic hybridization and cybridization. The mutant ß-tubulin gene of APM-resistant plants has not yet been identified or cloned, thereby somatic cell fusion technology was an ideal approach.

Asymmetric somatic hybrids

Asymmetric somatic hybrids are produced by fusing the fragmented donor genome with heavy dose of gamma radiation with the normal protoplasts of recipient parent. This approach would be ideal for partial genome transfer from donor to the recipient. Hall et al., (1992) suggested that ionizing radiation could be substituted with UV light for producing asymmetric somatic hybrids. UV light induces substantial physical fragmentation of DNA without producing immediate cytotoxicity. However, Jain et al., (1988) could not produce somatic cybrids by fusing protoplasts between tomato and UV irradiated black nightshade. By this approach, the partial genome can be transferred to the recipient parent by intergeneric, intrageneric, intratribal, intrafamiliar and interfamiliar protoplast fusion and the fusion product looks similar to the recipient parent with the complete genome. The main advantage of this approach is that asymmetric somatic hybrids have less of the undesirable donor traits and may require fewer or no backcrosses. In other words, it is possible to produce intergeneric somatic hybrids between any plant species, which are sexually incompatible, e.g. carrot × tobacco, barley × tobacco. There was an extensive chromosomal elimination in asymmetric somatic hybrids between carrot and tobacco and barley and tobacco. In 1993, Pehu research group, Finland produced asymmetric somatic hybrids between gamma-irradiated (300–500 Gy) protoplasts of virus-resistant *Solanum brevidens* and on irradiated protoplasts of dihaploid virus-susceptible *S. tuberosum*, and succeeded in transferring a virus resistance trait in the recipient *S. tuberosum*. The degree of elimination of alleles from the irradiated *S. brevidens* donor genome ranged from 10–65% in the asymmetric somatic hybrids analyzed (Xu et al., 1993).

Another approach for producing asymmetric somatic hybrids is by fusion of microprotoplasts with hypocotyls protoplasts in *Helianthus* species (Binsfeld et al., 2000). The cytological analysis of the metaphase cells of 16 hybrids revealed an addition of 2–8 extra chromosomes in these plants, and the phenotype of the most asymmetric somatic hybrids resembled the recipient *H. annuus* plants. These results provide evidence of the possible use of microprotoplasts for partial genome transfer.

CONCLUSIONS

Plant breeders require a wide range of genetic variation in improving their crops. This could be done either by using wide natural germplasm or by induced genetic variability. The term 'mutation breeding' is too narrow to describe all the possible applications of induced mutations, and has been replaced by another term 'Mutation Techniques' in plant breeding and research (Maluszynski 2001). Plants amenable to cell and tissue culture may provide a wide range of somaclonal variation not only in asexually-propagated species, but also in the seed propagated ones. In vitro selection of desirable mutants can be the greatest attribute of somaclonal variation for rapid selection of agronomical desirable traits having the cellular basis, e.g. salinity and metal tolerance, herbicide resistance. However, in vitro selected traits further require field and genetic stability testing. Somaclonal variation and induced mutations have resulted in the production of new genotypes with a limited change in the original genome. As a source of variation, somaclonal variation mimics induced mutations. However, it is desirable to test the genetic stability of somaclones before they are incorporated in a breeding program; otherwise somaclones that arise as a result of epigenetic changes will revert to normal in the subsequent sexual generations. Molecular markers would be ideal for identifying genetic and epigenetic somaclones. The use of mutagenesis with plant tissue culture and length of culture period, especially in cereals, could increase the genetic variability for exploitation by plant breeders. Somatic cell fusion technology is gradually getting attention for gene transfer from interspecific, intraspecific, and intergenera plants. The integration of new technologies with induced mutations would be needed before genetic engineering becomes a regular tool in plant breeding. The developing countries with high population growth can't wait before genetic engineering can reap high harvest. Thereby, genetic variability induced by a combination of tissue culture and mutagenesis can be useful to plant breeders in crop improvement.

REFERENCES

Afza R., Xie J., Shen M., Zapata F.A., Fundi H.K., Lee K.S., Bobadilla-Mucino E. and Kodym A., 2001, Detection of androclonal variation in anther-cultured rice lines using RAPDs. In Vitro Cell. Dev. Biol. Plant 37: 644–647.

Ahloowalia B.S., 1998, In vitro techniques and mutagenesis for the improvement of vegetatively propagated plants. In: S.M. Jain, D.S. Brar and B.S. Ahloowalia (Eds), Somaclonal Variation and Induced Mutations for Crop Improvement, pp. 293–309. Kluwer Academic Publishers, Dordrecht.

Al-Juboory K.H., Shibli R.A., Williams D.J., Nayfeh M.H. and Skirvin R.M., 2000, Effect of pulsed excimer laser radiation on shoot regeneration and reduction of contamination of date palm (*Phoenix dactylifera*) in vitro. Indian J. Agric. Sci. **70:** 482–483.

Arcioni S., Pezzotti M. and Damiani F., 1987, In vitro selection of alfalfa plants resistant to *Fusarium oxysporum* f. sp. *medicaginis*. Theor. Appl. Genet. **74:** 700–705.

Arroyo-Garcia R., Martinez-Zapater J.M., Fernandez-Prieto J.A. and Alvarez-Arbesu R., 2001, AFLP evaluation of genetic similarity among laurel population (*Laurus* L.). Euphytica **122:** 155–164.

Banerjee M.K. and Kalloo, 1989, Role of phenols in resistance to tomato leaf curl virus, *Fusarium* wilt and fruit borer in *Lycopersicon*. Current Sci. **58:** 575–576.

Barakat M.N. and Abdel-Latif T.H., 1996. In vitro selection of wheat callus tolerant to high levels of salt and plant regeneration. Euphytica **91:** 127–140

Bayliss M.W., 1980. Chromosomal variation in tissue culture. Intern. Rev. Cytol. Supple. **11A:** 113–144.

Bebeli P.J., Kaltsikes P.J. and Karp A., 1993, Field evaluation of somaclonal variation in rye lines differing in telomeric heterochromatin. J. Genet. & Breed. **47:** 15–22.

Behnke M., 1979, Selection of potato callus for resistance to culture filtrates of *Phytophthora infestans* and regeneration of resistant plants. Theor. Appl. Genet. **55:** 69–71.

Behnke M., 1980a, General resistance to late blight of *Solanum tuberosum* plants regenerated from callus resistant to culture filtrates of *Phytophthora infestans*. Theor. Appl. Genet. **56:** 151–152.

Behnke M., 1980b, Selection of dihaploid potato callus for resistance to culture filtrates of *Fusarium oxysporum*. Z. Pflanzenzucht **85:** 254–258.

Behl R.K., Nainawatee H.S. and Singh K.P., 1993, High temperature tolerance in wheat. In: International Crop Science 1. Crop Sci. Soc. America, Madison, Wisconsin, USA, pp. 349–355.

Benzion G. and Phillips R.L., 1988, Cytogenetic stability of maize tissue cultures A cell line pedigree analysis. Genome **30:** 318–325.

Bertin P., Bouharmont J. and Kinet J.M., 1997, Somaclonal variation and improvement of chilling tolerance in rice-changes in chilling-induced chlorophyll fluorescence. Crop Sci. **37:** 1727–1735.

Binding H., Jain S.M., Finger J., Mordhorst G., Nehls R. and Gressel J., 1982, Somatic hybridization of an atrazine resistant biotype of *Solanum nigrum* and *S. tuberosum*. 1. Clonal variation in morphology and I atrazine sensitivity. Theor. Appl. Genet. **63:** 273–277.

Binsfeld P.C., Wingender R. and Schnabl H., 2000, Characterization and molecular analysis of transgenic plants obtained by microprotoplast fusion in sunflower. Theor. Appl. Genet. **101:** 1250–1258.

Boscherini G., Muleo R., Montagni G., Cinelli F., Pellegrini M.G., Bernardini M. and Buiatti M., 1999, Characterisation of salt tolerant plants derived from a *Lycopersicon esculentum* Mill. somaclone. J. Plant Physiol. **155:** 613–619.

Bouman H. and De Klerk G.J., 1996, Somaclonal variation in biotechnology of ornamental plants. In: R. Geneve, J. Preece and S. Merkle (Eds), Biotechnology of Ornamental Plants, pp. 165–183. CAB International.

Bozorgipour R. and Snape J.W., 1997, An assessment of somaclonal variation as a breeding tool for generating herbicide tolerant genotypes in wheat (*Triticum aestivum* L.). Euphytica **94:** 335–340.

Brar D.S. and Jain S.M., 1998, Somaclonal variation: mechanism and applications in crop improvement. In: S.M. Jain, D.S. Brar and B.S. Ahloowalia (Eds), Somaclonal Variation and Induced Mutations in Crop Improvement, pp. 15–37, Kluwer Academic Publishers, Dordrecht.

Bressan R.A., Singh N.K., Handa A.K., Kononowicz A.K. and Hasegawa P.M., 1985, Stable and unstable tolerance to NaCl in cultured tobacco cells. In: M. Freeling (Ed.), Plant Genetics, pp. 755–769, New York: Liss.

Bressan R.A., Singh N.K., Handa A.K. and Mount R., 1987, Stability of altered genetic expression in cultured plant cells adapted to salt. In: L. Monti and E. Porceddu (Eds.) Drought Resistance in Plants, pp. 41–57, Brussels: Commission of the European Communities.

Broertjes C. and M. van Hartan A. (eds), 1978, Application of mutation breeding methods in the improvement of vegetatively propagated crops. Elsevier, Amsterdam, The Netherlands.

Busogoro J.P., Etame J.J., Lognay G., Van Custem P., Balatero C.H. and Lepoivre R.B., 2001, Selection for banana resistance to Black Leaf Streak Disease. IAEA Working Document, IAEA-312.D2RC.579, pp. 143–150.

Caretto S., Giardina M.C., Nicolodi C. and Mariotti D., 1993, In vitro cell selection: production and characterization of *Nicotiana tabacum* L. cell lines and plants resistant to the herbicide chlorosulfuron. J. Genet. Breed. **47:** 115–120.

Chaleff R.F., 1980, Further characterization of picloram-tolerant mutants of *Nicotiana tabacum.* Theor. Appl. Genet. **58:** 91–95.

Chandlee J.M., 1990, The utility of transposable elements as tools for the isolation of plant genes. Physiol Plant 79: 105–115.

Chawla H.S. and Wenzel G., 1987a, In vitro selection for fusaric acid resistant barley plants. Plant Breed. 99: 159–163.

Chawla H.S. and Wenzel G., 1987b, In vitro selection of barley and wheat for resistance against *Helminthosporium sativum.* Theor. Appl. Genet. **74:** 841–845.

Choi H.W., Lemaux P.G. and Cho M.J., 2000, Increased chromosomal variation in transgenic versus nontransgenic barley (*Hordeum vulgare* L.) plants. Crop Sci. **40:** 524–533.

Chowdhury M.K.U. and Vasil I.K., 1993, Molecular analysis of plants regenerated from embryogenic cultures of hybrid sugarcane cultivars (*Saccharum* spp.). Theor. Appl. Genet. **86:** 181–188.

Coury D.A. and Feldmann K.A., 1998, T-DNA insertion mutagenesis and the untagged mutants. In: Jain S.M., Brar D.S. and Ahloowalia B.S. (Eds), Somaclonal Variation and Induced Mutations in Crop Improvement, pp. 519–540. Kluwer Academic Publishers, Dordrecht.

Creissen S..S. and Karp A., 1985. Karyotypic changes in potato plants regenerated from protoplasts. Plant Cell Tiss. Org. Cult. 4: 171–182.

Crino P., Lai A., Bonito R.D. and Veronese P., 1994, Genetic variability in tomato plants regenerated from irradiated cotyledons. J. Genet. and Breed. 48: 281–290.

Davies P.A., Pallotta M.A., Ryan S.A., Scowcroft W.R. and Larkin, P.J., 1986, Somaclonal variation in wheat: Genetic and cytogenetic characterization of alcohol dehydrogenase 1 mutants. Theor. Appl. Genet. **72:** 644–653.

Davies Jr. F.T., Puryear J.F., Newton R.J., Egilla J.N. and Saraiva Grossi, J.A. 2001. Mycorrhizal fungi enhance accumulation and tolerance of chromium in sunflower (*Helianthus annuus*). J. Plant Physiol. **158:** 777–786.

Donini P. and Sonnino A., 1998, Induced mutation in plant breeding: Current status and future outlook.. In: S.M. Jain, D.S. Brar and B.S. Ahloowalia (Eds), Somaclonal Variation and Induced Mutations in Crop Improvement, pp. 255–291. Kluwer Aacdemic Publishers, Dordrecht.

Dörffling K., Dörffling H. and Lesselich G., 1993, In vitro selection and regeneration of hydroxyproline-resistant lines of winter wheat with increased proline content and increased frost tolerance. J. Plant Physiol. **142:** 222–225.

Dornelles A.L.C., Decarvalho F.I.F., Federizzi L.C., Sereno M.J.C.D., Handel C.L. and Mittelmann A., 1997, Somaclonal variation in aluminium tolerance and gibberellic acid sensibility in wheat. Pesquisa Agro. Brasileira **32:** 193–200.

Dugdale L.J., Collin H.A., Issac S. and Gill, J.J.B 1993, Leaf blight resistance in carrot somaclones. Acta Hort. **336:** 399–404.

Duncan D.R. and Widholm J.M., 1987, Proline accumulation and its implication in cold tolerance of regenerable maize callus. Plant Physiol. **83:** 703–708.

Duncan R.R., 1997, Tissue culture-induced variation and crop improvement. Adv. Agron. **58:** 201–240.

Elsen A., Lens K., Nguyet D.R.M., Broos S., Stoffelen R. and De Eaele D., 2001, Development of aseptic culture systems for *Radpholus similes* for in vitro studies. IAEA Working Document, IAEA-312.D2RC.579, pp. 55–62.

Escorial M.C., Sixto H., Garciabaudin J.M. and Chueca M.C., 1996, In vitro culture selection increases glyphosate tolerance in barley. Plant Cell. Tiss. Org. Cult **46:** 179–186.

Feldmann K.A., 1991, T-DNA insertion mutagenesis in *Arabidopsis*: Mutational spectrum. Plant L. **1:** 71–82.

Feldmann K.A., Marks M.D., Christianson M.L. and Quatrano R.S., 1989, A dwarf mutant of *Arabidopsis* generated by T-DNA insertional mutagenesis. Science **243:** 1351–1354.

Fourre J.L., Berger P., Niquet L. and Andre P., 1997, Somatic embryogenesis and somaclonal variation in Norway spruce: morphogenetic, cytogenetic and molecular approaches. Theor. Appl. Genet. **94:** 159–169.

Forster B.P., 2001, Mutation genetics of salt tolerance in barley: an assessment of Golden Promise and other semi-dwarf mutants. Euphytica **120:** 317–328.

Furosawa I., 1988, Production of disease resistant plants using somaclonal variation. Adv. Tiss. Cult. Tech. Plant Breed **113:** 6–13.

Gavazzi G., Tonelli C., Todesco G., Arreghini E., Raffaldi F., Vecchio F., Barbuzzi, G. Biasini M.G. and Sala F., 1987, Somaclonal variation versus chemically induced mutagenesis in tomato (*Lycopersicon esculentum*). Theor. Appl. Genet. **74:** 733–738.

Gonzalez A.I., Pelaez M.I. and Ruiz M.L., 1996, Cytogenetic variation in somatic tissue cultures and regenerated plants of barley (*Hordeum vulgare* L.). Euphytica **91:** 37–43.

Goral T. and Arseniuk E., 1997, Somaclonal variation in winter triticale for resistance to *Fusarium* head blight. Cereal Res. Commun. **25:** 741–742.

Gressel J., Ezra G. and Jain S.M., 1982, Genetic and chemical manipulation of crops to confer tolerance to chemicals. In: J.S. McLaren (Ed.) Chemical Manipulations of Crop Growth and Development, Butterworth, London, pp. 79–91.

Groose R.W. and Bingham E.T., 1986, An unstable anthocyanin mutation recovered from tissue culture of alfalfa. 1. High frequency of reversion upon reculture. 2. Stable nonrevertants derived from reculture. Plant Cell Rep. **5:** 104–110.

Grosser J.W., Ollitraukt P. and Olivares-Fuster O., 2000, Somatic hybridization in citrus: an effective tool to facilitate variety improvement. In Vitro Cell Dev. Biol. Plant. **36:** 434–449.

Guo W.W. and Deng X.X., 2001, Wide somatic hybrids of *Citrus* with its related genera and their potential in genetic improvement. Euphytica **118:** 175–183.

Gupta P.K., 1998, Chromosomal basis of somaclonal variation in plants. In: Jain S.M., Brar D.S. and Ahloowalia B.S. (Eds), Somaclonal Variation and Induced Mutations in Crop Improvement, pp. 149–168. Kluwer Academic Publishers, Dordrecht.

Gupta P.K., Varshney R.K., Sharma P.C. and Ramesh B., 1999. Molecular markers and their applications in wheat breeding. Plant Breed. **118:** 369–390.

Guthridge K.M., Dupal M.P., Kolliker R., Jones E.S., Smith K.F. and Forster J.W., 2001, AFLP analysis of genetic diversity within and between populations of perennial ryegrass (*Lolium perenne* L.). Euphytica **122:** 191–201.

Hall R.D., Krens F.A. and Rouwendal G.J.A., 1992, DNA radiation damage and asymmetric somatic hybridization: Is UV a potential substitute or supplement to ionizing radiation in fusion experiments? Physiol. Plant. **85:** 391–324.

Hammerschlag F.A., 1988, Selection of peach cells for insensitivity to culture filtrates of *Xanthomonas campestris* pv. *pruni* and regeneration of resistant plants. Theor. Appl. Genet. **76:** 865–869.

Hammerschlag F.A., 1992, Somaclonal variation. In: Hammerschlag F.A. and Litz R.E. (Eds). Biotechnology of perennial fruit crops. CAB International, Cambridge, pp. 35–55.

Hashmi G., Huettel R., Meyer R., Krusberg L. and Hammerschlag F., 1997, RAPD analysis of somaclonal variants derived from embryo callus cultures of peach. Theor. Appl. Genet. **16:** 624–627.

Heath-Pagliuso S., Pullman J. and Rappaport L., 1988, Somaclonal variation in celery: screening for resistance to *Fusarium oxysporum* f. sp. *apii*. Theor. Appl. Genet. **75:** 446–451.

Heinze D.J. and Mee G.W.P., 1971, Morphologic, cytogenetic and enzymatic variation in *Saccharum* species hybrid clones derived from callus culture. Amer. J. Bot. **58:** 257–262.

Heinze B. and Schmidt J., 1995, Monitoring genetic fidelity vs somaclonal variation in Norway spruce (*Picea abies*) somatic embryogenesis by RAPD analysis. Euphytica **85:** 341–345.

Henry R.J., 1998. Molecular and biochemical characterization of somaclonal variation. Jain S.M., Brar D.S. and Ahloowalia B.S. (Eds), Somaclonal Variation and Induced Mutations for Crop Improvement, pp. 487–501. Kluwer Academic Publishers, Dordrecht, Great Britain.

Hirochika H., Sugimoto K., Otsuki Y., Tsugawa H. and M. Kanda. 1996, Retrotransposons of rice involved in mutations induced by tissue culture. Proc. Natl. Acad. Sci. USA. **93:** 7783–7788.

Hou A., Geoffriau E. and Peffley, E.B. 2001, Esterase isozymes are useful to track introgression between *Allium fistulosum* L. and *A. cepa* L. Euphytica **122:** 1–8.

Infante R., Gonelli S., Rosatti P. and Mazzara M., 1996, Long-term cell suspension culture and regeneration of the single-leafed strawberry *Fragaria vesca monophylla*. J. Sci. Food Agric. **72:** 196–200.

Ilarslan R., Kaya Z., Tolun A.A. and Bretting P.K., 2001, Genetic variability among Turkish pop, flint and dent corn (*Zea mays* L. spp. *mays*) races: Enzyme polymorphism. Euphytica **122:** 171–179.

Ishii K., Thakur R. and Jain S.M., 1999, Somatic embryogenesis in *Quercus serrata*. In: Jain S.M., Gupta P.K. and Newton R.J. (Eds), Somatic Embryogenesis in Woody Plants, Vol. 4, pp. 403–414. Kluwer Academic. Publications, The Netherlands.

Jain R.K., Jain S., Nainawatee H.S. and Chowdhury J.B., 1990, Salt-tolerance in *Brassica juncea* L. 1. In vitro selection, agronomic evaluation and genetic stability. Euphytica **48:** 141–152.

Jain S.M., 1997a, Somaclonal variation and mutagenesis for crop improvement. Maatalouden tutkimuskeskuksen, Sirkka Immonen (Ed.), Vol. 18, pp. 122–132.

Jain S. M., 1997b, Creation of variability by mutation and tissue culture in improving plants. Acta Hort. **447:** 69–78.

Jain S.M., 1998, Induction of somaclonal variation and mutation in developing new improved cultivars. MIIT Pune J. pp. 23–31.

Jain S.M. 2001, Tissue culture-derived variation in crop improvement. Euphytica **118:** 153–166.

Jain S.M., 2002a. Feeding the world with induced mutations and biotechnology. Proceeding International Nuclear Conference 2002—Global trends and perspectives. Seminar 1: Agriculture and Bioscience. MINT, Bangi, Malaysia. pp. 1–14.

Jain S.M. 2002b. A review of induction of mutations in fruits of tropical and subtropical regions. Acta Hort. 575: 295–302.

Jain S.M. and Newton R.J., 1988. Proto-variation in protoplast-derived *Brassica napus* plants. In: Puite K.J., Dons J.J.M., Huizing H.J., Kool A.S.J., Koornneef M. and Krens F.A. (Eds), Progress in Plant Protoplast Research. pp. 403–404. Kluwer Academic Publishers, Dordrecht.

Jain S.M. and Newton R.J., 1989, Evaluation of protoclonal variation versus chemically induced mutagenesis in *Brassica napus* L. Current Sci. **58:** 176–180.

Jain S.M., Shahin E.A. and Sun S., 1988, Interspecific protoplast fusion for the transfer of atrazine resistance from *Solanum nigrum* to tomato (*Lycopersicon esculentum* L.). Plant Cell Tiss. Org. Cult. **12:** 189–192.

Jain S.M., Oker-Blom C., Pehu E. and Newton R.J., 1992, Genetic engineering: an additional tool for plant improvement. Agric. Sci. Finland **1:** 323–338.

Jain S.M., Brar D.S. and Ahloowalia B.S. (eds), 1998. Somaclonal variation and induced mutations in crop improvement. Kluwer Academic Publishers, Great Britain.

Jaligot E., Rival A., Beule T., Dussert S., and Verdeil J., 2000, Somaclonal variation in oil palm (*Elaeis guineensis* Jacq.): the DNA methylation hypothesis. Plant Cell Rept. **19:** 684–690.

Jan V.V., Demacedo C.C., Kinet J.M. and Bouharmont J., 1997, Selection of Al-resistant plants from a sensitive rice cultivar using somaclonal variation, in vitro and hydroponic cultures. Euphytica **97:** 303–310.

Jayasankar S. and Litz R.E., 1998, Characterization of resistance in mango embryogenic cultures selected for resistance to *Colletotrichum gloeosporioides* culture filtrate and phytotoxin. Theor. Appl. Genet. **96:** 823–831.

Jayasankar S., Litz R.E., Gray D.J. and Moon P.A., 1999, Responses of mango cultures and seedling bioassays to a partially purified phytoxin produced by a mango leaf isolate of *Colletotrichum gloeosporioides* Penz. In Vitro Cell Develop. Biol. Plant. **35:** 475–479

Jayasankar S., Litz R.E., Schnell R.J. and Cruz-Hernandez A., 1998, Embryogenic mango cultures selected for resistance to *Colletotrichum gloeosporioides* culture filtrate show variation in random amplified polymorphic DNA (RAPD) markers. In Vitro Cell Develop. Biol. Plant **34:** 112–116.

Jayasankar S., Li Z. and Gray D.J., 2000. In vitro selection of *Vitis vinifera* "Chardonnay" with *Elsinoe ampelina* culture filtrate is accompanied by fungal resistance and enhanced secretion of chitinase. Planta **211:** 200–208.

Kaeppler S.M., Phillips R.L. and Olhoft P., 1998, Molecular basis of heritable tissue culture-induced variation in plants. In: S.M. Jain, D.S. Brar and B.S. Ahloowalia (Eds), Somaclonal Variation and Induced Mutations in Crop Improvement, pp. 467–486. Kluwer Academic Publishers, Dordrecht.

Kaeppler S.M. and Phillips R.L., 1993, DNA methylation and tissue culture-induced variation in plants. In Vitro Cell. Dev. Biol. **29P:** 125–130.

Karp A., Steele S.H., Parmar S., Jones M.G.K., Shewry P.R. and Breiman A., 1987. Relative stability among barley plants regenerated from cultured immature embryos. Genome **29:** 405–412.

Kawchuk L.M., Lynch D.R., Martin R.R., Kozub GC and Ferries B., 1997, Field resistance to the potato leafroll luteovirus in transgenic and somaclone potato plants reduces tuber disease symptoms. Can. J. Plant Pathol. **19:** 260–266.

Keskitalo M., Angers P., Earle E. and Pehu E., 1999, Chemical and genetic characterization of calli derived from somatic hybridization between tansy (+L.) and pyrethrum [(*Tanacetum cinerariifolium* (Trevir.) Schultz-Bip.]. Theor. Appl. Genet. **98:** 1335–1343.

Kirti P.B., Hadi S., Kumar P.A. and Chopra V.L., 1991, Production of sodium-chloride-tolerant *Brassica juncea* plants by *in vitro* selection at the somatic embryo level. Theor. Appl. Genet. **83:** 233–237.

Kocabek T., Rakousky S., Ondej M., Repkova J. and Relichova J., 1999, Identification and mapping of a T-DNA induced flower mutation in *Arabidopsis thaliana*. Biol. Plant, **42:** 349–359.

Kuksova V.B., Piven N.M. and Gleba Y.Y., 1997, Somaclonal variation and in vitro induced mutagenesis in grapevine. Plant Cell. Tiss. Org. Cult. **49:** 17–27.

Kumar A. and Hirochika H., 2001, Applications of retrotransposons as genetic tools in plant breeding. Trends in Plant Sci. **6:** 127–134.

Laferriere L.T., Helgeson J.P. and Allen C., 1999, Fertile *Solanum tuberosum* + *S. commersonii* somatic hybrids as sources of resistance to bacterial wilt caused by *Ralstonia solanacearum*. Theor. Appl. Genet. **98:** 1272–1278.

Lambe P., Mutambel H.S.N., Fouche J.G., Deltour R., Foidart J.M. and Gaspar T., 1997, DNA methylation as a key process in regulation of organogenic totipotency and plant neoplastic progression. In vitro Cell. Develop. Biol. Plant. **33:** 155–162.

Larkin P.J. and Scowcroft S.C., 1981, Somaclonal variation—a novel source of variability from cell culture for plant improvement. Theor. Appl. Genet. **60:** 197–214.

Larkin P.J., Banks P.M., Brettell R.I.S., Davies P.A., Ryan S.A., Scowcroft W.R., Spindler L.H. and Tanner G.J., 1989, From somatic variation to variant plants: mechanisms and applications. Genome **31:** 705–711.

Lee M. and Phillips R.L., 1987, Genome rearrangements in maize induced by tissue culture. Genome **29:** 122–128.

Levall M.W. and Bornman J.F., 1993, Selection in vitro for UV-tolerant sugar beet (*Beta vulgaris*) somaclones. Physiol. Plant **88:** 37–43.

Levall M.W., Bengtsson K., Nilsson N.O., Hjerdin A. and Hallden C., 1994. Molecular characterization of UV-treated sugar beet somaclones using RFLP markers. Physiol. Plant. **90:** 216–220.

Lineberger R.D., Pogany M., Malinich T., Druckenbrod M. and Warne A., 1993, Genotype segregation and chimeral rearrangements in tissue culture. A potential source of new ornamental cultivars. In: Proc. XVIIth Eucarpia Symposium, Creating Genetic Variation in Ornamentals, T. Schiva and A. Mercuri (Eds), Sanremo, Italy, pp. 83–92.

Locy R.D., Chang C.C., Nielsen B.L. and Singh N.K., 1996, Photosynthesis in salt-adapted heterotrophic tobacco cells and regenerated plants. Plant Physiol. **110:** 321–328.

Maddock S.E., 1986, Somaclonal variation in wheat. In: J. Semal (Ed), Somaclonal Variation and Crop Improvement, pp. 127–137. Martinus Nijhoff Publishers, Dordrecht.

Maluszynski M., 2001, Officially released mutant varieties—The FAO/IAEA database. Plant Cell, Tiss. Org. Cult. **65:** 175–177.

Maluszynski M., Ahloowalia B.S. and Sigurbjönsson B., 1995. Application of in vivo and in vitro mutation techniques for crop improvement. Euphytica **85:** 303–315.

Maluszynski M., Nichterlein K., Van Zanten L. and Ahloowalia B.S., 2000, Officially released mutant varieties—the FAO/IAEA database. Mutation Breed. Revs. 12:84. http://www-INFOCRIS.iaea.org/MVD/.

Maluszynski M., Ahloowalia B.S., Nichterlein K., Jain S.M., Lee S., Nielen S., Zapata Arias J., Roux N., Afza R. and Kodym A., 2001, Mutating genes to meet the challenge for crop improvement and food security. AgBiotechNet Vol. 3, August ABN 071.

Mandal A.B., Pramanik S.C., Chowdhury B. and Bandyopadhyay A.K., 1999, salt-tolerant Pokkali somaclones: performance under normal and saline soils in Bay Islands. Field Crops Res. **61:** 13–21.

Martelli G., Greco I., Mezzetti B. and Rosatti P., 1993, Isozymic analysis of somaclonal variation among regenerants from apple root stock leaf tissue. Acta Hort. **336:** 381–387.

Micke A., Donini B. and Maluszynski M., 1990. Induced mutations for crop improvement. Mutation Breed rev., FAO/IAEA, Vienna No. 7, pp. 1–41.

Mohamed M.A.H., Harris P.C.J. and Henderson J., 2000, In vitro selection and characterization of a drought tolerant clone of *Tagetes minuta*. Plant Sci. **159:** 213–222.

Moon D.H., Ottoboni L.M.M., Souza A.P., Sibov S.T., Gasper M. and Arruda P., 1997, Somaclonal-variation-induced aluminium-sensitive mutant from an aluminium-inbred maize tolerant line. Plant Cell Rep. **16:** 686–691.

Nakajima K., 1991, Biotechnology for crop improvement and production in Japan. Paper presented at the Regional Expert Consultation on the Role of Biotechnology in Crop Production, FAO Regional Office for Asia and the Pacific, Bangkok, June 18–21, 1991, 21 pp.

Nehra S.N., Chibber R.N., Kartha K.K., Datla R.S.S., Crosby W.L. and Stushnoff C., 1990, Genetic transformation of strawberry by *Agrobacterium tumefaciens* using a leaf disk regeneration system. Plant Cell Rept. **9:** 293–298.

Novak F.J., Afza R., van Duren M. and Omar M.S., 1990, Mutation induction by gamma irradiation of *in vitro* cultured shoot tips of banana and plantain (*Musa* cvs.) Trop. Agric. (Trinidad) **67:** 21–28.

Osipova E.S., Kokaeva Z.G., Troitskj A.V., Dolgikh Yu. I., Shamina Z.B. and Gostimskij S.A., 2001. RAPD analysis of maize somaclones. Russian J. Genetics **37:** 80–84.

Peraza-Echeverria S., Herrera-Valencia V.A. and James-Kay A., 2001, Detection of DNA methylation changes in micropropagated banana plants using methylation sensitive amplification polymorphism (MSAP). Plant Sci. **161:** 359–367.

Peschke V.M. and Phillips R.L., 1992, Genetic implications of somaclonal variation in plants. Adv. Genet. **30:** 41–75.

Phillips R.L., Kaeppler S.M. and Peschke V.M., 1990, Do we understand somaclonal variation? In: Proceedings of the 7th International Congress on Plant Tissue Cell Culture, pp. 131–141.H.J.J. Nijkamp, L.H.W. van der Plas and J. van Aartrijk (Eds), Kluwer Academic Publishers, Dordrecht.

Phillips R.L., Kaeppler S.M. and Olhoft P., 1994, Genetic instability of plant tissue cultures: breakdown of normal controls. Proc. Natl. Acad. Sci. USA **91:** 5222–5226.

Piccioni E., Barcaccia G., Falcinelli M. and Standardi A., 1997, Estimating alfalfa somaclonal variation in axillary branching propagation and indirect somatic embryogenesis by RAPD fingerprinting. Intern J. Plant Sci. **158:** 556–562.

Pontaroli A.C. and Camadro E.L., 2001, Increasing resistance to *Fusarium* crown and root rot in asparagus by gametophyte selection. Euphytica **122:** 343–350.

Powell W., Morgante M., Andre C., Hanafey M., Vogel J., Tingey S. and Rafalski A., 1997, The comparison of RFLP, RAPD, AFLP, and SSR (microsatellite) for germplasm analysis. Mol. Breed. **2:** 225–238.

Racchi, M.L., Rebecchi M., Todesco G., Nielsen E. and Forlani G., 1995. Glyphosate tolerance in maize (*Zea mays* L.) 2. Selection and characterization of a tolerant somaclone. Euphytica **82:** 165–173.

Rahman M.H. and Rajora O.P., 2001, Microsatellite DNA somaclonal variation in micropropagated trembling aspen (*Populus tremuloides*). Plant Cell Repts. **20:** 531–536.

Ramos Leal R., Maribona R.H., Ruiz A., Konerva S. et al., 1996. Somaclonal variation as a source of resistance to eyespot disease of sugarcane. Plant Breed. **115:** 37–42.

Ramgareeb S., Watt M.P., Marsh C. and Cooke J.A. 1999. Assessment of Al^{+3} availability in callus culture media for screening tolerant genotypes of *Cynodon dactylon*. Plant Cell Tiss. Org. Cult. **56:** 65–68.

Rani V., Parida A. and Raina S.N., 1995, Random amplified polymorphic (RAPD) markers for genetic analysis in micropropagated plants of *Populus deltoides* Marsh. Plant Cell Repts **14:** 459–462.

Rani V., Singh K.P., Shiran B., Nandy S., Goel S., Devarumath R.M., Sreenath H.L. and Raina S.N., 2000, Evidence for new nuclear and mitochondrial genome organizations among high-frequency somatic embryogenesis-derived plants of allotetraploid *Coffea arabica* L. (Rubiaceae). Plant Cell Repts **19:** 1013–1020.

Reinbothe C., Tewes A. and Reinbothe S., 1992, Altered gene expression during somatic embryo genesis in *Nicotiana plumbagifolia* and *Digitalis lanata*. Plant Sci. **82:** 47–58.

Remotti P.C., 1998. Somaclonal variation and in vitro selection for crop improvement. In: Jain S.M., Brar D.S. and Ahloowalia B.S. (Eds). Somaclonal Variation and Induced Mutations in Crop Improvement, pp. 169–201. Kluwer Academic Publishers, Dordrecht.

Remotti P.C., Loffler H.J.M. and Vanvlotendoting L., 1997, Selection of cell-lines and regeneration of plants resistant to fusaric acid from *Gladiolus* × *grandiflorus* cv. Peter pears. Euphytica **96:** 237–245.

Rietveld R.C., Hasegawa P.M. and R. Bressan A., 1991, Somaclonal variation in tuber disc-derived populations of potato. I. Evidence of genetic stability across tuber generations and diverse locations. Theor. Appl. Genet. **82:** 430–440.

Rosati P., Menzzetti B., Anchenari M., Foscolo S., Predieri S. and Foscolo F., 1990. In vitro selection of apple rootstock somaclones with *Phytophthora cactorum* culture filtrate. Acta Hort. **280:** 409–416.

Roth R., Ebert I. and Schmidt J., 1997, Trisomy associated with loss of maturation capacity in a long-term embryogenic cultures of *Abies alba*. Theor. Appl. Genet **95:** 353–358.

Rout G.B., Samantaray S. and Das P., 2001, Aluminum toxicity in plants: a review. Agronomie **21:** 3–21.

Rugienius R. and Stanys V., 2001, In vitro screening of strawberry plants for cold resistance. Euphytica **122:** 269–277.

Rus A., Yokoi S., Sharkhuu A., Reddy M., Lee B.H., Matsumoto T.K., Koiwa H., Zhu J.K., Bressan R.A. and Hasegawa P.M., 2001, AtHKT1 is a salt tolerance determinant that controls Na+ entry into plant roots. PNAS **98:** 14150–14155.

Russell J.R., Fuller J.D., Macaulay M., Hatz B.G., Jahoor A., Powell W. and Waugh R., 1997. Direct comparison of level of genetic variation among barley accessions detected by RFLPs, AFLPs, SSRs and RAPDs. Theor. Appl. Genet. **95:** 714–722.

Sabir A., Newbury H.J., Todd G., Catty J. and Ford-Lloyd B.V., 1992, Determination of genetic stability using isozymes and RFLPs in beet plants regenerated in vitro. Theor. Appl. Genet. **84:** 113–117.

Sabry S.R.S., Smith L.T. and Smith G.M., 1995, Osmoregulation in Spring wheat under drought and salinity stress. J. Genet. and Breed. **49:** 55–60.

Samantaray S., Rout G.R. and Das P., 1999, In vitro selection and regeneration of zinc tolerant calli from *Setaria italica* L. Plant Sci. **143:** 201–209.

Sato F., Shigematsu Y. and Yamada Y., 1988, Selection of an atrazine-resistant tobacco cell line having a mutant *psb*A gene. Mol. Gen. Genet. **2134:** 358–360.

Saunders J.W., Acquaah G., Renner K.A. and Doley W.P., 1992, Monogenic dominant sulfonylurea resistance in sugarbeet from somatic cell selection. Crop Sci. **32:** 1357–1360.

Shahin E.A. and Spivey R., 1986, A single dominant gene for *Fusarium* wilt resistance in protoplast-derived tomato plants. Theor. Appl. Genet. **73:** 164–169.

Sharma J.R., Lal R.K., Misra H.O., Naqvi A.A. and Patra D.D., 1999, Combating opium-linked global abuses and supplementing the production of edible seed and seed oil: A novel non-narcotic var. *Sujata* of opium poppy. Curr. Sci. **77:** 1584–1589.

Shenoy V.B. and Vasil I.K., 1992, Biochemical and molecular analysis of plants derived from embryogenic tissue cultures of napier grass (*Pennisetum purpureum* K. Schum). Theor. Appl. Genet. **83:** 947–955.

Shepard, J.F., Bidney D. and Shahin E., 1980, Potato protoplasts in crop improvement. Science **208:** 17–24.

Singer S.R. and McDaniel C.N., 1984, Selection of amitrole tobacco calli and the expression of this tolerance in regenerated plants and progeny. Theor. Appl. Genet. **67:** 427–432.

Skirvin R.M. and Janick J., 1976. Tissue culture-induced variation in scented *Pelargonium* spp. J. Amer. Soc. Hort. Sci. **101:** 281–290.

Skirvin R.M., Norton M. and McPheeters K.D., 1993, Somaclonal variation: Has it proved useful for plant improvement. Acta Hort. **336:** 333–340.

Sowik I., Bielenin A. and Michalczuk L., 2001, In vitro testing of strawberry resistance to *Verticillium dahliae* and *Phytophthora cactorum*. Sci. Hort. **88:** 31–40.

Stephens P.A., Nickell C.D. and Widholm J.M., 1991, Agronomic evaluation of tissue culture-derived soybean plants. Theor. Appl. Genet. **82:** 633–635.

Stewart G.R. and Lee J.A., 1974, The role of chlorine accummulation in halophytes. Planta **20:** 279–289.

Sunderland N., 1977, Nuclear cytology. In: H.E. Street (Ed.). Plant Tissue and Cell Culture, pp. 177–205. Oxford: Blackwell.

Taylor P.W.J., Geijskes J.R., Ko H.L., Fraser T.A., Henry R.J. and Birch R.G., 1995, Sensitivity of random amplified polymorphic DNA analysis to detect genetic changes in sugarcane during tissue culture. Theor. Appl. Genet. **90:** 1169–1173.

Thakur R.C., Goto S., Ishii K. and Jain S.M., 1999. Monitoring genetic stability in *Quercus serrata* Thunb in somatic embryogenesis using RAPD markers. J. For. Res. **4:** 157–160.

Thanuthong P., Furosawa I. and Yamamoto M., 1983, Resistant tobacco plants from protoplast-derived calluses selected for their resistance to *Pseudomonas* and *Alternaria* toxins. Theor. Appl. Genet. **66:** 209–215.

Thomas C.M., Vos P., Zabeau M., Jones D.A., Norcott K.A., Chadwick B.P. and Jones J.D.G. 1995. Identification of amplified restriction fragment polymorphism (AFLP) markers tightly linked to the tomato Cf-9 gene for resistance to *Cladosporium fluvum*. The Plant J. **8:** 785–794.

Toker C., Gorham J. and Cagirgan M.I., 1999, Assessment of response to drought and salinity stresses of barley (*Hordeum vulgare* L.) mutants. Cereal Res. Comm. **27:** 411–426.

Tomilov A.A., Tomilova N.B., Ogarkova O.A. and Tarasov V.A., 1999, Insertional mutagenesis in *Arabidopsis thaliana*: presonication of germinating seeds increases the efficiency of their transformation. Russian J. Genetics **35:** 1043–1049.

Turner S., Krauss S.L., Bunn E., Senaratna T., Dixon K., Tan B. and Touchell D., 2001. Genetic fidelity and viability of *Anigozanthos viridis* following tissue culture, cold storage and cryopreservation. Plant Sci. **161:** 1099–1106.

Van den Bulk R.W. and Dons J.J.M., 1993, Somaclonal variation as a tool for breeding tomato for resistance to bacterial canker. Acta Hort. **336:** 347–355.

Veilleux, R.E., Shen L.Y. and Paz, M.M. 1995, Analysis of the genetic composition of anther-derived potato by randomly amplified polymorphic DNA and simple sequence repeats. Genome **38:** 1153–1162.

Winicov I., 1991, Characterization of salt tolerant alfalfa (*Medicago sativa* L.) plants regenerated from salt tolerant cell lines. Plant Cell Rep. **10:** 561–564.

Winicov I., 1996, Characterization of rice (*Oryza sativa* L.) plants regenerated from salt-tolerant cell lines. Plant Sci. **13:** 105–111.

Wolff K., Zietkiewicz E. and Hofstra H., 1995, Identification of chrysanthemum cultivars and stability of DNA fingerprint patterns. Theor. Appl. Genet. **91:** 439–447.

Xing W. and Rajashekar C.B., 2001. Glycine betaine involvement in freezing tolerance and water stress in *Arabidopsis thaliana*. Environ. Expt. Bot. **46:** 21–28.

Xiong L.Z., Xu C.G., Saghai Maroof M.A. and Zhang Q., 1999, Patterns of cytosine methylation in an elite rice hybrid and its parental lines, detected by a methylation-sensitive amplification polymorphism technique. Mol. Gen. Genet. **26:** 439–446.

Xu Y S., Murto M., Dunckley R., Jones M.G.K. and Pehu E., 1993, Production of asymmetric somatic hybrids between *Solanum tuberosum* and irradiated *S. brevidens*. Theor. Appl. Genet. **86:** 754–760.

Yamazaki M., Tsugawa H., Miyao A., Yano M., Wu J., Yamamoto S., Matsumoto T., Sasaki T. and Hirochika H., 2001, The rice retrotransposon *Tos17* prefers low-copy-number sequences as integration targets. Mol. Gen. Genet. **265:** 336–344.

Yanofsky M.F., Ma H., Bowman J.L., Drews G.N., Feldmann K.A. and Meyerowiz E.M., 1990. The protein encoded by the *Arabidopsis* homeotic gene agamous resembles transcription factors. Nature **346:** 35–39.

Yemets A.I., Kundel'Chuk O.P., Smertenko A.P., Solodushko V.G., Rudas V.A., Gleba Y.Y. and Blume Y.B., 2000, Transfer of amiprophosmethyl resistance from a *Nicotiana plumbaginifolia* mutant by somatic hybridization. Theor. Appl. Genet. **100:** 847–857.

11

Current Strategies in Molecular Breeding to Increase Food Production

Ahmad S. Islam[1], Gregory Clark[1] and Zeba I. Seraj[2]

[1] *Molecular Cell and Development Biology, University of Texas, Austin, TX 78713, USA;*

[2] *Department of Biochemistry and Molecular Biology, Dhaka University, Bangladesh*

ABSTRACT

This chapter outlines the recent advances in genetically-modified plants that may lead to better production of food and increased nutrition in food for the growing human population. Aluminium toxicity limits the growth of many crops in acidic soils, but recent findings on aluminium tolerant crops suggest that certain genes help confer tolerance and have led to the production of the first transgenic plants with increased aluminium tolerance. Advances in producing legume seeds and cereal grains with low phytic acid content, while still maintaining good seed/grain yield, have also been discussed. The chapter also brings into focus the advantages of incorporating desirable genes into chloroplasts over production of transgenic plants by insertion of genes into the nuclear genome. One possible advantage of this technology is in the production of vaccines in food crops, and the prospects for its use have been examined. The importance of early flowering genes, discovered recently in *Arabidopsis*, has been discussed in relation to utilizing them for hastening flowering time in perennial fruit plants; the strategy increases the prospect of development of early bearing fruit trait in otherwise late-bearing plant species. Recent results in bioengineering of drought- and salt-tolerance as well as disease resistance in plants have also been reviewed.

INTRODUCTION

By 2025, the human population is predicted to increase to a staggering number of more than 8 billion from the present figure of 6 billion. In the last 50 years, the increase has been doubled, the increment almost entirely confined to the developing countries. Due to human interference, the planet's ecological carrying capacity has been steadily shrinking. The two trends combined have made food security more precarious than ever before in human history. Besides food security, malnutrition due to lack of balanced food is another formidable problem. According to a recently published statistics by FAO, nearly 800 million of the world populations suffer from malnutrition. According to the same source* of information, the gap between intake of food and malnutrition will substantially decrease within the next decade but there will still be 680 million (13%) people suffering from chronic malnutrition. In the backdrop of the existing situation, the obvious solution lies in the production of more food—both in quantity and quality, utilizing the inhospitable landmass, now considered to be unsuitable for any crop cultivation. This chapter discusses the manner in which some of recent technologies developed in the post-genomic era may contribute to ease the global food situation, providing food not only to the increasing populace but also save them from under-nourishment and immunizing them against deadly diseases such as hepatitis B through genetically-modified (GM) food containing vaccines.

Aluminium Tolerance in Plants

In most of the world's tropical areas, where soil is acidic ranging from 4.5 to 5, aluminium (Al^{3+}) toxicity limits the productivity of crop plants. At this soil pH, between 4.5 and 5 or lower, there is an inhibition of root growth impairing its function. J.F. Ma et al. (2001) report that certain crop cultivars secrete organic acids such as citrate, oxalate and malate from their roots in response to Al^{3+} treatment. These organic acid anions form sufficiently strong complexes with Al^{3+} in order to protect the plant roots. Only the root apical cells secrete these organic acids to create a zone of detoxification that shields the root, allowing its growth through the acidic soil.

Ma et al. (2001) cite examples such as rye roots, which show a 30% increase only in citrate synthase activity, following an exposure to Al^{3+}. In the Al^{3+}-tolerant species *Paraserianthes falcataria*, mitochondrial citrate synthase gene expression and its activity showed an increase in response to Al^{3+}. Overexpression of carrot mitochondrial citrate synthase in *Arabidopsis* increased citrate synthase activity, enhanced production of citrate, and showed a 60% increase in citrate efflux.

Some species naturally accumulate high levels of Al^{3+} in aboveground tissues. The authors cite *Melastoma malabathricum* as an example of an extremely tolerant tropical rainforest species. Interestingly, its growth is stimulated by the presence of Al^{3+} in soil. Since the binding affinity of Al^{3+} for ATP is almost 10.7 times stronger than that of Mg^{2+}, this species must have mechanisms for detoxification of intracellular Al^{3+}. Other Al^{3+}-accumulating species—they quote in their article—are: wheat, buckwheat, canola, oats, radish, rye, triticale, taro and hydrangea.

---

* FAO Food, Agriculture and Food Security: The Global Dimension. (Document WFS 96/Tech 1), Rome 1996, p. 27

Ma et al. (2001) consider that the formation of stable complexes between Al^{3+} and organic acids lead to the reduction of the level of Al^{3+} in the cytosol, thus reducing cytotoxic effects. The intake of Al^{3+} by the root is followed by the formation of a 1:3 Al^{3+}-oxalate complex in the root cells. Al^{3+} is, however, transported through the xylem as Al^{3+}-citrate. Al^{3+}-citrate is converted back to Al^{3+}-oxalate in leaf cells before storage of the oxalate complex in the vacuole. This is illustrated by GM tobacco and papaya that secrete citric acid from their roots. The secreted citric acid forms complex with Al^{3+} ions, preventing their entry into the root. This special trait enables the GM tomato and papaya to grow well in aluminium concentration, 10 times higher than those tolerated by control plants.

The authors also note that many crop plants have no Al^{3+}-tolerant cultivars. Creation of transgenic lines utilizing genes from Al^{3+}-tolerant species may allow cultivation of such crops in Al^{3+}-contaminated soil. As Al^{3+} toxicity remains a significant problem in agriculture, the authors advocate both proper land use strategies to avoid Al soil contamination and development of new cultivars for use on marginal soil.

The International Rice Research Institute (IRRI) has in its germplasm a collection of some aluminium-tolerant rice lines. Transfer of genes for Al^{3+} tolerance across the species barrier is now possible through genetic engineering. IRRI has developed a number of molecular markers to accelerate selection process. Molecular markers help identify a trait at the seedling stage. Molecular marker for aluminium tolerance is one of them and its use to determine aluminium tolerance at the seedling stage will doubtless accelerate the breeding programme and minimize land use and labour cost required for field testing of adult individuals.

Production of Cultivars with Low Phytic Acid Content is Now Possible

Merely maximizing food production will not solve the problem of feeding 8 billion people within the next 25 years. Along with boosting food yield, scientists need to change the phosphorus status of the seed storage protein. Otherwise, in spite of the plenteous food, the poor, who do not have the means to buy meat, eggs and milk, will suffer from malnutrition. Phytic acid constitutes up to about 75% of the total seed phosphorus content. However, most of this acid is excreted because of the inability of humans and non-ruminant animals (poultry, pig and fish) to absorb phosphorus. As a result, excreta of poultry, pig and fish, containing dietary phytic acid, creates health hazard by polluting water bodies. Phosphorus pollution contributes to algal blooms and fish kills. The reduction of phosphorus loads in the manure minimizes leaching into the groundwater, preventing pollution to a considerable extent. In the developing countries, excretion of dietary phytic acid contributes to public health problems, one of which is iron and zinc deficiency.

In a recent article Dr. Victor Raboy (2001) at U.S. Department of Agriculture, Idaho, reviews the current knowledge about phytic acid and the genetic strategies available for reducing its levels in legume seeds and cereal grains without reducing the total phosphorus content.

One of the ways to solve the problem partially is by supplementing feed with phytase enzymes produced industrially, using fungal phytase genes. Phytase supplements break down up to 50% of grain-based feed, releasing normally unavailable P for use by livestock, thereby reducing the amount of phosphorus waste. Plant genes encoding phytases have already been identified.

Dr. Raboy cites examples of certain approaches employed by researchers to enhance mineral contents in legume seeds and grain crops. In one study, the expression of a *Phaseolus* phytoferritin gene in rice was found to double the iron content of endosperm iron. In another study, cysteine-rich peptides that facilitate absorption of iron in the gut have been expressed in rice. In a third study, rice was transformed with a thermo-tolerant fungal phytase in an effort to reduce seed phytate content.

The author advocates the use of both classical and molecular genetics to disrupt the biosynthesis of phytic acid during the seed development. The main problem that breeders are confronted with is the direct relationship between seed/grain yield and the low phytate content. Recently, in maize, two loss-of-function *InsP6* mutants showing decoupling of this relationship have been found. The two mutants are: low phytic acid 1-1 *(Ipa1-1)* and low phytic acid 2-1 *(Ipa2-1)*. Both the mutants were reported to produce seeds with low phytic acid content without reduction in the normal levels of total phosphorus. The InsP6 reduction was about two-thirds in *Ipa1-1* mutant line with a molar equivalent increase in inorganic phosphorus. In the second mutant line *Ipa2-1*, the reduction was by ~50%. The number of maize *Ipa1* mutants has exceeded 20 and in homozygous condition, the reduction in the levels of InsP6 has been found to range from 50–95%. Similar *Ipa1* mutants showing a reduction in the amount of InsP6 by 50 to 95% have been reported in barley, rice and soybean. However, some of the homozygous mutants with low levels of InsP6 do not produce any grains, a factor that poses a big challenge to breeders for restoring normal grain formation in the lack-of-grain types. The author suggests a crossing programme: non-mutant types X mutant lines with ~75% reduction in the amount of InsP6. Individuals showing a yield range from 5-15% compared to wild type may serve as a forerunner for developing commercial varieties with an acceptable yield combined with a low phytic acid level.

The results of recent experiments in which the normal feed of maize or/and barley to animals was substituted by a low 'maize phytate' show that the excreta contained considerably less waste phosphorus (10–85%). According to the author, results of a small-scale clinical trials with human beings, using a phytase-low maize food, indicate that absorption of iron and zinc by the body system increased by ~50 -75%. Volunteers, who ate corn-flour made tortillas from the low phytate corn, showed less iron deficiency than the control group.

Dr. Raboy suggests that by making adjustments in seed metabolism such as targeting the desired change in gene expression to the appropriate seed tissue and decoupling starch accumulation and inorganic phosphorus concentration, it may be possible to restore seed weight and yield. In support of this approach, the author says that the levels of InsP6, typical of normal seeds, are not essential to seed function. This is evident from the fact that nearly all InsP6 are synthesized in seeds. Since they occur in storage tissue, they have apparently no active role in cellular metabolism or signal transduction. InsP6, present in the storage tissue, has been reported to play a very important role in the survival of species in the wild. Unlike wild species, cultivated crops are protected from herbivores due to human intervention and as such there is no particular need for InsP6-rich 'seed storage' tissue for its survival. Thus, for the benefit of human and animal health, programmes to breed crops with low phytate content can be undertaken without environmental risks.

Low-Phytate Non-GE Rice Developed in the USA: True to the expectation of Raboy, breeders have been recently successful to evolve a rice variety with low-phytate content without the help of bioengineering techniques. The strategy adopted by breeders for downsizing phytic acid production in cereal seeds was to insert a gene i.e., myo-inositol phosphate synthase (*MIPS*) for inactivating enzymes that are involved in the formation of phytic acid. For more details about *MIPS* see the most recent publication of Raboy (2003). Molecular biologists collaborating with breeders bio-engineered a modified form of the *MIPS* gene. Its defective product gave rise to a phenotype with lower phytate content (LPA). As a result, the defective product introduced into the system was destroyed by the plant's natural defense system. During the process the natural *MIPS* product was also eliminated giving rise to a phenotype with hardly any or a lower content of phytate. According to a recent report*** published in USDA

**(a) A tobacco-produced IgG-IgA antibody against a surface antigen of *Streptococcus mutans* designed to prevent tooth decay; (b) a soybean-produced antibody against herpes simplex virus (HSV) found to prevent transmission of HSV in a mouse model; (c) a rice-and wheat-produced antibody against carcinoembryonic antigen used for cancer diagnostic and treatment techniques; and (d) a transiently-expressed antibody produced by a plant virus vector in a tobacco system used for the treatment of lymphoma.

****http://www.ars.usda.gov/is/AR/archive/sep02/rice0902.htm*

News Bulletin, J. Neil Rutger together with Victor Raboy successfully used this technique to produce rice grains with a lower amount of phytic acid. The new rice variety with a lower phytic acid content has already been released to breeders and researchers. It is expected to be marketed shortly. The unique property of this rice variety will go a long way to provide nutrition to the world's malnourished people in Asia, Africa and Latin American countries. This would also mean more nutritious animal feed, and better water quality because of far less contamination by manure, which is one of the main sources of water pollution.

There has been a recent report in which Pilu et al. (2003) screened some maize mutants for lower phytic acid content in a population subjected to chemical mutagenesis. They isolated a viable recessive mutant, *lpa 241* that showed about 90% reduction in phytic acid amount and about a tenfold increase in seed-free phosphate content. Seed germination rate was 30% lower compared to the wild type. However, the lower germination rate did not seem to affect the development of mutant plants. An analysis of their results showed that the mutation affected the *MIPS1S* gene that codes for the first enzyme of the phytic acid biosynthetic pathway.

Bioengineering Plants via Chloroplasts

Production of transgenic plants through the insertion of genes in the nuclear genome has heralded a new chapter in the field of plant breeding. This technique enables breeders to transfer genes of interest across species or even of higher order barrier. However, soon it was realized that it is not as potent a method as it was originally envisaged. Among the major disadvantages, one is the chimerical nature of transformation, prolonging the breeding time to release a commercial variety. The other disadvantage is the indiscriminate insertion of the transgene of interest in the host genome. Random insertion of the transgene in the nuclear genome creates position effects, which simply means that the foreign-gene-controlled trait is not faithfully expressed. The third disadvantage is its weakness, when it comes to breeding for multiple traits, e.g. yield or tolerance to salt.

The problem has been addressed by incorporating genes of interest via chloroplasts. Professor Lawrence Bogorad (2000) at Harvard University discusses the advantages of transforming plants via chloroplasts over the current use of T-DNA or microprojectiles to bring about transformation. The author points out that transformation via chloroplasts gives a definite advantage over the old method in that it ensures introduction of a foreign gene at a specific location on a particular chromosome. For instance, Knoblauch and his associates were able to demonstrate that the DNA that was injected directly into individual tobacco leaf chloroplasts expressed itself in the photosynthetic leaf cells.

This technique also enables a breeder to introduce a block of genes controlling a character such as nitrogen or carbon dioxide fixation. Compared to the nuclear-cytoplasmic compartment, plastids and mitochondria provide more favourable environments for certain biochemical reactions and for accumulating large amounts of some gene- and enzyme products. In addition, these two organelles have the great advantage of having a smaller genomic size. Chloroplasts with 60 copies of a single circular chromosome have only 120 to150 genes. Since many chloroplast genes occur in operons, it may be possible to introduce blocks of foreign genes into a single operon. Furthermore, since the same kind of promoter is involved in transcription, the degree of expression of a foreign gene is the same, regardless of its location in the circular chloroplast chromosome. Transformation via chloroplasts has another great advantage. Since the chloroplasts are transmitted via ovules, the pollen of transformants would not contain any transgene, keeping the environment safe, thereby allaying the fears of adversaries of GM crops.

The author is of the opinion that with the techniques now available, it may be possible to bioengineer chloroplasts for nitrogen fixation in non-leguminous plant species. The present difficulty that prevents this seemingly difficult task is the sensitivity of nitrogenase to oxygen, which is released as a result of photosynthesis. Devising a method in higher plants similar to the one used by Cyanobacteria may solve this problem. Through the help of heterocysts, this class of microorganisms keep the two processes, namely, photosynthesis and nitrogen fixation separate temporally. The latter is a kind of thick-walled asexual spores in which there is no oxygen flow to the cells where nitrogen fixation occurs. Watching the trend of current research in gene transfer via organelles, the author envisions that it may be possible to convert C3- to C4 plants by manipulating bundle sheath chloroplasts and their association with Rubisco.

Hou et al. (2003) at the Institute of Genetics, Chinese Academy of Sciences have recently reported that by means of biolistic delivery, they have successfully inserted two genes, *aadA* and *cry1Aa10* into the chloroplast genomes of some oilseed rape varieties. Chloroplast transformants were resistant to second instar *Plutella xylostera* larvae in that 47% of the larvae died after consuming transgenic oilseed rape leaves. The weight of the surviving larvae was significantly lower than the control. The authors believe that the demonstration of chloroplast transformation, giving rise to partially insect resistant oil seed rape varieties, opens up avenues for breeding resistant crop varieties without transmitting harmful genes to the environment because pollen does not transmit any maternally inherited traits via chloroplasts to the progeny.

Transformation via Chloroplasts for Medical Molecular Farming

Daniell et al. (2001) discuss the superiority of the method of producing transformants via chloroplast in a recently published review article in *Trends in Plant Science.* Citing some examples of the production of antibodies in plants, the authors emphasize the need to reduce the quantity of plant tissue constituting a vaccine dose to a practical size. At present, due to the low level of antibodies in transgenic plants, those seeking immunity against a disease through plant vaccine, would be required to eat a large amount of the transformed plant material. This is inconvenient compared to a small vaccine dose. It is in this context that Professor Henry Daniell advocates switchover targeting chloroplasts for transformation rather than using nuclear DNA.

A combination of the two factors—small size of the vaccine dose and better expression of surface antigen—will prove to be more effective for conferring immunity. There are about 10,000 copies of chloroplast genomes per cell. This facilitates introduction of 10,000 copies of foreign genes per transformed cell. Such an increase in gene dosage results in several hundred-fold increase in gene expression as compared to nuclear transgenic plants. Such a dramatic increase has been demonstrated with the human somatotropin expressed via chloroplast- or nuclear genomes. Success in this direction will lead to 10-20 times cost reduction in the production of such oral vaccines compared to cell culture expression systems. Furthermore, a chloroplast transformation approach may prove to be more effective as an oral vaccine because of the uniformity of foreign gene expression among transgenic lines. The authors suggest to accomplish this by integration of transgenes into the spacer region at a precise location of the chloroplast genome. This procedure will eliminate position effects, frequently observed in nuclear transgenic plants due to random insertion of transgenes. In addition, gene silencing is not observed in chloroplast genetic engineering.

The toxicity of transgenic pollen containing pharmaceuticals to non-target insects or animals is a matter of great concern. To prevent the formation of superweeds that contain transgenic pollen, the authors underscore the importance of generating apomictic plants, creating incompatible genomes, suicide genes and male sterility. Targeting genes of interest to chloroplasts for the production of biopharmaceuticals is another powerful way to contain harmful genes because transgenes are not transmitted via chloroplasts. Such a procedure has been successfully demonstrated in tobacco and potato, suggesting that it may be possible to extend the same method to other crop plants.

According to Professor Henry Danielle, the state of the art currently available to produce plant vaccines in terms of optimization of protein quantity, environment biosafety and market size, is such that undertaking projects for the safe production of pharmaceuticals for clients requiring them, holds a great promise for the future.

Citrus Trees Transformed with *LEAFY* or *APETALA* Gene from *Arabidopsis* Produce Flowers in the First Year

Following the determination of base sequence of *Arabidopsis thaliana*, molecular geneticists have concentrated their efforts to determine the traits that the different base sequences constituting about 29,000 genes in the model plant, control. In the post-genomic era, a number of such genes and their functions have come to light. Some of these genes control characters of agronomic importance. Although they have been isolated from the model plant, they can be transferred to the genome of plant species of interest. Recently, two *Arabidopsis* genes, *LEAFY (LFY)* and *APETALA1 (AP1)* have been cloned. Both the genes induce early flowering in *Arabidopsis*. Dr. Jose Martinez-Zapater and his associates at the Instituto Valenciano de Investigaciones Agrarias, Spain were interested in using these genes in order to hasten flowering time in citrus trees. Normally, juvenile citrus trees do not flower and produce fruit until they are at least 6–20 years old (depending upon the species). The authors report in the March issue of *Nature Biotechnology,* that they produced two types of transgenic plants: one transformed with the *LEAFY* gene and the other with *APETALA* gene. The two foreign genes had a dramatic effect on citrus trees in that the transformed plants flowered in their first spring and continued to do so in consecutive years.

Their data revealed that the integration of the alien gene for flowering was stable. Moreover, their environmental response was similar to the control plants. The authors believe that their results open new possibilities for domestication, genetic improvement, and experimental research in citrus and other woody species.

There are many fruit plants such as mango, date palm, coconut, etc., which do not flower and bear fruit until they are several years old. It will be a worthwhile exercise for breeders to use the above two genes, *LEAFY* and *APETALA* to advance fruiting time in the plants of their interest. The quantum of fruit production in the first few years will be small. Nonetheless, early fruit-bearing ability of these species will be welcome, particularly by home gardeners as they will not have to wait long to harvest fruits from their own garden.

Prospects of Engineering Starches for Quality and Quantity

Starch in its unmodified form has limited industrial uses. Therefore, there is an urgent need to produce modified starches in order to enhance its multipurpose uses in the industry, where it serves as a raw material in the food, paper, adhesive and livestock industries. The limited reactivity of glucose as a building block makes it difficult to modify starch molecules. Recently, Slattery et al. (2001), at the Institute of Biological Chemistry, Washington University, Pullman, reviewed the progress of research on starch synthesis and its regulation. The bioengineering objective could be achieved by introducing glucose residues with reactive side-chains or charged groups. A variety of strategies are discussed concerning the creation of novel starch molecules. Some enzymes that can be used for generating novel starches are: starch-branching enzyme (SBE), debranching enzyme (DBE), granular-bound starch synthase (GBSSI) and starch synthase (SS). For example, there is an increased amount of amylose production, when SBE- and SS genes are activated. Similarly, the activation of ADP-glucose pyrophosphorylase (AGPase) and GBSS lead to the production of an increased amount of amylopectin. Starch yield is enhanced with an increase in the activity of both *SBE-* and *DBE* genes in conjunction with ATP/ADP transporter and AGPase. The fact, that the antisense *AGPase* gene in transgenic potatoes reduces starch formation, suggests that it may be possible to accelerate starch production by increasing the activity of this enzyme. Some experimental data support this view. When the potato variety, Russett Burbank was transformed with a mutant *E. coli AGPase* gene, there was a 30% increase in its starch content.

The authors propose that insertion of the *AGPase* gene could be achieved by combining the two subunits of this enzyme, large and small in different proportions. The larger subunit plays a regulatory role, while the smaller one, a catalytic one. The different levels of *AGPase* also have profound effects on starch structure. For instance, reduced levels of *AGPase* lower the amylose content of starch and higher levels enhance short-chain amylopectin.

Studies on starch synthase isoforms reveal that by manipulating this enzyme, it may be possible to generate novel starch molecules by extending glucan chains, or changing their thermosensitivity or even by producing chimeric proteins through fusion of an SBE isoform with a GBSS isoform. The authors also suggest that a DNA shuffling approach, in which divergent species and isoforms of SBE are used, may be applied to bioengineer new enzymes for different catalytic turnover and chain lengths.

Drought- and Salt-tolerant Plants without Yield Loss

Extensive research work was carried out at several institutions to produce salt tolerant crops. In most cases, salt tolerance was claimed not for enhanced yield under stress, but for some measure of survival. Some of these genes code for one of four classes of proteins, namely, proteins of ion homeostasis, protective proteins, osmoprotectants or regulatory factors. Genes that affected yield minimally, and yet conferred salt tolerance, have been reported by scientists working with the vacuolar H^+ pyrophosphatase gene *(AVP)* or the vacuolar Na^+/H^+ antiporter gene *(NHX1).* One of the most detailed analyses is that of Zhang et al. (2001), who engineered salt tolerance in *Brassica* plants using the Na^+/H^+ antiporter of *Arabidopsis,* that promotes vacuolar sodium accumulation. When transgenic plants were grown in 200mM NaCl, sodium accumulated to 6% of the dry weight of the plants but growth was only marginally affected. Neither seed yield nor seed oil quality was affected.

Zhang Hong-Xia and Eduardo Blumwald (2001) have reported bioengineering of a salt tolerant tomato line by inserting and overexpressing the same vacuolar Na^+/H^+ antiporter gene. These GM tomato plants were able to grow, flower and produce fruits in the presence of 200 mM NaCl. The fruit size of the salt tolerant line was slightly smaller than that of the control. While the transgenic leaves accumulated Na^+ to almost 1% of their dry weight, the fruits displayed only a marginal increase in Na^+ content and a 25% increase in K^+ content. The results of their experiments show clearly that transgenic tomato plants can utilize salty water for growth. Based on their positive results, it is suggested that it may be possible to produce salt-tolerant crops by introducing a fewer genes into the host genome than had been hitherto envisaged.

Roberto et al. (2001) at MIT, Harvard University, University of Connecticut and University of California, San Diego have shown that transgenic plants over expressing the vacuolar H^+-pyrophosphatase gene are much more salt- and drought-tolerant than wild type plants. The leaf tissue of the transgenic plants accumulates Na^+ and K^+ to higher levels than in the control. A comparison of isolated vacuolar membrane vesicles demonstrated that the *AVP1* transgenic plants exhibit enhanced cation uptake. Capacity for solute accumulation and water retention in the transgenic plants is enhanced, possibly as a result of an enhanced vacuolar proton gradient. The sequestration of excess Na^+ in the vacuoles of these plants makes high levels of this cation less toxic to the plant.

There may be opportunities to enhance salt tolerance by use of a master gene that regulates a cascade of stress tolerance genes and the use of a promoter that gives enhanced expression of the master gene in response to stress. Suitable genes could be transcription factors such as *DREB1A.* When *DREB1A* was expressed from the stress-inducible rd29A promoter, plants looked normal and showed enhanced tolerance to salinity, drought or cold (Kasuga et al., 1999). Another suitable gene is the one encoding a specific mitogen-activated protein kinase kinase kinase (MAPKKK), designated *ANP1* in *Arabidopsis* leaf cells. Kovtun et al. (2000) reported that transgenic tobacco plants that expressed a constitutively active *ANP1* orthologue, *NPK1,* display enhanced tolerance to multiple stress conditions without activating drought, cold, and abscissic acid signaling pathways described by other groups, including Kasuga and coworkers (1999). Winicov (2000) showed that root-specific transcription factor *Alfin1* is an essential gene for root growth and that its overexpression in transgenic plants confers a many-fold increase in root growth under normal and saline conditions and significantly increases plant growth and salt tolerance. *Alfin1* over-expressing transgenic plants showed salinity tolerance comparable to NaCl-tolerant plants (Winicov and Bastola, 1999). *Alfin1* recognizes the *cis*-element GNGGTG or GTGGNG that is found in a salt-inducible proline-rich protein MsPRP2.

Possibility of Evolving Drought-tolerant Plant Varieties through Manipulation of the Gene Regulating Stomatal Aperture

Dr. Alan Jones, Dr. H. Ullah in Alan Jones' Lab and their collaborators in Dr. Sarah Assmann's laboratory (Wang XQ et al., 2001) demonstrated that the water loss in *Arabidopsis gpa1* mutants is greater than its wild counterpart because of its insensitivity to inhibition by abscisic acid (ABA). It may be mentioned here that ABA prevents water loss from terrestrial plants by controlling the dimension of stomatal pores via inward K^+ channels. In the *gpa1* mutant, ABA activation does not occur, rendering anion channels non-functional and insensitive to environmental fluctuation. Although preliminary, the above finding is extremely important as it has shown the manner is which future stomatal apertures in plants may be controlled, leading to the development of drought-tolerant plants.

Possibility of Controlling Crown Gall Formation in Walnut

During infection of plant hosts, *Agrobacterium tumefaciens* transfers oncogenes such as *iaaM* and *ipt* to the host genome. This causes tumour formation via *de novo* synthesis of auxin and cytokinin. This mechanism of infection is well conserved, and *iaaM* and *ipt* show approximately 90% DNA sequence identity in various *A. tumefaciens* strains. Recently, Dr. Abhaya M. Dandekar and his associates at the University of California, Davis (Escobar et al., 2001) report producing transgenic tomato *(Lycopersicon esculentum)* and *Arabidopsis thaliana* lines containing constructs encoding two self-complementary RNA fragments. The two fragments inactivated the action of *iaaM* and *ipt* through RNA interference (RNAi) Tumor initiation upon infection with *A. tumefaciens* was greatly reduced in the transgenic lines. Unlike plant resistance traits conferred by highly specific receptor-ligand binding mechanisms, the operational mechanism in these transgenic lines is based on mRNA sequence homology. The authors suggest that the immunity may be durable, and express hope that RNAi oncogene silencing will be used for improvement of ornamental rootstocks and tree crops, if this technique proves to be workable under field conditions. Dandekar's team is currently transforming walnut, where crown gall is a serious problem. Although his team is well on the way to providing a practical solution for the industry, Dandekar (personal communication) thinks commercialization will be years away. The technique devised by Dandekar may be tried to combat the incidence of tumour formation in other tree cultivars where it is a serious problem.

Never before were so many powerful molecular tools available to plant breeders to tailor plants to feed the hungry mouths of the global population as well as to customize plants to the needs of consumers' special needs. Now has come the time, when plant breeders in collaboration with FAO, CGIAR and those who determine policy in National Governments, need to prove to the world community that by applying these new devices, they are in a position to meet the challenge facing the undernourished world's population, who often do not have two square meals a day to satiate their hunger, what to talk of obtaining balanced nutritious meals.

REFERENCES

Bogorad L., 2000, Engineering chloroplasts: An alternative site for foreign genes, proteins, reactions and products. Trends Biotechnol. **18(6):** 257-63.

Daniell H., Streatfield S.J., Wycoff K., 2001, Medical molecular farming: Production of antibodies, biopharmaceuticals and edible vaccines in plants. Trends Plant Sci. **6(5):** 219-26.

Escobar M.A., Civerolo E.L., Summerfelt K.R., Dandekar A.M., 2001, RNAi-mediated oncogene silencing confers resistance to crown gall tumorigenesis. Proc. Natl Acad. Sci. USA **98(23):** 13437-42.

Hou B.K., Zhou Y.H, Wan .H, Zhang Z.L., Shen G.F., Chen Z.H., Hu Z. M., 2003, Chloroplast transformation in oilseed rape. Transgenic Res. **12**:111-4.

Kasuga M., Liu Q., Miura S., Yamaguchi-Shinozaki K. and Shinozaki K., 1999, Improving plant drought, salt and freezing tolerance by gene transfer of a single stress-inducible transcription factor. Nature Biotechnol. **17**:287-291.

Kovtun Y., Chiu W. L., Tena G., and Sheen J., 2000, Functional analysis of oxidative stress activated mitogen-activated protein kinase cascade in plants. Proc. Natl. Acad. Sci. USA. **97**: 2940-5.

Ma J.F., Ryan P.R., Delhaize E., 2001, Aluminium tolerance in plants and the complexing role of organic acids. Trends Plant Sci. **6**: 273-8.

Pena L., Martin-Trillo M., Juarez J., Pina J.A., Navarro L., Martinez-Zapater J.M., 2001, Constitutive expression of *Arabidopsis LEAFY* or *APETALA1* genes in citrus reduces their generation time. Nat. Biotechnol. **19**: 263-7.

Pilu R., Panzeri D, Gavazzi G., Rasmussen S.K., Consonni G., Nielsen E., 2003, Phenotypic, genetic and molecular characterization of a maize low phytic acid mutant *(lpa241)*. Theor Appl Genet.**107**:980-7.

Raboy V., 2001, Seeds for a better future: 'low phytate' grains help to overcome malnutrition and reduce pollution Trends Plant Sci. **6**:458-62.

Raboy V., 2002, Progress in Breeding Low Phytate Crops, J. Nutr. 132:503S-505S presented in the Plant Breeding Symposium: A New Tool for Fighting Malnutrition.

Raboy V., 2003, myo-Inositol-1,2,3,4,5,6-hexakisphosphate. Phytochemistry **64(6)**: 1033-43.

Roberto A., Gaxiola, Jisheng Li., Soledad Undurraga, Lien M. Dang, Gethyn J. Allen, Seth L. Alper and Gerald R. Fink, 2001, Drought- and salt-tolerant plants result from overexpression of the AVP1 H+-pump. Proc. Natl. Acad. Sci. USA **98:** 11444-49.

Slattery C.J., Kavakli I.H., Okita T.W., 2000, Engineering starch for increased quantity and quality. Trends Plant Sci. **5**: 291-8.

Wang X.Q., Ullah H., Jones A.M., Assmann S.M., 2001, G-protein regulation of ion channels and abscisic acid signaling in Arabidopsis guard cells. Science **292:** 2022-3.

Winicov I., 2000, *Alfin1* transcription factor overexpression enhances plant root growth under normal and saline conditions and improves salt tolerance in alfalfa. Planta **210**: 416-422.

Winicov I. and Blastola D. R., 1999, Transgenic overexpression of the transcription factor *Alfin1* enhances expression of the endogenous MsPRP2 gene in alfalfa and improves salinity tolerance of the plants. Plant Physiol. **120**: 473-480.

Zhang H. X., and Blumwald E., 2001, Transgenic salt-tolerant tomato plants accumulate salt in foliage but not in fruit. Nature Biotech. **19:**765-8.

Zhang H. X., Hodson J. N., Williams J. P. and Blumwald E., 2001, Engineering salt tolerant *Brassica* plants: characterization of yield and seed oil quality in transgenic plants with increased vacuolar sodium accumulation. Proc. Natl. Acad. Sci. USA **98**:12382-36.

12

Genetic Engineering to Develop Resistance against Potato Virus Y

S.B. Ghosh, T.R. Ganapathi, and V.A. Bapat*
Plant Cell Culture Technology Section, Nuclear Agriculture and Biotechnology Division, Bhabha Atomic Research Centre, Trombay, Mumbai 400 085, India

ABSTRACT

Potato is the fourth most important food crop cultivated and consumed all over the world. The improvement of potato for agronomic qualities through conventional breeding is time consuming and laborious due to its biology. Many of the good cultivars are susceptible to virus infection; fungal and bacterial pathogens also attack potato cultivars, resulting in considerable yield loss. Recent developments in genetic engineering in potato have opened up many avenues for improvement of potato cultivars with desirable traits. Several genes for disease and pest resistance have been incorporated into potato genome. Among the viruses infecting potato cultivars, potato Virus Y (PVY), Potato Virus X (PVX) and Potato Leaf Roll Virus (PLRV) are the important ones. PVY, in particular, can cause yield loss up to 30–40% worldwide. In this chapter, different strategies employed in imparting resistance to PVY—the problem and prospects have been described.

INTRODUCTION

Potato (*Solanum tuberosum* L.) is an economically-important food crop cultivated and consumed all over the world. In total food production, potato ranks fourth after wheat, maize and rice (Heling 1998). Potato varieties with higher yield, better agronomic qualities and resistance to diseases are of immense importance for increased production (Naik and Sarkar 2000). However, the process of varietal improvement through conventional breeding is time consuming as well as complicated and laborious due to tetrasomic segregation pattern and incomplete fertility in many tetraploid cultivars. Many good cultivars are susceptible to virus infection, fungal and bacterial diseases such as late blight, ringrot, black leg, bacterial wilt, etc., leading to major yield losses. Among viruses, Potato Virus Y (PVY), Potato Virus X (PVX) and Potato Leaf Roll Virus (PLRV) are the major viral pathogens, while Potato Virus S, Potato Virus

*Corresponding author. Email: vabapat@magnum.barc.ernet.in

A and Potato Virus V comprise the other notable ones (Khurana 1992). Potyviruses (PVY, PVA, PVV) cause major damage to potato crop and 30 to 40% yield loss is quite common (Rykbost et al., 1999). Infection of tobacco with PVY strains indigenous to southeastern United States resulted in a yield reduction ranging from 10 to 100% (Gooding and Tolin 1973). A synergistic effect involving a dual infection of PVY and PVX can cause serious damage to the crop in the field (Pruss et al., 1997).

Viral diseases in crop plants are specially important because of the fact that any other microbial attack on a crop plant can be controlled to a great extent by pesticides or other agents, whereas there is no chemical or physical agent which can be used in a mass scale to cure viral infection in the field. Viral infection can only be prevented by cultivating resistant varieties. Availability of conventional resistance in related wild species is not well known as also not dependable and most resistances can be broken down either by virulent strains or by higher level of viral inoculum (Brigneti et al., 1997; Hamalainen et al., 1998; Hinrichs et al., 1999).

Potato Virus Y is a type member of the potyvirus group of plant viruses (de Bokx and Huttinga 1981; Shukla et al., 1994; Mathews 1991). It is a flexous helical rod-shaped virus of approximately 700-nm length, 12-nm diameter and 3.4-nm helical pitch. It has a single stranded positive sense RNA of approximately 10-Kb length, encapsidated by about 2000 copies of a single coat protein of approximately 30 kD size. Reports are available on the complete genome sequence of PVY (Thole et al., 1993; Singh and Singh 1996; Robaglia et al., 1989). PVY infects potato and many other solanaceous crops, including tobacco, tomato, etc., but major economical damage is caused in potato. Two loci imparting resistance against PVY in homozygous recessive condition in tomato were studied but it has not yet been possible to transfer the genes between these two distant solanaceous species (Legnani et al., 1995). Acceptable sources of resistance have been rarely found in tobacco (Wernsman and Rufty 1980; Burk et al., 1982). One of the sources of resistance or tolerance in tobacco has been reported in 'NC 744' breeding line, which carries a *va* homozygous recessive gene conferring resistance to PVY derived from 'Virgin A mutant' (VAM) (Sudarsono et al., 1996; Burk et al., 1982; Werns-man and Rufty, 1980; Gooding and Kennedy, 1985; Gupton and Burk, 1973). However, some of the negative attributes have been associated with the tobacco genotypes carrying the *va* gene in recessive homozygous condition, the most notable among them being the hypersensitivity to blue mold (*Peronospora tabacina*) (Gooding et al., 1985).

The recent progress in genetic engineering has opened up many avenues to develop virus and disease resistant crop varieties. Sanford and Johnston (1985) first proposed the concept of pathogen-derived resistance; since then, structural and non-structural viral genes have been used to engineer virus resistance against many virus groups, including potyviruses (Lindbo et al., 1993; Dinant et al., 1993; Baulcombe, 1996). The coat protein (CP) encoding region of PVY has been successfully used to engineer resistance to PVY in potato cultivars 'Russet Burbank', 'Russet Norkotah', 'Bintje' and 'Flova' (Kaniewski et al., 1990; Lawson et al., 1990; Farinelli et al., 1992; Malnoe et al., 1994; Smith et al., 1995) (Table 12.1). A high level of resistance against PVY has also been reported by transferring other nonstructural genes like RNA-dependant RNA polymerase gene (NIb), nuclear inclusion proteinase gene (NIa) and P1 proteinase gene, both in sense and antisense orientation (Maki-Valkama et al., 2000a,b; Audy et al., 1994; Pehu et al., 1995; Vardi et al., 1993).

Table 12.1 Reports on potato virus Y resistance in different potato cultivars

Cultivar	*Strategy adopted*	*Reference*
1. 'Kufri Jyoti	Coat protein	Ghosh et al., 2003
2. 'Pito'	P1 antisense	Maki-Valkama et al., 2000
3. 'Russet Burbank', 'Shepody' and 'Norchip'	PVY coat protein under PVX CP leader sequence	Hefferon et al., 1997
4. "Flova"	Coat protein	Okamoto et al., 1996
5. "Favorita", "Tiger Head" and "K4"	Coat protein	Song et al., 1996
6. 'Pito'	P1 in sense orientation	Pehu et al., 1995
7. 'Russet Norkotah'	Coat protein	Smith et al., 1995
8. "Bintje"	Coat protein	Farinelli et al., 1992 Malnoe et al., 1994
9. "Russet Burbank"	Coat protein	Lawson et al., 1990

In our laboratory, a collaborative research program with the Central Potato Research Institute (CPRI), Shimla, India, to develop PVY resistance in Indian potato cultivars using coat protein mediated resistance is being pursued. PVY coat protein gene from an Indian isolate (PVY^{o}) has been cloned and sequenced. The virus has been multiplied in the greenhouse on 'Havana 425' variety of tobacco. Subsequently, the virus was isolated from the infected plants and the protein coat was removed in order to isolate genomic RNA. cDNA was synthesized, using the viral genomic RNA as template and the PVY coat protein gene was PCR amplified, using a pair of specific primers. This was sequenced and sub-cloned into a plant expression vector with a strong constitutive promoter. Initially, as a model system tobacco leaf discs were transformed and transgenic plants were regenerated. Molecular analyses such as PCR and dot-blot hybridization confirmed the integration of this gene in tobacco genome. The transgenic plants were hardened in the greenhouse and were inoculated with virus inoculum. The virus titre was determined through ELISA in both control as well as in transgenic plants. It was clearly indicated that the virus titre was less in transgenic plants compared to control non-transformed ones (Ghosh et al., 2002). The same construct was used to transform Indian elite potato cultivar 'Kufri Jyoti'. The transgenic nature of the transformed plantlets that grew in the selection media were confirmed by molecular analysis. Seven such transgenic lines are under evaluation trial in the greenhouse (Ghosh et al., 2003).

GENOME ORGANIZATION OF POTATO VIRUS Y

The PVY genome organization is depicted in Fig. 12.1. The genome acts as mRNA, and upon uncoating of coat protein in the host cell, the whole genome is translated into a single polyprotein which is self-cleaved so as to generate individual functional proteins. The length of the genome, the precise position of the initiator ATG codon and the length of the 5'UTR and 3'UTR, etc., vary from strain to strain. The PVY genomic RNA codes for a single polyprotein of approximately 3060 amino acids which is self cleaved by *cis*-acting proteinase domains of the polyprotein to give individual viral proteins. From N to C terminus of the polyprotein, the order of the final products is: P1, HC-Pro (helper component - proteinase), P3, $6K_1$, CI (cylindrical inclusion protein), $6K_2$, VPg part of NIa (nuclear inclusion protein), Protease part of NIa, NIb (large nuclear inclusion protein) and CP (coat protein).

Two activities of P1 protein have so far been reported: (a) *cis*-cleavage of P1-HC-Pro) precursor performed by the chymotrypsin like serine protease domain of P1; and (b) activation of Viral RNA replication, which probably involves the P1 specific RNA-binding activity. HC-Pro is a multifunctional protein having a papain-like cystein protease domain located at the C-terminal. It cleaves the polyprotein in *cis* at the boundary between HC-Pro and P3. The HC domain participates in the aphid transmission of the virus and the central region of HC-Pro promotes genome amplification and virus spread through the plant tissue. The function of P3 and $6K_1$ proteins are still unclear; however they are reported to contain some of the pathogenicity determinants of Potyviruses. CI protein has ATPase-helicase domain and has RNA helicase activity in vitro and is also essential for cell to cell movement of the virus. $6K_2$ protein associates with large vesicles derived from endoplasmic reticulum by an interaction of its central hydrophobic domain with the membranes. $6K_2$ is speculated to target replication complexes (primarily NIa Protein) to membranous sites of replication. NIa is an essential replication protein

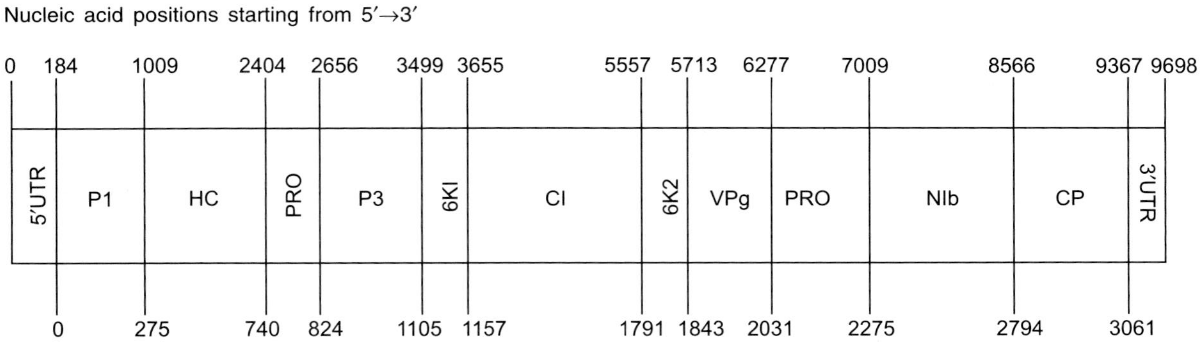

Figure 12.1 Schematic representation of PVY genomic RNA. Figures at the top indicate the nucleotide position, whereas those at the bottom indicates amino acid position in the polyprotein (as adapted from Singh and Singh, 1996).
5′UTR = 5′ untranslated region
3′UTR = 3′ untranslated region followed by poly a tail.

and a stretch of 188 amino acid is cleaved off from its N-terminal to generate VPg protein, which is found to be covalently linked to the 5′ end of the viral genomic RNA in the mature virus particle through a tyrosine residue. The remaining portion of NIa is called NIa proteinase, a chymotrypsine-like cystein protease which cleaves both in *cis* and *trans* at the seven sites in the polyprotein in the following stretch: P3-6K_1-CI-6K_2-Vpg-NIa(Pro)-NIb-CP. NIa proteinase also interacts with the NIb to form replication complex. NIb protein is the potyviral RNA dependent RNA polymerase (RdRp) and it functions by specific interaction with NIa, coat protein and possibly with VPg. Potyviral CP encapsidates the genomic RNA in mature virion particles and the conserved tripeptide DAG located at the N terminus is a determinant of Aphid transmission. Two types of functional mutations of the CP have been recorded. Some mutations affect the internal regions, causing a defect in assembly while others affect terminal regions, causing impaired cell-to-cell movement without any effect on virus assembly (Morozov S. and Soloveyev A. 1999).

There are two proteases in the PVY genome, the 27 kDa NIa-Pro protein is one protease whereas the 9 kDa fragment which initially remains attached to the HC protein and self cleave itself to generate mature HC protein is the other.

COAT PROTEIN MEDIATED RESISTANCE (CPMR)

Coat protein mediated resistance to viruses is one of the successful methods employed for virus resistance in crop plants (Beachy et al., 1990). After the theoretical concept of pathogen-derived, resistance as postulated by Sanford and Johnston (1985), Powell-Abel et al., (1986) first reported that transgenic tobacco plants expressing the tobacco mosaic virus (TMV) coat protein had exhibited resistance to TMV infection. Since then, this type of resistance has been utilized against more than 12 different groups of plant viruses. Several major crop plants such as tobacco, tomato, potato, squash, etc., have been transformed by different viral coat protein genes to generate resistance against different viral infections. A number of reports are available on the transfer of potato virus Y CP gene into potato or tobacco plants which imparted a high level of resistance against PVY under different agroclimatic conditions (Kaniewski et al., 1990; Farinelli et al., 1992; Heling, 1998). As early as 1989, Stark and Beachy showed that the transfer of coat protein gene of one potyvirus, Soybean Mosaic Virus (SMV) to tobacco plants can impart resistance against two other potyviruses such as Potato Virus Y (PVY) and Tobacco Etch Virus (TEV). Although SMV, PVY and TEV all belong to potyvirus group, they are antigenically distinguished from each other. This is one of the early reports of heterologous virus resistance.

MECHANISM OF COAT PROTEIN MEDIATED RESISTANCE

The expression of viral coat protein genes in transgenic plants can lead to different phenotypes of resistance (Beachy et al., 1990). Sometimes, transgenic plants escape infection completely and do not accumulate virus or develop symptoms. In other cases, local and systemic virus accumulation and development of systemic infection proceed at a rate slower than in non-transgenic plants. In transgenic plant lines, the proportion of plants that develop symptoms after inoculation is frequently lower than in control lines. It has also been shown that transgenic plants can become locally infected and accumulate virus in the inoculated leaf, but do not support systemic infection. The different phenotypes of resistance suggests that there is probably not one common mechanism by which virus infection is affected in transgenic plants, but different steps of virus infection are inhibited in different host-virus combination.

Powell-Abel et al., (1986) first demonstrated the presence of coat protein mediated resistance (CPMR) or coat protein mediated protection (CPMP) in tobacco against tobacco mosaic virus (TMV) and the degree of resistance correlated with the level of CP accumulation in the transgenic plants. Those plant lines that accumulated higher levels of CP were more resistant to TMV than those with low accumulation. The exact mechanism of CP mediated resistance is still unknown. Initially, it was thought that the existence of the preformed coat protein in the host cell cytoplasm prevents the disassembly of the virus particle or other early events of the virus life cycle. In support of this hypothesis, resistant transgenic tobacco plants were inoculated with naked viral RNA and the resistance was overcome (Nelson et al., 1987). On the other hand, inoculation with PVX genomic RNA did not overcome CP mediated resistance against PVX (Hemenway et al., 1988).

However, subsequently in many reports of CP mediated virus resistance and particularly against PVY, the level of coat protein in the host plant did not correlate with the extent of resistance (Beachy et al., 1990; Lawson et al., 1990). Plants with low CP levels or even with no detectable CP accumulation were shown to be highly resistant. This indicated that viral CP transgene can lead to virus resistance in different ways and that in some cases, the transcript of the transgene—rather than the protein itself—was effective. Since high protein levels are usually associated with high transcript levels, the correlation of CP levels with the degree of resistance suggests, but does not prove, that CP, rather than RNA is effective. The only sure way to exclude a protein effect is to transform plants with a gene construct that encodes a nontranslatable CP gene and test the resulting plants for resistance. This has indeed been demonstrated against two different potyviruses—Potato Virus Y and

Tobacco Etch Virus (Lindbo and Dougherty 1992a; Farinelli et al., 1992). The PVY^N coat protein encoding region, supplemented with a translational ATG start signal (CP^{+ATG}) or a CP cistron devoid of translational start codon (CP^{-ATG}), were expressed in tobacco (Van der Vlugt 1992). A high level of viral CP specific mRNA in all transgenic plants were obtained, but the coat protein production was undetectable. Virus protection analysis in the progeny lines of self-pollinated transgenic plants showed that the levels of over 80% immunity to mechanical inoculation with PVY^N were observed in plant lines transformed with either CP^{+ATG} or CP^{-ATG} suggesting that the resistance conferred was predominantly due to the viral CP gene transcript and not the accumulation of viral CP. A similar result was reported by Lee et al., (1998) and they could develop potato virus Y resistant tobacoo plants by introduction of untranslatable coat protein cDNA.

NON-STRUCTURAL GENE-MEDIATED RESISTANCE

Non-structural genes in virus code for proteins are required for viral replication but are not part of the assembled particles. PVY has 2 structural genes: VPg and CP. A single molecule of the VPg protein is covalently attached to 5' end of the genomic RNA in the virion particle, whereas almost 2000 copies of the CP encapsidates the 10 kb long RNA genome. All the other remaining genes of the potyvirus are non-structural genes.

A high level of resistance against PVY has been achieved by transforming tobacco plants with the RNA dependent RNA polymerase (NIb) gene or the nuclear inclusion proteinase gene (NIa-PRO) (Audy et al., 1994; Vardi et al., 1993). The use of antisense oriented viral genes to engineer resistance to RNA viruses has usually resulted in weaker resistance, expressed as a delayed symptom development, slower virus accumulation and reduced virus titres (van den Elzen et al., 1989; Hull and Davies, 1992). However, in case of potyviruses, there are many reports where antisense oriented viral genes have imparted a similar resistance as the one imparted by sense oriented viral genes. A transgenic potato line expressing PVY CP gene in antisense orientation has been reported to be resistant against PVY infection (Smith et al., 1995). Further effective resistance against bean yellow mosaic virus (BYMV) has been reported in two transgenic tobacco lines that express the carboxy-terminal portion of the BYMV CP gene together with its complete 3' non-coding sequence in antisense orientation (Hammond and Kamo 1995). In addition to these, the potyvirus P1 protein, also termed as N-pro, is produced from the N-terminus of the potyvirus encoded polyprotein. It has non-specific RNA binding activity and an accessory or regulatory role in stimulating potyviral genome amplification. A high level of resistance was obtained by expressing the PVY P1 gene in sense orientation in potato (Pehu et al., 1995). Resistance to PVY was also achieved by expressing PVY P1 in transgenic potato in antisense orientation (Maki-Valkama et al., 2000b). Specificity and strength of the resistance conferred by antisense P1 transgene was similar to that expressed by the transgenic plants of the same potato variety transformed with the P1 gene in sense orientation, and was also effective following graft inoculation, in four transgenic lines.

ANTI VIRAL PROTEINS

A new development in the engineering of virus-resistant plants is to express genes of various antiviral proteins of non-viral origin. Tavladoraki et al., (1993) reported that transgenic plants expressing a functional single chain Fv antibody are specifically protected from artichoke mottle crinkle virus. This was the first report of resistance to a pathogen by expressing transgenic antibodies, recognizing a structurally important epitope of viral coat protein. This approach of generating virus resistant transgenic crop varieties has an important implication. If the transgenic antibodies recognize the conserved sequence of viral replicase protein, it would confer a broad spectrum of viral resistance. On the other hand, pathogen-derived resistance is a comparatively much more specific, imparting resistance against a particular virus or other closely related viruses.

Another successful application of antiviral protein in transgenic research is 2′-5′ oligoadenylate synthetase. Potato plants expressing mammalian 2'-5' oligoadenylate synthetase are reported to be resistant to PVX infection (Truve et al., 1993). This enzyme produces a key component of the interferon mediated antiviral activity in the animal system, which in turn suggests the possibility of existence of interferon-like molecules in a plant antiviral defense mechanism. Resistance against PVY, PVX and cauliflower mosaic virus has been achieved by expression of Pokeweed Antiviral Protein (PAP) in transgenic tobacco plants (Lodge et al., 1993). This protein is located in the cell wall and released into the cell when the plasma membrane is broken by physical damage (Reddy et al., 1986). Once released into the cytoplasm, it inactivates the ribosome by highly efficient deadenylation of ribosomal RNAs and thus acts as a cellular "suicide" protein. In addition to these a 26 amino acid amphipathic peptide called mellitin confers resistance against tobacco mosaic virus (TMV) (Marcos et al., 1995). It is known to inhibit Human Immuno-deficiency Virus-1 (HIV-1) multiplication. It also shares a sequence and structural similarity to a region of TMV CP known to be critical for protein-protein and protein-RNA interaction. This concept is extremely valuable to be applied against potyviruses if a suitable natural or synthetic peptide with similar properties could be found.

HOMOLOGY DEPENDENT AND RNA MEDIATED RESISTANCE (RMR)

There are a few reports indicating that the expression of coat protein is not necessary for engineering resistance (Van der Vlugt et al., 1992; Lindbo and Dougherty, 1992b). It was demonstrated that very strong resistance is due to the presence of untranslatable forms of coat protein gene in transgenic plants through a sense RNA mediated process. The mechanism of this RNA mediated resistance is not well known, but it is clear that the resistance is based on sequence homology in the transgene and the inoculated virus. This homology need not be in the coat protein gene or in any sequence that gives protein product (Baulcombe, 1994). A high level of accumulation of the transgene RNA probably activates this resistance mechanism. It was considered that this type of resistance is extremely useful for its non-dependence on a particular gene and its specificity.

Smith et al., (1994) clearly demonstrated the transgenic PVY resistance by untranslatable sense RNAs. He and his co-workers transformed haploid tobacco leaf tissue with several constructs containing cDNA of potato virus Y coat protein open reading frame (ORF). The various constructs containing PVY CP ORF produced: (1) the expected mRNA and CP product; (2) an mRNA rendered untranslatable by introduction of a stop codon immediately after the start codon; or (3) an antisense RNA that was untranslatable as a result of the incorrect orientation of the PVY CP ORF behind the transcriptional promoter. Homozygous double haploid plants were produced and selfed progenies from these plants were examined. Resistance was virus specific and functioning only against PVY. An inverse correlation between transgene derived PVY transcript steady state levels and resistance was generally noted with lines expressing the untranslatable sense version of the PVY CP ORF. A collection of double haploid lines derived from a single transformation event of a common haploid plant and isogenic for the PVY transgene expressing untranslatable sense RNA, displayed different levels of PVY resistance. Lines with actively transcript, methylated transgene sequences had low steady state levels of transgene transcript and a virus resistant phenotype.

As mentioned in the previous sections, resistance against potyviruses could be obtained by transferring and subsequent transcriptional expression of different genes and non-translatable fragments of the viral genome in transgenic host plants. For most viruses, antisense RNAs are also capable of conferring virus resistance, but this does not seem to be applicable for all plant viruses. Possibly some target viral RNAs are somehow protected and are not capable of interacting with transgenic RNA. Considering all the available data on RMR, it seems that, at least for the majority of viruses, the polarity of the transgenic transcripts is not important for their ability to confer resistance. This leads to the hypothesis that RMR and antisense inhibition of gene expression might operate in a similar or even identical mechanisms. The schematic representation of a possible mechanism of RMR is depicted in Figure 12.2. It is mediated by a cytoplasomic factor which can be conceived as a RdRp-RNase complex. The transgenic mRNAs are recognized by this host factor which synthesizes a short complementary RNA and subsequently degrades the single strand portion of the mRNA. With the help of the remaining short duplex RNA, the complex can tag and degrade any RNA sequence homologous or complementary to the transgenic mRNAs, and offer virus resistance.

All transgenes, including the antisense ones are always cloned under strong promoter like CaMV 35S, resulting in a high level of constitutive expression. RdRp-RNase complex might be targeting only those RNase species whose number increases beyond a threshold level and which has no role in that tissue at that point of time and developmental state. A number of observations support this model. Many eukaryotic RNA polymerases and RNases contain small nucleic acid fragments and host encoded RdRp has been found in virus infected plant tissues. An in vitro experiment to show that an RdRp complex from a virus-resistant transgenic plant can degrade genomic RNA from same virus or closely-related ones but cannot degrade a non-specific RNA will strengthen this model immensely.

CONCLUSIONS

Viruses infecting the crop plants are the most important and devastating pathogens. They are widely distributed and their prevention by chemical means is not possible. Breeding for resistance to viruses through classical methods has been reported in many crop varieties. However, the slight availability of genetic resistance sources and their breakdown are the major limitations in employing this strategy. In this context, the recent advances in genetic engineering, specially transfer of virus-resistance genes and pathogen-derived resistance are promising alternatives. The viral coat protein and replicase mediated resistance is being used widely against the majority of viruses.

Potato virus Y is a RNA virus which infects solanaceous crops, in particular potato, causing considerable crop loss throughout the world. Resistant potato cultivars have been obtained through different approa-

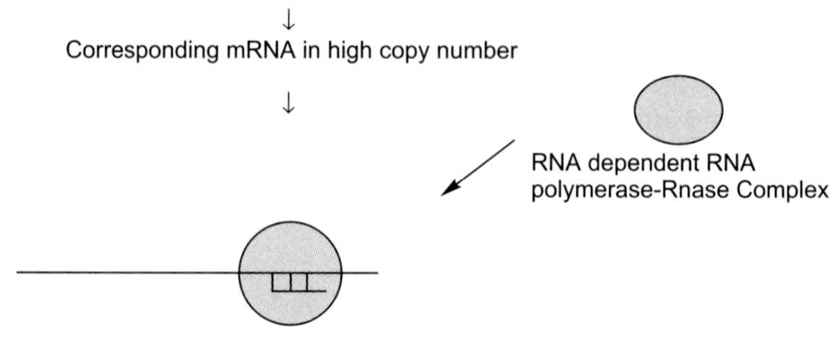

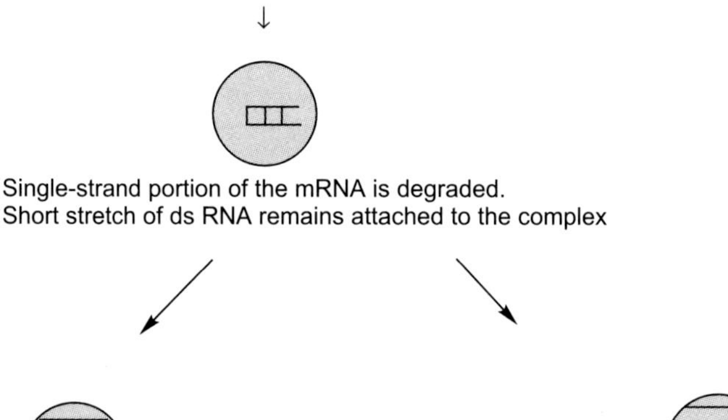

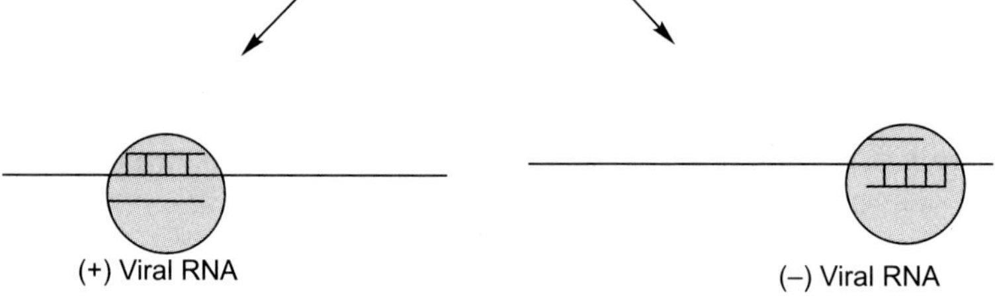

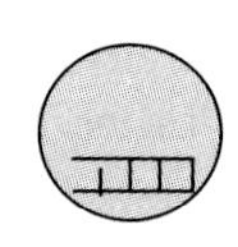

Fig. 12.2 Model to Explain RMR. The transcribed transgene mRNA comes out into the cytoplasm and comes in contact with the RNA dependant RNA polymerase (RdRp)-RNase complex. The complex synthesizes a short complementary RNA molecule. The transgenic mRNA molecule are rapidly degraded and the RdRp-RNase complex uses the small stretch of ds RNA molecule as the molecular memory, and binds to incoming viral RNA of either polarity and subsequently degrades them (adapted from de Haan, 1998).

ches, the most promising to date are coat protein and replicase mediated resistance. Considering the crop loss due to viral infection, the transgenic resistant cultivars will probably prove to be the most practical approach particularly for crops like potato, where, the classical breeding is difficult to perform owing to its polyploid nature and the poor availability of natural resistance.

The exact mechanism of resistance imparted through coat protein, P1 or replicase genes is still not clear, despite various theories having tried to explain it. The plethora of reports indicating a high level of resistance against potyviruses imparted by untranslatable viral transgenes in the host plants strengthens the possibility of RNA mediated resistance. The RNA mediated resistance based on the coat protein transgene transcript might be preferred over resistance mechanism based on the coat protein itself. In the former strategy, there is no coat protein accumulation in the plant and this eliminates the risk of transencapsidation with other viral RNAs. Further research on the understanding the mechanism of resistance, newer sources of resistance and the biosafety of the transgenic resistant cultivars are warranted.

REFERENCES

Audy P., Palukaitis P., Slack S.A., and Zaitlin M., 1994, Replicase-mediated resistance to potato virus Y in transgenic tobacco plants. Mol. Plant Microbe Interactions, **7:** 15-22.

Baulcombe D.C., 1994, Novel strategies for engineering virus resistance in plants. Current Opinion in Biotechnology. **15:** 117-124.

Baulcombe D.C., 1996, Mechanism of pathogen-derived resistance to viruses in transgenic plants. The Plant Cell, **8:** 1833-1844.

Beachy R.N., Loesch-Fries S., and Tumer N.E., 1990, Coat protein-mediated resistance against virus infection. Annual Review of Phytopathology **28:** 451-474.

Brigneti G., Garcia-Mas J., and Baulcombe D.C., 1997, Molecular mapping of the potato virus Y resistance gene *Rysto* in potato. Theor. Appl. Genet., **94:** 198-203.

Burk L.G., Gooding G.V. Jr., and Chaplin J.F., 1982, Reaction of *Nicotiana* species and cultivars of breeding lines of *Nicotiana* to three strains of potato virus Y. Tob. Sci., **26:** 85-88.

de Bokx J.A., and Huttinga H., 1981. Potato virus Y., CMI/AAB Description of plant viruses 242.

de Haan P., 1998, Mechanism of RNA-mediated resistance to plant viruses. In: G.D. Foster, S.C. Taylor (Eds.). Methods in Molecular Biology, Vol. 81: Plant Virology Protocols: From Virus Isolation to Transgenic Resistance. pp. 533-546.

Dinant D., Blaise F., Kusiak C., Astier-Manifacier S., and Albouy J., 1993, Heterologus resistance to potato virus Y in transgenic tobacco plants expressing the coat protein gene of lettuce mosaic potyvirus. Phytopathology. 83, 818-824.

Farinelli L., Malnoe P., and Collet G.F., 1992, Heterologus encapsidation of potato virus Y strain O (PVY^O) with the transgenic coat protein of PVY strain N (PVY^N) in *Solanum tuberosum* cv Bintje Bio/Technology, **10:** 1020-1025.

Ghosh S.B., Nagi L.H.S., Ganapathi T.R., Paul Khurana S.M., and Bapat V.A., 2002, Cloning and sequencing of potato virus Y coat protein gene from an Indian Isolate and development of transgenic tobacco for PVY resistance. Current Science, 82: 855-859.

Ghosh S.B., Nagi L.H.S., Ganapathi T.R., Paul Khurana S.M., and Bapat V.A., 2003, Development of coat protein gene mediated resistance against potato virus Y in Indian potato cultivar 'Kufri Jyoti'. Plant Cell Rep. (Communicated).

Gooding G.V. Jr, and Tolin S.A., 1973, Strains of potato virus Y affecting fluecured tobacco in the southeastern United States. Plant Dis. Rep., **57:** 200-204.

Gooding G.V. Jr, and Kennedy G.G., 1985, Resistance in tobacco breeding line NC 744 to potato virus Y and inoculation by aphids. Plant Dis. **69:** 396-397.

Gooding G.V. Jr, Wernsman E.A., and Rufty R.C., 1985, Reaction of *Nicotiana tabacum* L. cultivar Havana 307 to potato virus Y, tobacco vein mottling virus, tobacco etch virus, and *Peronospora tabacina*. Tob. Sci. **29:** 32-35.

Gupton C.L., and Burk L.G., 1973, Location of the factor for resistance to potato virus Y in tobacco. J. Hered. **64:** 289-290.

Hamalainen J.H., Sorri V.A., Watanabe K.N., Gebhardt C., and Valkonen J.P.T., 1998, Molecular examination of a chromosome region that controls resistance to potato Y and A potyviruses in potato. Theor. Appl. Genet. **96(8):** 1036-1043.

Hammond J., and Kamo K., 1995, Effective resistance to potyvirus infection conferred by expression of antisense RNA in transgenic plants. Molecular Plant Microbe Interactions **8:** 674-682.

Hefferon K.L., Khalilian H., and Abouhaidar M.G., 1997, Expression of the PVY^O coat protein (CP) under the control of the PVX CP gene leader sequence: protection under greenhouse and field conditions against PVY^O and PVY^N infection in three potato cultivars. Theor. Appl. Genet., **94(2):** 287-292.

Heling Z., 1998, Genetic engineering for resistance to potato viruses. In: S.M. Paul Khurana, Ramesh Chandra and Mahesh D. Upadhya (Eds), Comprehensive Potato Biotechnology. Malhotra Publishing House, New Delhi 110 018, India, pp. 269-303.

Hemenway C., Fang R.X., Kaniewski J.J., Chua N.H., and Tumer N.E., 1988, Analysis of the mechanism of protection in transgenic plants expressing the potato virus x coat protein or its antisense RNA. EMBO J., **7:** 1273-1280.

Hinrichs-Berger J., Harfold M., Berger S., and Buchenauer H., 1999, Cytological response of susceptible and extremely resistant potato plants to inoculation with potato virus Y. Physiol. Mol. Plant Pathol., **55(3):** 143-150.

Hull R. and Davies J.W., 1992, Approaches to nonconventional control of plant virus diseases. Critical Rev. In Plant Sci., 11(1): 17-33.

Kaniewski W., Lawson G., Sammons B., Haley L., Hart J., Delanny X., and Tumer N.E., 1990, Field resistance of transgenic Russet burbank potato to effects of infection by potato virus X and potato virus Y. Bio/Technology, **8:** 750-754.

Lawson C., Kaniewski W., Haley L., Rozman R., Newell C., Sanders P., and Tumer N.E., 1990, Engineering resistance to mixed virus infection in a common potato cultivar: resistance to potato virus X and potato virus Y in transgenic Russet Burbank. Bio/Technology, **8:** 127-134.

Lee Yung Ge, Lee Cheong Ho, Kang Shin Woong, Park Seong Weon, Kim Sang Seock, Park Eunkyung, 1998, Development of potato virus Y resistant tobacco plants by introduction of untranslatable coat protein cDNA. Korean Jour. Genet. 20: 1–9

Legnani R., Selassie K.G., Womdim R.N., Gognalons P., Moretti A., Laterrot H., and Mmarchoux G., 1995, Evaluation and inheritance of the *Lycopersicon hirsutum* resistance against potato virus Y. Euphytica, **86:** 219-226.

Lindbo J.A., and Dougherty W.G., 1992a. Untranslatable transcripts of Tobacco Etch Virus coat protein gene sequence can interefere with TEV replication in transgenic plants and protoplasts. Virology, **189:** 725-733.

Lindbo J.A., and Dougherty W.G., 1992b, Pathogen derived resistance to a Potyvirus: immune and resistant phenotypes in transgenic tobacco expressing altered forms of a potyvirus coat protein nucleotide sequence. Molecular Plant Microbe Interactions, **5:** 144-153.

Lindbo J.A., Silva-Rosales L., and Dougherty W.G., 1993. Pathogen derived resistance to Potyviruses: working but why? Semin. Virol., **4:** 369-379.

Lodge J.K., Kaniewski W.K., and Tumer N.E., 1993, Broad-spectrum virus resistance in transgenic plants expressing pokeweed antiviral protein. Proc. Natl. Acad. Sci. USA, **90:** 7089-7093.

Maki-Valkama T., Valkonen J.P.T., Kreu J.F., and Pehu E., 2000a, Transgenic resistance to PVY^{O} associated with post-transcriptional silencing of P1 transgene is overcome by PVY-N strains that carry highly homologous P1 sequences and recover transgene expression at infection. Mol. Plant Microbe. Interactions **13(4):** 366-373.

Maki-Valkama T., Pehu T., Santala A., Valkonen J.P.T., Koivu K., Lehto K., and Pehu E., 2000b, High level of resistance to potato virus Y by expressing P1 sequence in antisense orientation in transgenic potato. Molecular Breeding, **6:** 95-104.

Malnoe P., Farinelli L., Collet G.F., and Reust W., 1994, Small-scale field tests with transgenic potato, cv Bintje, to test resistance to primary and secondary infections with potato virus Y. Plant Mol. Biol., **25:** 963-975.

Marcos J.E., Beachy R.N., Houghton R.A., Blondelle S.E., and Perez-Paya E., 1995, Inhibition of plant virus infection by analogs of melittin. Proc. Natl. Acad. Sci. USA, **92:** 12466-12469.

Mathews R.E.F., 1991, Plant Virology (Third Edition). Academic Press Inc. pp. 197-202.

Morozov S. and Soloveyev A., 1999, Genome Organization in RNA viruses. In: Molecular Biology of Plant Viruses; (Eds.). C.L. Mandahar. Kluwer Academic Publisher, USA. pp. 49-51.

Naik P.S., and Sarkar D., 2000, *In vitro* propagation and conservation of genetic resources in potato. In: K.L., Chadda, P.N., Ravindran, and L., Sahijram (Eds.). Biotechnology in Horticulture and Plantation Crops. Malhotra Publishing House, India, pp. 369-406.

Nelson R.S., Powell-Abel P., and Beachy R.N., 1987, Lesions and virus accumulation in inoculated transgenic tobacco plants expressing the coat protein gene of tobacco mosaic virus. Virology, **158:** 126-132.

Okamoto D., Nielsen S.V.S., Albrechsten M., and Borkhardt B., 1996, General resistance against potato Virus Y introduced into a commercial potato cultivar by genetic transformation of PVYN coat protein. Potato Research, **39:** 271-282.

Paul. Khurana S.M., 1992, Potato viruses and viral diseases. Technical Bulletin No. 35, Central Potato Research Institute, Shimla, India.

Pehu T.M., Maki-Valkama T.K., Valkonen J.P.T., Koivu K.T., Lehto K.M., and Pehu E.P., 1995, Potato plants transformed with a potato virus Y P1 gene sequence are resistant to PVY-O. Am. Potato J., **72:** 523-532.

Powell-Abel P., Nelson R.S., De B., Hoffman N., Roger S.G., Fraley R.T., and Beachy R.N., 1986, Delay of disease development in transgenic plants that express the tobacco mosaic virus coat protein gene. Science, **232:** 738-743.

Pruss G., Ge X., Shi X.M., Carrington J.C., and Vance V.B., 1997, Plant Viral synergism; the potyviral genome encodes a broad-range pathogenicity enhancer that transactivates replication of heterologous viruses. Plant Cell, **9:** 859-868.

Robaglia C., Durand-Tardif M., Tronchet M, Bouzadin G., Astier-Manifacier S., and Casse-Delbart F., 1989, Nucleotide sequence of potato virus Y (N Strain) genomic RNA. J. Gen. Virology, **70:** 935-947.

Reddy M.P., Brown D.T., and Roberts J.D., 1986, Extracellular localization of pokeweed antiviral protein. Proc. Natl. Acad. Sci. USA, **83:** 5053-5056.

Rykbost K.A., Hane D.C., Hamm P.B., Voss R., and Kirby D., 1999, Effects of seedborne Potato Virus Y on Russet Norkotah performance. Am. J. Potato Res. **76(2):** 91-96.

Sanford J.C., and Johnston S.A., 1985, The concept of parasite derived resistance: deriving resistance genes from parasite's own genome. J. Theor. Biol., **113:** 395-405.

Shukla D.D., Colin W.W. and Brunt A.A., 1994, The Potyviridae. Cambridge University Press, Cambridge, UK.

Singh M., and Singh R.P., 1996, Nucleotide sequence and genome organization of a Canadian isolate of the common strain of the Potato Virus 'Y' (PVY^O). Canadian Journal of Plant Pathology. **18:** 209-224.

Smith H.A., Powers H., Swaney S., Brown C., and Dougherty W.G., 1995, Transgenic potato virus Y resistance in potato: evidence for an RNA mediated response. Phytopathol. **85:** 864-870.

Smith H.A., Swaney S.L., Parks T.D., Warnsman E.A., Dougherty W.G., 1994, Transgenic plant virus resistance mediated by untranslatable sense RNAs: Expression, regulation and fate of non-essential RNAs. Plant Cell **6:** 1441-1453.

Song Y.R., Ma Q.H., Hou L.L., Zhang L.Z., Yang W.Y., Peng X.X., and Wang H.Y., 1996, Transgenic potato with PVY coat protein gene and its small scale field test. Acta Botanica Sinica, **38(9):** 711-718.

Sudarsono, Hellman G.M., Lommel S.A., and Weissinger A.K., 1996, Transgenic tobacco with resistance to various isolates of potato virus Y. Asia Pacific Journal of Mol. Biol. and Biotechnology, **4(2):** 74-84.

Tavladoraki P., Benvenuto E., Trinca S., De Martinis D., Cattaneo A., and Galeffi P., 1993, Transgenic plants expressing a functional single-chain Fv antibody are specifically protected from virus attack. Nature, **366:** 469-472.

Thole V., Dalmay T., Burgyan J., and Balazs E., 1993, Cloning and sequencing of potato virus Y (Hungarian Isolate) genomic RNA. Gene **123:** 149-156.

Truve E., Aaspollu A., Honkanen J., Puskj R., Mehto M., Hasi A., Teeri T.H., Kelve M., Seppanen P., and Saarma M., 1993, Transgenic potato plants expressing mammalian 2′-5′ oligoadenylate synthetase are protected from potato virus X infection under the field conditions. Bio/Technology, **11:** 1048-1052.

Van den Elzen P.J.M., Husman M.J., Willink D.P., Jongedijk E., Hoekema A. and Cornelissen B.J.C. (1989). Engineering virus resistance in agricultural crops. Plant Mol. Biol. **13:** 337-346.

Van der Vlugt R.A.A., Ruiter R.K., and Goldbach R., 1992, Evidence for sense RNA mediated protection to PVI^{N} in tobacco plants transformed with the viral coat protein cistron. Plant Mol. Biol., **20:** 631-639.

Vardi E., Sela I, Edelbaum O., Livneh O., Kuznetsova L., and Stram Y., 1993, Plants transformed with a cistron of potato virus Y protease (NIa) are resistant to virus infection. Proc. Natl. Acad. Sci. USA, **90,** 7513-7517.

Wernsman E.A., and Rufty R.C., 1980, Tobacco. In : W. Fehr and H.H. Hadley (Eds) Hybridization of Crop Plants. pp. 669-698. Crop. Sci. Soc. Amer., Madison, WI, USA.

13

Direct DNA Transfer Using Particle Gun Bombardment—A Potential Tool for Crop Improvement

Susan Eapen
Plant Biotechnology and Secondary Products Section,
Nuclear Agriculture and Biotechnology Division,
Bhabha Atomic Research Centre, Mumbai 400 085, India.

ABSTRACT

Crop improvement through the development of transgenic plant research depends on the availability of powerful techniques to introduce genes into plant cells and particle gun technology is the most widely used technique for development of transgenic crop plants. Particle bombardment uses high velocity microprojectiles to penetrate plant cells in order to introduce genetic material into living cells. The introduced genes express in transformed cells and such cells can be regenerated into plants, which again express the trait. Utilization of the technology opens the way for transformation of our important crop plants—which are not accessible to genetic modification—using other techniques. Development of transgenic plants for disease and pest resistance and other agronomic characters using this technique is a major breakthrough towards the improvement of the crop plants. In this chapter, the technology is summarized and discussed in terms of present and future applications.

INTRODUCTION

Ever since the dawn of civilization, man has depended on plants as a source of food and feed. Dramatic improvements in crop yield were obtained in the previous century as a result of intensified use of land, fertilizers and pesticides and by planting improved varieties developed through different plant-breeding techniques based on genetic principles. However, there are indications that "The Green Revolution" is slowing down and there is an urgent need to have a "Gene revolution" to meet the increase in demand for food. In future, major increases in cropped area are unlikely. Pesticide use is on the decline due to increasing concerns about their harmful effects on the environment and human health. Hence, it is essential to produce more food from less land with lesser pesticides, lesser labour and lesser water. Of the different strategies for increasing crop productivity, one of them is to genetically engineer plants so as to increase their productivity. In order to apply genetic engineering techniques for crop improvement, one should be able to identify and isolate the agronomically-useful genes, modify them according to the requirements, transfer them into cells/tissues and ultimately recover them in mature transgenic plants that can be used in breeding programmes. Genes for production of transgenic plants can be accessed from diverse sources such as plants, bacteria, animals, fungi or viruses. Targets for genetic transformation will include:

Resistance to herbicides
Resistance to diseases and pests
Improving protein content
Changing oil quality

The genes to be introduced have to be specifically modified for expression in specific tissues/organs at particular stages of plant development.

Transgenic plant technology depends on the availability of procedures for plant transformation. Two types of methods are employed for obtaining plant transformation—the *Agrobacterium*-mediated gene transfer and direct DNA transfer. Among the different direct DNA transfer techniques, microprojectile-mediated transformation is the most commonly-used method for production of transgenic plants.

Genetic material (DNA) can be introduced into living cells using microprojectile bombardment which employs high velocity metal particles to deliver the DNA. This technique has circumvented the limitations of *Agrobacterium* host range and requirement for plant regeneration from protoplasts when other direct DNA methods are employed. On the issue of particle gun bombardment, due to the physical nature of the process, there is no biological limitation to the actual DNA delivery. Combining the relative ease of DNA introduction into plant cells with an efficient regeneration protocol avoiding protoplasts or callus culture or somatic embryogenesis, it is an optimal system for development of transgenic crops. Monocotyledonous crops which are difficult to be transformed by *Agrobacterium*-mediated methods are amenable to genetic transformation using particle gun bombardment. Direct DNA transfer methods requiring protoplast regeneration systems are ineffective in monocots for reasons such as recovery of sterile regenerants, genotype specificity for tissue culture response and somaclonal variation. Experiments with wheat (Lonsdale et al., 1990; Chibbar et al., 1991; Shimada et al., 1991; Vasil et al., 1992, 1993); barley (Lee et al., 1991; Knudson and Müller 1991), rice (Christou et al., 1991, 1992) and pearl millet (Taylor and Vasil 1991) have demonstrated the usefulness of genetic transformation using the technique of particle gun bombardment for crop improvement in graminaceous crops.

HISTORY OF PARTICLE GUN BOMBARDMENT

The use of high velocity DNA coated microprojectiles for gene transfer has been referred to as microprojectile bombardment, particle gun bombardment, particle acceleration, particle gun method or gene gun method. The first microprojectile technique developed in the 1960s was used by plant virologists for wounding plant cells to facilitate entry of viral nucleic acids into plant cells (Mac Kenzie et al., 1966). However, it was only in the late 1980s that key steps were taken to develop a microprojectile technique by John Sanford and his group at Cornell University, who developed techniques to accelerate microprojectiles (1–4 μm in diameter) to velocities of approximately 250 m/s, sufficient enough to penetrate plant cell walls and membranes. TMV RNA was incorporated into onion epidermal cells using tungston microprojectile bombardment and the cells remained viable and developed crystalline inclusions, indicating the expression of viral nucleic acids (Sanford et al., 1987). Following the demonstration of transient expression of reporter gene in onion epidermal cells and maize cells (Klein et al., 1987; 1988 b,c), the first stable transformation was reported in soybean (Christou et al., 1988) and tobacco (Klein et al., 1988a). Immature embryos of soybean were bombarded using an electric discharge instrument and protoplasts isolated, cultured and stably transformed callus and roots obtained (Christou et al., 1988). Following this, the same group reported transgenic soybean plants after bombardment of intact soybean meristems isolated from seeds (Mc Cabe et al., 1988 b). Recovery of transformed tobacco (Klein et al., 1988a), cotton (Finer and Mc Mullen 1990) and maize (Gordon-Kamm et al., 1990; Fromm et al., 1990) validated the general applicability of this technique for plant transformation. Later, this technique was adopted and modified successfully by numerous research groups worldwide and today, particle gun technology is used not only for production of transgenic crop plants, but also for introduction of DNA into bacteria, fungi, yeast and animal cells.

PRINCIPLES OF PARTICLE GUN DESIGN

Microprojectiles are coated with DNA and accelerated at velocities of 300 m/second to penetrate intact plant cell walls. The accelerating force can be supplied with gun-powder detonation, electric discharge through a water droplet or a gas tube. The DNA-coated microprojectile is spread on a macroprojectile, which is normally high molecular polyethylene plate and it is stopped by a stopping screen, thus enabling the stoppage of the macroprojectile from falling on the target tissue.

Particle bombardment should result in transient expression of the introduced DNA in several hundred cells where it is hit. Only a small portion of penetrated cells will be stably transformed and will be capable of plant regeneration. Some of the normal transient and stable selectable markers are listed in Table 13.1.

The important factors which affect the frequency of transient expression of bombarded tissues, include:

type and size of microprojectile
attachment of DNA to microprojectile prior to bombardment
velocity of microprojectile/DNA complex
degree of tissue damage

Type and Size of Microprojectiles

The microprojectiles should be chemically inert to be able to reduce the possibility of explosive oxidation of the microprojectile powder. A microprojectile should be 1 – 2 μm in diameter, should not react with DNA and other precipitating chemicals and should not be toxic to the bombarded tissues.

Tungsten and gold satisfy the above criteria and are the most commonly-used microprojectiles. However, tungsten is known to cause toxicity in some plant species. Gold is a better microprojectile, but is expensive. Stably-transformed tobacco colonies produced was four-fold higher when gold was used instead of tungsten (Russell et al., 1992).

Method of Attachment of DNA Particles to the Microprojectiles

DNA has to be effectively bound to the surface of the microprojectile to achieve effective genetic transformation. DNA can be bound to the microprojectile using Ca Cl_2/spermidine co-precipitation method of Klein et al., (1988b). DNA can also be precipitated onto the surface of gold particles with ethanol and dried onto the carrier surface before acceleration (Mc Cabe et al., 1988b). Precipitation with PEG and Ca Cl_2/ spermidine co-precipitation followed by ethanol wash can also be used (Ye et al., 1990; Christou et al., 1991).

Velocity of Microprojectile/DNA Complex

Velocity of microprojectiles can be controlled by altering the acceleration force (gun-powder discharge, gas pulse pressure, electrical discharge energy, etc.), the vacuum in the target chamber or the distance travelled by the microprojectile.

Tissue Damage during Bombardment

Macroprojectiles and stop plate devices are also important in biolistic experiments, since in some cases, cell death occurs in target tissue due to

Table 13.1 Commonly used selectable and screenable markers for transformation experiments

Gene	*Enzyme*	*Selective agent*	*Reference*
uid A	Beta-glucuronidase	....	Jefferson et al., 1987
npt II	Neomycin phosphotransferase II	Kanamycin, G 418	Beck et al., 1982
		Paromomycin	Torbert et al., 1995
cat	Chloramphenicol acetyl transferase	Chloramphenicol	Gorman et al., 1982
nos	Nopaline synthase	....	Depicker et al., 1982
lux	Luciferase		Ow et al., 1986
bar	Phosphinothricin acetyl transferase	Phosphinothricin	De Block et al., 1987
aad A	3' adenyltransferase	Spectinomycin	Svab and Maliga 1993
		Streptomycin	
gfp	Green fluorescent protein gene	....	Haseloff and Amos 1995
man A	Phosphomannose isomerase	Mannose	Joersbo et al., 1998

collateral damage (see Birch and Bower 1994). Birch and Bower (1994) found that nylon 66 and ultra high molecular weight polyethylene macroprojectiles cause less fragmentation than many other plastics including PVC, teflon, polycarbonates and polyurethanes. Tissues are normally bombarded in a vacuum chamber to maintain particle velocity over a greater distance. Vacuum levels higher than 27" Hg are most effective to maintain the velocity of particles of 1-μm diameter (Klein et al., 1992).

Another important factor to be taken into account is the optimum size of the microprojectile. Klein et al., (1988b) found the best transient expression in maize cells when tungsten particles were having a diameter of 1 – 2 μm as compared to 0.6 or 2.4 μm particles.

PORTABLE, HAND-HELD GENE GUN

While the particle gun targeted tissues are held in a vacuum, some target tissues are sensitive to conditions created by an evacuated chamber. Besides, experiments can be conducted only under laboratory conditions and not on plants growing in the field. Hence, there is need for a gun which could be hand held and used without a vacuum. The Helios gun (Bio-Rad, Hercules, CA) is a hand held device which uses low pressure helium to accelerate DNA particles from the interior of a small plastic cartridge towards the target tissue. The Helios gun can accommodate upto 12 loaded cartridges at once, thus allowing multiple firing of the device before relocating and can be taken to field for gene introduction into plants.

INTRODUCTION OF DNA INTO PLANT CELLS

The gene to be introduced can be in a cloning vector in *Escherichia coli* with appropriate regulatory sequences. The gene to be introduced need not be arranged between any T-DNA or viral sequences. Increased transformation with linearized plasmid DNA has been reported by some authors (Sautter et al., 1991), but others found no difference between linearized and supercoiled plasmids (Armaleo et al., 1990). There are indications that large plasmids (>10 kbp) may be subjected to fragmentation during particle gun bombardment (Mendel et al., 1989; Fitch et al., 1990). However, Mullen et al., (1998) showed that large insert DNA on 80 kb YAC can be transferred and integrated intact into the genome of tobacco cells using microprojectile bombardment.

For optimization of bombardment conditions, it is essential to study the transient expression frequency. The rate of conversion of transiently-expressing cells to stably-transformed cells has been shown to vary at least from 0.9 to 9 % (Russell et al., 1992).

TARGET TISSUE AND REGENERATION

Target tissue for particle gun bombardment should be highly regenerable to produce transgenic plants. One can use organized meristems from immature/mature embryos, cell suspension cultures, callus, tissue explants or somatic embryos as target tissues. An advantage of the meristematic target tissue is that it can be excised and differentiated into plants for virtually all species with minimal time in tissue culture, so as to minimize somaclonal variation. The disadvantage is that a high proportion of transformed regenerants may be chimeric as observed with soybean and cotton. When chimeric plants are obtained, it may be possible to derive uniformly-transformed progeny from lateral buds through proliferation in vitro. Regeneration in presence of selectable marker permits the selection of transformed plants as in case of maize, rice, sugarcane and wheat (Gordon-Kamm et al., 1990; Christou et al., 1991; Bower and Birch 1992; Vasil et al., 1992).

The first report on recovery of intact transgenic plants was in soybean using particle gun bombardment, where meristems excised from immature soybean seeds were used (Mc Cabe et al., 1988 b) as the target material. Approximately 2% of shoots derived from these meristems were chimeric for *gus* gene and one of the plants transferred the introduced gene to the progeny in a Mendelian fashion. Two years later, the same authors reported on the transformation frequencies of the order of 15% for all events with a germline transformation frequency of 0.5–0.1% based on the number of bombarded explants. These experiments resulted in the recovery of transgenic soybean plants from 30 commercially important cultivars (Christou et al., 1990) in a procedure that is presently used for commercial production. Meristems have been used for development of transgenic *Phaseolus* (Russell et al., 1993) and peanut (Brar et al., 1992).

Embryogenic suspensions were used to obtain transgenic plants in cotton (Finer et al., 1990), maize (Gordon-Kamm et al., 1990; Fromm et al., 1990), wheat (Vasil et al., 1992) and oats (Somers et al 1992). In monocotyledonous plants, immature embryos were used to obtain transgenic plants as in wheat (Weeks et al., 1993) and barley (Wan and Lemaux 1994), while in rice, scutellar tissues of embryos (Christou et al., 1991) were used. Christou et al., (1992) reported a variety independent gene transfer for rice capable of introducing any gene into any variety at a high frequency (Christou et al., 1992) using particle gun bombardment.

Embryogenic callus is readily produced in many plant species and provides an ideal target for production of transgenic plants via particle gun bombardment. It is important to minimize the duration in culture in order to reduce the probability of undesirable somaclonal variation.

When organized tissues are used, it is essential to know from which particular cell layer the plants are originating. Although Christou et al., (1990, 1992), obtained a large number of transiently-expressing cells in rice and soybean, they failed to produce any stably-transformed plants because DNA-coated microprojectile failed to penetrate the cells that had given rise to regenerated plants.

One can produce transgenic seeds through fertilization with pollen that has been bombarded with DNA-coated microprojectiles. Pollen grains of tobacco, maize and tradescantia have been shown to express reporter genes (Twell et al., 1989; Hamilton et al., 1992). This approach aims to produce large numbers of transformants without the need to tissue culture manipulations. A patent application was filed for a method to produce transgenic seeds through fertilization with pollen that has been bombarded with DNA-coated microprojectiles (Mc Cabe et al., 1988a).

TRANSFORMATION OF ORGANELLES

Plant cells contain important organelles such as chloroplasts and mitochondria. Organelle transformation using particle gun bombardment, although possible, is less studied in comparison with transformation of the nucleus. It is essential to introduce organellar gene control sequences and also efficient selection markers. In higher plants, due to the presence of a large number of organelles in cells, different selection strategies may be required to recover stable transformants.

INTEGRATION PATTERN

An integration pattern of the target DNA is analogous to those resulting from other direct DNA transfer techniques. Majority of transgenic lines produced by microprojectile bombardment have complex transgene loci composed of multiple copies of whole, truncated and delivered DNAs frequently organised as direct and inverted repeats (Pawlowski and Somers 1998, Svitashev and Somwrs 2002). Complete sequence analysis of transgene loci in transgenic plants obtained by bombardment should rearranged delivered DNA and flunking sequence (Makarevitch et al 2003). The pattern of integration may vary from one individual to the other. Transformed plants will have one to many copies of genes inserted into the genome. Integration pattern may be influenced by both the genetic construct and DNA load per bombardment (Finer and Mc Mullen 1991; Drummond et al., 1991). In many cases, integrated DNA remains stable and is inherited in a Mendelian fashion. However, upto 25% of the transformed lines have been reported to show instability or non-Mendelian inheritance (Christou et al., 1990; Spencer et al., 1992; Walters et al., 1992).

When individual transformants are analyzed after repeated subculture, the integration patterns are usually unchanged indicating that no genomic rearrangements have occurred subsequent to transformation (Kindle et al., 1989; Day et al., 1990; Gordon-Kamm et al., 1990; Christou et al., 1992). Hence, the concatemerization and rearrangement of foreign DNA occur prior to the integration event. When different plasmids are co-precipitated and co-transformed, many of them are found to be linked in the genome (see Day et al., 1990; Gordon-Kamm et al., 1990; Spencer et al., 1992; Christou et al., 1990, 1992; Walters et al., 1992) of the transgenic plants.

In microprojectile-bombardment mediated gene transfer experiments, stable transformants could be generated with two marker genes derived from the same or different constructs. Patterns of co-transformation and co-expression for microprojectile-mediated gene transfer is similar to that of other direct gene transfer techniques.

APPLICATION TO CROP PLANTS

Transgenic plants using particle gun bombardment have been produced in most of the crop plants such as rice, wheat, sorghum, maize, soybean, peanut, tapioca, etc., and the list is increasing every day. A few selected examples of successful particle gun bombardment in crop plants have been listed in Table 13.2. Herbicide resistance (Kim et al., 1999), virus resistance (Sivamani et al., 1999), insect resistance (Altpeter et al., 1999), improved baking qualities of wheat (Barro et al., 1997) bacterial blight and aphid resistance (Tang et al., 1999) have been introduced into the crop plants using this technique (Table 13.3).

The potential of particle gun bombardment has been clearly proven by the long list of species already transformed by this method (see Varsha et al., 1997). A major strength of the technique is that it is the only transformation method in which gene transfer is virtually species-and tissue-independent, a fact which has made a new generation of transgenic plants possible.

CONCLUSION

Transformation of crop plants through particle gun bombardment is a major milestone in the history of plant biotechnology. It is also the most important method of choice for introducing useful genes into monocotyledonous plants, to which the major cereals and millets belong

Table 13.2 Selected examples of transgenic plants recovered through particle gun bombardment

Plant species	*Explant used*	*Reference*
I Cereals and millets		
1. *Avena sativa*	Embryogenic callus	Somers et al., 1992
	-do-	Torbert et al., 1995
	Highly regenerative, green tissues	Cho et al., 1999
2. *Hordeum vulgare*	Embryonic callus	Wan and Lemaux 1994
	Suspension culture derived from embryos	Ritala et al., 1993
	Immature embryos	Hagiou et al., 1995
	Suspension culture cells	Stiff et al., 1995
3. *Oryza sativa*	Zygotic embryos	Christou et al., 1991, 1992
	Embryogenic callus	Cao et al., 1992
	Immature embryos	Christou and Ford 1995 a
	Primary and axillary meristems	Christou and Ford 1995 b
	Immature embryos	Kohli et al., 1999
	-do-	Kim et al., 1999
	Immature embryos and embryogenic callus	Sivamani et al., 1999
	Seed-derived callus	Tang et al., 1999
4. *Secale cereale*	Young embryogenic cultures	Castillo et al., 1994
5. *Sorghum bicolor*	Immature zygotic embryo	Casas et al., 1993
6. *Tritico secale*	Immature embryos	Zimny et al., 1995
7. *Triticum aestivum*	Embryogenic callus	Vasil et al., 1992
	Immature embryos	Vasil et al., 1993
	Zygotic embryo callus	Weeks et al., 1993
	Scutellar tissue of immature embryos	Becker et al., 1994
	Somatic embryos	Nehra et al., 1994
	Immature embryos	Altpeter et al., 1999

Contd.

Table 13.2 Contd.

Plant species	*Explant used*	*Reference*
8. *Zea mays*	Embryogenic suspension	Gordon-Kamm et al., 1990
	-do-	Fromm et al., 1990
	Immature embryos	Koziel et al., 1993
	Suspension culture	Hunold et al., 1994
	Suspension culture cells	Rasmussen et al., 1994
II Oil crops		
1. *Arachis hypogaea*	Meristem	Brar et al., 1992, 1994
	Embryonic callus	Ozias-Akins et al., 1993
2. *Brassica napus*	Microspore-derived secondary embryoids	Chen and Beversdorf 1994
3. *Glycine max*	Meristem	Mc Cabe et al., 1988 b
	Embryonic axes	Christou et al., 1990
	Embryogenic suspension	Finer and Mc Mullen 1991
	-do-	Sato et al., 1993
4. *Helianthus annuus*	Apical meristem	Bidney et al., 1992
	Half shoot apices	Knittel et al., 1994
	Cotyledons	Hunold et al., 1995
III Sugar-yielding plants		
1. *Beta vulgaris*	Embryo/seedling explants	Mahn et al., 1995
2. *Saccharum officinarum*	Embryogenic callus	Bower and Birch 1992
IV Pulse crops		
1. *Phaseolus vulgaris*	Meristem	Russell et al., 1993
2. *Vigna unguiculata*	Leaf tissue	Finer et al., 1992
3. *Vigna aconitifolia*	Hypocotyl	Kamble et al., 2003

Table 13.3 A few selected examples of useful genes transferred to crop plants by particle gun bombardment

Trait	*Plant*	*Reference*
Bacterial blight resistance	Rice	Tang et al., 1999
Sap sucking insect resistance		
Aphid resistance	Wheat	Stoger et al., 1999
Bread-making properties	Wheat	Barro et al., 1997
Insect resistance	Wheat	Altpeter et al., 1999
Wheat scab resistance	Wheat	Chen et al., 1999
Herbicide resistance	Rice	Kim et al., 1999
Antiviral resistance		
Rice tungro virus resistance	Rice	Sivamani et al., 1999

to. In less than ten years, particle bombardment technology allowed the creation and evaluation of transgenic plants in a variety of crops, elite cultivars and advanced breeding lines. It made possible to transfer a number of genes of agronomic interest including pest, disease and herbicide resistance. One of the criticisms of particle gun bombardment is that it requires sophisticated instruments and laboratories. The criticism is slowing down since it is being replaced by inexpensive devices. Evaluation of large number of transgenic plants generated through particle gun bombardment showed that they contained only 1 – 3 copies of transgenes, which is similar to that reported by Agrobacterium (Hiei et al., 1994). Issues of biosafety, public perception and commercialization of transgenic plants and engineering of crop plants for multigene traits are areas receiving attention at present.

REFERENCES

Altpeter, F., Diaz, I., McAuslane, H., Gaddour, K., Carbonero, P., and Vasil. I.K., 1999, Increased insect resistance in transgenic wheat stably expressing trypsin inhibitor CMe. Molecular Breeding, **5:** 53-63.

Armaleo, D., Ye, G.N., Klein, T.M., Shark, K.B., Sanford, J.C., and Johnston, S.A., 1990, Biolistic nuclear transformation of *Saccharomyces cerevisiae* and other fungi., Curr. Genet. **17:** 98-103.

Barro, F., Rooke, L., Bekes, F., Gras, P., Tatham, A.S., Fido, R., Lazzeri, P.A., Shewry, P.R., and Barcelo, P., 1997, Transformation of wheat with high molecular weight subunit genes results in improved functional properties. Nature Biotech., **15:** 1295-1299.

Becker, D., Brettschneider, R., and Lörz, H., 1994, Fertile transgenic wheat from microprojectile bombardment of scutellar tissue. Plant J., **5:** 299-307.

Beck, E., Ludwig, E.A., Reiss, B., and Schaller, A., 1982, Nucleotide sequence and exact localization of the neomycin phosphotransferase gene from transposon Tn 5. Gene. **19:** 324-336.

Bidney, D., Scelonge, C., Martich, J., Burrus, M., Sims, L., and Huffman, G., 1992, Microprojectile bombardment of plant tissues increases transformation frequency by *Agrobacterium tumefaciens*. Plant Mol. Biol. **18:** 301-313.

Birch, R.G., and Bower, R., 1994, Principles of gene transfer using particle bombardment. In: Yang, N.S., and P. Chrostou (Eds). Particle Bombardment Technology for Gene Transfer. UWBC Biotech. Res. Series. New York-Oxford. Oxford University Press, pp. 3-37.

Bower, R., and Birch, R.G., 1992, Transgenic sugar beet plants via microprojectile bombardment. Plant J. **2:** 409-416.

Brar, G.S., Cohen, B.A., and Vick, C.L., 1992, Germline transformation of peanut (*Arachis hypogaea* L.) utilizing electric discharge particle acceleration (ACCELL) technology. Proc. of Amer. Peanut Research and Education Soc. Inc. Norfolk Virginia **24:** 21.

Brar, G.S., Cohen, B.A., Vick, C.L., and Johnson, G.W., 1994, Recovery of transgenic peanut (*Arachis hypogaea* L.) plants elite cultivars utilizing ACCELL ® technology. Plant J. 5: 745-753.

Bruce, W.B., Christensen, A.H., Klein, T., Fromm, M., and Quail, P.H., 1989, Photoregulation of a phytochrome gene promoter from oat transferred into rice by particle bombardment. Proc. Natl. Acad. Sci. USA, **86:** 9692-9696.

Cao, J., Duan, X., McElroy, D., and Wu. R., 1992, Regeneration of herbicide resistant transgenic rice plants following microprojectile mediated transformation of suspension culture cells. Plant Cell Rep. **11:** 586-591.

Casas, A.M., Kononowicz, A.K., Zehr, U.B., Tomes, D.T., Axtell, J.D., Butler, I.G., Bressan, R.A., and Hasegawa, P.M., 1993, Transgenic sorghum plants via microprojectile bombardment. Proc. Natl. Acad. Sci. USA. **90:** 11212-11216.

Castillo, A.M., Vasil, V., and Vasil, I.K., 1994, Rapid production of fertile transgenic plants of rye (*Secale cereale* L.) Bio/Technology **12:** 1366-1371.

Chen, J.L., and Beversdorf, W.D., 1994, A combined use of microprojectile bombardment and DNA inhibition enhances transformation frequencies of canola (*Brassica napus* L.). Theor. Appl. Genet. **88:** 187-192.

Chibbar, R.N., Kartha, K.K., Leung, N., Quereshi, J., and Caswell, K., 1991, Transient expression of marker genes in immature zygotic embryos of spring wheat (*Triticum aestivum*) through microprojectile bombardment. Genome **34:** 453-460.

Cho, M.-J., Jiang, W., and Lemaux, P.G., 1999, High-frequency transformation of oat via microprojectile bombardment of seed-derived highly regenerative cultures. Plant Sci. **148:** 9-17.

Christou, P., Mc Cabe, D.E., and Swain, W.F., 1988, Stable transformation of soybean callus by DNA coated gold particles. Plant Physiol. **87:** 671-674.

Christou, P., McCabe, D., Martinell, B.J., and Swain, W.F., 1990, Soybean genetic engineering—commercial production of transgenic plants. Trends Biotech. **8:** 145-151.

Christou, P., Ford, T., and Kofron, M., 1991, Production of transgenic rice (*Oryza sativa* L.) plants from agronomically important indica and japonica varieties via electric discharge particle acceleration of exogenous DNA into immature zygotic embryos. Bio/Technology **9:** 957-962.

Christou, P., Ford, T.L., and Kofron, M., 1992, The development of a variety independent gene transfer method for rice. Trends Biotech. **10:** 239-246.

Christou, P. and Ford, T.I., 1995a, Parameters influencing stable transformation of rice immature embryos and recovery of transgenic plants using electric discharge particle acceleration. Ann. Bot. **75:** 407-413.

Christou, P. and Ford, T.I., 1995b, Recovery of chimeric rice plants from dry seeds using electric discharge particle acceleration. Ann. Bot. **75:** 449-454.

Day, A., Debuchy, R., van Dillewijn, J., Purton, S., and Rochaix, J.D., 1990, Studies on the maintenance and expression of cloned DNA fragments in the nuclear genome of the green alga *Chlamydomonas reinhardii.* Physiol. Plant. **78:** 254-260.

De Block, M., Botterman, J., Vandewiele, M., Dockx, J., Thoen, C., and Gossele, V., Rao Movva, N., Thompson, C., Van Montagu, M., and Leemans, J., 1987, Engineering herbicide resistance in plants by expression of a detoxifying enzyme. EMBO. J. **6:** 2513-2518.

Depicker, A., Stachel, S., Dhaese, P., Zambryski, P., and Goodman, H.M., 1982, Nopaline synthase transcript mapping and DNA sequence. J. Mol. Appl. Gen. 1: 561.

Drummond, B., Beck, P., Martich, J., Lane, S., Ross, M., Blair, D., Fox, T., and Bowen, B., 1991, The amount of input DNA affects phenotypic variation in tobacco transformed by the particle gun. Abstracts 3rd Int. Conf. Plant Mol. Biol. Tucson. 1809.

Finer, J.J., and McMullen, M.D., 1990, Transformation of cotton (*Gossypium hirsutum* L.) via particle bombardment. Plant Cell Rep. **8:** 586-589.

Finer, J.J., and McMullen, M.D., 1991, Transformation of soybean via particle bombardment of embryogenic suspension culture tissue. In Vitro Cell Dev. Biol. **27P:** 175-182.

Finer, J.J., Vain, P., Jones, M.W., and Mc Mullen, M.D., 1992, Development of particle inflow gun for DNA delivery to plant cells. Plant Cell. Rep. **11:** 323-328.

Fitch, M.M.M., Manshardt, R.H., Gonsalves, D., Slightom, J.L., and Sanford, J.C., 1990, Stable transformation of papaya via microprojectile bombardment. Plant Cell Rep. **9:** 189-194.

Fromm, M.E., Morrish, F., Armstrong, C., Williams, R., Thomas, J., and Klein, T.M., 1990, Inheritance and expression of chimeric genes in the progeny of transgenic maize plants. Biotechnology. **8:** 833-844.

Gordon-Kamm, W.J., Spencer, T.M., Mangano, M.L., Adams, T.R., Daines, R.J., Start, W.J., O'Brien, J.V., Chambers, S.A., Adams, W.R., Willetts, N.G., Rice, T.B., Mackey, C.J., Krueger, R.W., Kaush, A.P., and Lemaux, P.G., 1990, Transformation of maize cells and regeneration of fertile transgenic plants. Plant Cell. **2:** 603-618.

Gorman, C.M., Moffat, L.F., and Howard, B.H., 1982, Recombinant genomes which express chloramphenicol acetyl transferase in mammalian cells. Mol. Cell. Biol. **2:** 1044-1051.

Hagio. T., Hirabayashi, T., Machii, H., and Tomotsune, H., 1995, Production of fertile transgenic barley (*Hordeum vulgare* L.) plant using hygromycin resistance marker. Plant Cell Rep. **14:** 329-334.

Hamilton, D.A., Roy, M., Rueda, J., Sindhu, R.K., Sanford, J., and Mascarenhas, J.P., 1992, Dissection of a pollen specific promoter by transient transformation assays. Plant Mol. Biol. **18:** 211-218.

Haseloff, J., and Amos, B., 1995, GFP in plants. Trends in Genetics. **11:** 328-329.

Hiei, Y., Ohta, S., Komari, T., and Kumashiro, T., 1994, Efficient transformation of rice (*Oryza sativa* L.) mediated by *Agrobacterium* and sequence analysis of boundaries of T-DNA. Plant J. **6:** 271-282.

Hunold, R., Bronner, R., and Hahne, G., 1994, Early event in microprojectile bombardment, cell viability and particle location. Plant J. **5:** 593-604.

Hunold, R., Burrus, M., Bonner, R., Durret, J.P., and Hahne, G.C., 1995, Transient gene expression in sunflower (*Helianthus annuns* L.) following microprojectile bombardment. Plant Sci. **105:** 95-109.

Jefferson, R.A., Kavanagh, T.A., and Bewan, M.W., 1987, GUS fusions: β-glucuronidase as a sensitive and versatile gene fusion marker in higher plants. EMBO. J. **6:** 3901-3907.

Joersbo, M., Donaldson, I., Kreiberg J., Peter sen S.G., Brun Stedt J., Okkela F.T., 1998. Analysis of mannose selection used for transformated of sugarbeet. Mole Breed **4:** 111 - 117

Kamble, S., Misra, H.S., Mahajan, S.K., and Eapen, S. 2003. A protocol for efficient biolistic transformation of mothbean *Vigna aconitifolia* L. Jacq. Marechal. Plant Molecular Biology Reporter (In Press).

Kindle, K.L., Schnell, R.A., Fernandez, E., and Lefefvre, P.A., 1989, Stable nuclear transformation of Chlamydomonas using the *Chlamydomonas* gene for nitrate reductase. J. Cell Biol. **109:** 2589-2601.

Kim, J.K., Duan, X., Wu, R., Seok, S.J., Boston, R.S., Jang, I.C., Eun, M.Y., and Nahm, B.H., 1999, Molecular and genetic analysis of transgenic rice plants expressing the maize ribosome inactivating protein b-32 gene and the herbicide resistance *bar* gene. Mol. Breed. **5:** 85-94.

Klein, T.M., Wolf, E.D., Wu, R., and Sanford, J.C., 1987, High velocity microprojectiles for delivering nucleic acids into living cells. Nature **327:** 70-73.

Klein, T.M., Harper, E.C., Svab, Z., Sanford, J.C., Fromm, M.E., and Maliga, P., 1988a, Stable transformation of intact *Nicotiana* cells by the particle gun bombardment process. Proc. Natl. Acad. Sci. USA. **85:** 4305-4309.

Klein, T.M., Gradziel, T., Fromm, M.E., and Sanford, J.C., 1988b, Factors influencing gene delivery into *Zea mays* cells by high velocity microprojectiles. Bio/Tehnology **6:** 559-563.

Klein, T.M., Fromm, M.E., Weissinger, A., Tomes, D., Schaaf, S., Sletten, M., and Stanford, J. C., 1988c, Transfer of foreign genes into intact maize cells with high velocity microprojectiles. Proc. Natl. Acad. Sci. USA. **85:** 4305-4309.

Klein, T.M., Arentzen, R., Lewis, P.A., and Fitzpatrick-Mc Elligot, S., 1992, Transformation of microbes, plants and animals by particle bombardment. Bio/Technology **10:** 286-291.

Knittel, N., Gruber, V., Hahne, G., and Lence, P., 1994, Transformation of sunflower (*Helianthus annus* L.): a reliable protocol. Plant Cell Rep. **14:** 81-86.

Knudsen, S., and Müller. M., 1991, Transformation of the developing barley endosperm by particle bombardment. Planta **185:** 330-336.

Kohli, A., Gahakwa, D., Vain, P., Laurie, D.A., and Christou, P., 1999, Transgene expression in rice engineered through particle bombardment: molecular factors controlling stable expression and transgene silencing. Planta **208:** 88-97.

Lee, B.T., Murdoch, K., Topping, J., Jones, M.G.K., and Kreis, M., 1991, Transient expression of foreign gene introduced into barley endosperm protoplasts by PEG mediated transfer or into intact endosperm tissue by microprojectile bombardment. Plant Science. **78:** 237-246.

Londsale, D., Onde, S., and Cuming, A., 1990, Transient expression of exogenous DNA in intact, viable wheat embryos following particle bombardment. J. Exper. Bot. **41:** 1161-1165.

Mac Kenzie, D.R., Anderson, P.M., and Wernham, C.C., 1966, A mobile air-blast inoculator for plot experiments with maize dwarf mosaic virus. Plant Dis. Rep. **50:** 363-367.

Mahn, A., Matzk, A., Sautter, C., and Schiemann. J., 1995, Transient gene expression in shoot apical meristems of sugarbeet seedlings after particle bombardment. J. Exp. Bot. **46:** 1625-1628.

Makarevitch, I., Svitashev, S.K. and Somers, D.A. 2003, Complete sequence analysis of transgene loci from plants transformed via microprojectile bombardment. Plant Molec. Biology 52: 421–432.

Mc Cabe, D.E., Swain, W.F., and Martinell, B.J., 1988a. Pollen-mediated plant transformation (to Agracetus) Application. USA 9388570. 12 May 1986.

Mc Cabe, D.E., Swain, W.F., Martinell, B.J., and Christou, P., 1988b, Stable transformation of soybean (*Glycine max*) by particle acceleration. Bio/Technology **6:** 923-926.

Mendel, R.R., Müller, B., Schulze, J., Kolesnikov, V., and Zelenin, A., 1989, Delivery of foreign genes to intact barley cells by high velocity microprojectiles. Theor. Appl. Genet. **78:** 31-34.

Mullen, J., Adam, G., Blowers, A., and Earle, E., 1998, Biolistic transfer of large fragments to tobacco cells using YACs retrofilled for plant transformation. Mol. Breed. **4:** 449-457.

Nehra, N.S., Chibbar, R.N., Leung, N., Caswell, K., Mallard, C., Steinhauer, L., Baga, M., and Kartha, K.K., 1994, Self fertile transgenic wheat plants regenerated from isolated scutellar tissues following microprojectile bombardment with two distinct gene constructs. Plant J. **5:** 285-297.

Ow, D.W., Wood, K.V., Deluca, M., De Wet, J.R., Helinski, D.R., and Howell, S.H., 1986, Transient and stable expression of firefly gene in plant cells and transgenic plants. Science **234:** 856-859.

Ozias-Akins, P., Schnall, J.A., Anderson, W.F., Singsit, C., Clemente, T.E., Adang, M.J., and Wessinger, A.K., 1993, Regeneration of transgenic peanut plants from stably transformed embryogenic callus. Plant Sci. **93:** 185-194.

Pawlowski, W.P., and Somers, D.A. 1996. Transgene inheritance in plants genetically engineered by microprojectile bombardment Mol Biotechnol **6:** 17–30.

Rasmussen, J.L., Kikkert, J.R., Roy, M.K., and Sanford, J.C., 1994, Biolistic transformation of tobacco and maize suspension cells using bacterial cells as microprojectiles. Plant Cell Rep. **13:** 212-217.

Ritala, A., Mannonen, L., Aspegren, K., Salmenkallio-Marttila, M., Kurten, U., Hannus, R., Mendez Lozano, J., Teeri, T.H., and Kaupinen, V., 1993, Stable transformation of barley tissue culture by particle bombardment. Plant Cell Rep. **12:** 435-440.

Russell, J.A., Roy, M.K., and Stanford, J.C., 1992, Physical trauma and tungsten toxicity reduce the efficiency of biolistic transformation. Plant Physiol. **98:** 1050-1056.

Russell, D.R., Wallace, K.M., Bathe, J.M., Martinell, B.J., and Mc Cabe D.C., 1993, Stable transformation of *Phaseolus vulgaris* via electric discharge mediated particle acceleration. Plant Cell Rep. **12:** 165-169.

Sanford, J. C., Klein, T. M., Wolf, E. D., and Allen, N., 1987, Delivery of substances into cells and tissues using a particle gun bombardment process. Partic. Sci. Technol. 5: 27-37.

Sato, S., Newell, C., Kolacz, K., Tredo, L., Finer, J., and Hinchee, M., 1993, Stable transformation via particle bombardment in two different soybean regeneration systems. Plant Cell Rep. **12:** 408-413.

Sautter, C., Waldner, H., Neuhaus-Url, G., Galli, A., Neuhaus, G., and Potrykus, I., 1991, Micro-targeting: High efficiency gene transfer using a novel approach for the acceleration of microprojectiles. Bio/Technology. **9:** 1080-1085.

Shimada, T., Seki, M., and Morikawa, H., 1991, Transient expression of GUS in wheat pollen embryos via microprojectile bombardment. Wheat Inf. Service **72:** 106-108.

Sivamani, E., Huet, H., Shen, P., Ong, C.A., De Kochko, A., Fauquet, C., and Beachy, R.N., 1999, Rice plant (*Oryza sativa* L.) containing Rice tungro spherical virus (RTSV) coat protein transgenes are resistant to virus infection. Mol. Breed. **5:** 177-185.

Somers, D.A., Rines, H.W., Gu, W., Kaeppler, H.F., and Bushnell, W.R., 1992, Fertile transgenic oat plants. Bio/Technology **10:** 1589-1594.

Spencer, T.M., O'Brien, J.V., Start, W.G., Adams, T.R., Gordon-Kamm, W.J., and Lemaux, P.G., 1992, Segregation of transgenes in maize. Plant Mol. Biol. **18:** 201-210.

Stiff, C.M., Kilian, A., Zhou, H., Kudrna, D.A., and Kleinhops, A., 1995, Stable transformation of barley callus using biolistic ® particle bombardment and bar gene. Plant Cell Tissue Organ. Cult. **40:** 243-248.

Stoger, E., Williams, S., Christou, P., Down, R.E., and Gatehouse, J.A., 1999, Expression of the insecticidal lectin from snowdrop (*Galanthus nivalis* agglutinin, GAA) in transgenic wheat plants: effects of predation by grain aphid *Sitobion avenae*. Molec. Breed. **5:** 65-73.

Svab, Z., and Maliga, P., 1993, High frequency plastid transformation in tobacco by selection for a chimeric *aad A* gene. Proc. Natl. Acad. Sci. USA. **90:** 913-917.

Svitashev S.K., Somers D.A. 2002. Characterizatoin of transgene loci in plants using Fish: A picture is worth on thousand words. Plant Cell, Tissue Organ. Cult. **69:** 205 – 214.

Tang, K., Tinjuangjun, P., Xu, Y., Sun, X., Gatehouse, J.A., Ronald, P.C., Qi, H., Lu, X., Christou, P., and Kohli, A., 1999. Particle bombardment mediated co-transformation of elite rice cultivars with genes conferring resistance to bacterial blight and sap sucking insect pests. Planta **208:** 552-563.

Taylor, M.G., and Vasil, I.K., 1991, Histology and physical factors affecting transient GUS expresssion in pearl millet (*Pennisetum glaucum*) embryos following microprojectile bombardment. Plant Cell Rep. **10:** 120-125.

Torbert, K.A., Rines, H.W., and Somers, D.A., 1995, Use of paromomycin as a selective agent for oat transformation. Plant Cell Rep. **14:** 635-640.

Twell, D., Klein, T.M., Fromm, M.E., and McCormick, S., 1989, Transient expression of chimeric genes delivered into pollen by microprojectile bombardment. Plant Physiol. **91:** 1270-1274.

Varsha, R.L., Dubey, R.K., Srivastava, A.K., and Kumar, S., 1997, Microprojectile mediated plant transformation: A bibliographic search. Euphytica **95:** 269-294.

Vasil, V., Castillo, A.M., Fromm, M.E., and Vasil, I.K., 1992, Herbicide resistant fertile transgenic wheat plants obtained by microprojectile bombardment of regenerable embryogenic callus. Bio/Technology. **10:** 667-674.

Vasil, V., Srivastava, V., Castillo, A.M., Fromm, M.E., and Vasil, I.K., 1993. Rapid production of transgenic wheat plants by direct bombardment of cultured immature embryos. Bio/Technology **11:** 1553-1558.

Ye, G. N., Daniell, H., and Sanford, J.C., 1990, Optimization of delivery of foreign DNA into higher plant chloroplasts. Plant Mol. Biol. **15:** 809-819.

Walters, D.A., Vetsh, C.S., Potts, D.E., and Lundquist, R.C., 1992, Transformation and inheritance of a hygromycin phosphotransferase gene in maize plants. Plant Mol. Biol. **18:** 189-200.

Wan, Y., and Lemaux, P.G., 1994. Generation of large numbers of independently transformed fertile barley plants. Plant Physiol. **104:** 37-48.

Weeks, T.J., Anderson, O.D., and Blechl, A.E., 1993, Rapid production of multiple independent lines of fertile transgenic wheat (*Triticum aestivum*). Plant Physiol. **102:** 1077-1084.

Zimny, J., Becker, D., Brettschneider, R., and Lörz, H., 1995, Fertile transgenic Triticale (× *Triticale wittmack*). Mol. Breed. **1:** 155-164.

14

Comparative Assessment of Somatic Embryogenesis and Plantlet Regeneration Potential in Salt-tolerant and Salt-susceptible *indica* Varieties

A.B. Mandal and Aparna Maiti
Biotechnology Laboratory, Central Agricultural Research Institute, Port Blair 744 101, India

ABSTRACT

Somatic embryogenesis and plantlet regeneration of a salt-tolerant tall traditional *indica* variety Pokkali were compared with a salt-susceptible variety PTB 10 following four different culture routes. MS, LS and N_6 media were used for callus induction and plantlet regeneration. N_6-N_6 route was found to be the best in facilitating increased somatic embryogenesis and plantlet regeneration in both the varieties. Differential genotypic response in respect of somatic embryogenesis and plantlet regenerability was largely influenced by the type of medium used in in vitro culture. Maximum plantlet regeneration of 68% was observed in PTB 10 following N_6-N_6 route. Calli from individual seeds in PTB 10 produced on an average of 6.65 plants, while Pokkali produced only 3.83 plants across successful treatments in MS, LS and N_6 (A-a and B-b routes) media. This advocates less potential of the latter in plantlet regeneration with the media and hormones used in this study. Somatic embryo derived plantlets were found to be fertile and they set seeds normally. However, the extent of spikelet sterility in in vitro developed plantlets was higher in comparison to parental control.

INTRODUCTION

Cell technological approaches could be an effective alternative in genetic modulation of crop plants. Somatic embryogenesis (SE) holds enormous potential especially due to its key role as a recipient of alien genes in genetic transformation to develop transgenics (Potrykus 1990); in micropropagation and production of synthetic seeds. Nevertheless, conventional plant breeding has been the major route in increasing crop productivity. Genetic improvement through the incorporation of gene/s responsible for resistance to disease and insect pests, adverse environment and energy intensive management practices has efficiently exploited the available germplasms through better partitioning of biomass into products of economic use and with improved stability of performance of the cultivars (Chopra and Narasimhulu 1990).

Tissue culture involves aseptic culture of plant parts. During in vitro culture, a number of physico-chemical and biological factors such as type and levels of plant hormones, nutrient composition of the medium, duration of culture, source and physiological condition of explants and the degree of departure from organized growth have been known to induce genetic variation. Such a variation, namely, somaclonal variation (SV) constitutes an important source of variability for crop improvement. This is important because SV could be induced in already improved high-yielding variety background (Chopra and Narasimhulu 1990). Rice somatic embryo formation from diverse explants, viz., mature and immature seeds (Henk et al., 1978; Abe and Futsuhara 1986; Kavikishor and Reddy 1986; Tsukahara and Hirosawa 1992; Chowdhury et al., 1993; Singh et al., 1993; Rueb et al., 1994; Bajaj and Raham 1995; Suprasanna et al., 1995; Dey et al., 1995; Suprasanna et al., 1996; Khanna and Raina 1997) has already been reported. The present work is aimed at studying the in vitro culture response of two *indica* genotypes contrasting in respect of salt tolerance. Three standard media were used with mature seeds as the starting material to work out an efficient route for optimum somatic embryogenesis and plantlet regeneration. Efficient embryogenesis is expected to boost up genetic transformation work through the use of biolistics in contrast to robust embryogenic cell suspension (ECS) development for protoplast-mediated genetic transformation.

MATERIALS AND METHODS

One tall, traditional salt-tolerant photosensitive cultivar Pokkali obtained from International Rice Research Institute (IRRI), Manila, Philippines and a salt-susceptible variety PTB 10 obtained from Directorate of Rice Research (DRR), Hyderabad were employed as test material for this study. Healthy, mature seeds were dehusked and cleaned manually, shaked with aqueous teepol (Qualigens, India) detergent solution (5%, v/v) at 80 rpm for 5 min. followed by washing in distilled water. The seeds were surface sterilized in freshly prepared 0.1% (w/v) $HgCl_2$ solution with two drops of tween 20 as wetting agent for 10 min. under laminar air flow and washed thrice in sterile distilled water. The seeds were dried on sterile tissue paper and inoculated onto callus induction medium (CIM) consisting of MS (Murashige and Skoog 1962), LS (Linsmaier and Skoog 1965), and N_6 (Chu et al,. 1975) media with 2 mg L^{-1} 2,4-D as control set (A). Three other treatments involving different organic adjuvants, viz., coconut water (CW) 10% (v/v) (B); casein hydrolysate (CH) 250 mg L^{-1} (C) and yeast extract (YE) 250 mg L^{-1} (D) were also used (Table 14.1). The pH of the media was adjusted to 5.8 using 0.1 N HCl or KOH before autoclaving, 0.8% (w/v) agar (SRL, India) was added to solidify the media. About 20 ml media were dispensed into 25 × 150 mm (60 ml capacity) culture tubes (Borosil) and plugged with non-absorbent cotton plugs wrapped in one layer of fine cotton cloth. The inoculated culture tubes were kept in the dark at 25 ± 2°C to obtain optimum callus. Each treatment was constituted of 50 replicates and each experiment was repeated three times. After 10 days, small calli appeared on the swollen junction of mesocotyl and radicle were excized and kept in the same tube. Twenty-one-day-old calli were subcultured onto callus maintenance medium (CMM) constituted of the same medium with half dose of 2,4-D of CIM. After a passage of 28-day duration, the calli were transferred onto different regeneration media (RM) (Table 14.1). The calli from individual seeds were maintained separately. RM were supplemented with 2% sucrose and 3% sorbitol (a), 6 % sucrose (b), 0.5 mg L^{-1} IAA and 2 mg L^{-1} kinetin (c), 0.5 mg L^{-1} IAA, 1 mg L^{-1} BAP and 250 mg L^{-1} CH (d) (Table 14.1). The cultures were kept in culture room at 25±2°C under 16 h photoperiod. The light was received from cool, white fluorescent incadescent tubes (Philips make) with irradiance of ~ 130μ Einstein m^{-2} s^{-1}. After 10–15 days, green spots were visible seldom with emerging miniature shoots. Gradually, plants recorded well-developed roots and within 40–45 days, attained a height of 6–8 cm. These plants were transplanted onto plastic pots filled with wet soil. The pots were covered with polyethylene bags for initia1 3–4 days and watered regularly. After 15 days, the plants were shifted to big cement pots. The plants were grown and managed under experimental net house condition and harvested individually at maturity.

RESULTS

Cultured seeds started germinating within 4–5 days in both the varieties. Bright white calli appeared on the swollen junction of mesocotyl and radicle of growing seedlings within 10–12 days of cultures in PTB 10 and 13-15 days in Pokkali. Callus-forming ability in Pokkali was 98.0% on MS control whereas it was 88.6% in PTB 10. A comparative study of the callus forming ability on three media revealed that N_6 was the most appropriate among others in inducing the maximum calli (Table 14.1). Among different combinations of organic adjuvants in CIM, both the varieties produced maximum calli on N_6. The callusing response on MS, N_6 and LS fortified with CW was 76.5, 89.9 and 81.4%, respectively, in PTB 10. However, LS with CW was found to slightly inhibit callusing in Pokkali (66.0%) than MS (80.0%) and N_6 (88.0%). Table 14.1 indicates that LS enriched with CH was more suitable for Pokkali (90.0%) than PTB 10. In case of PTB 10, N_6 supplemented with CH was more responsive than other two media supplemented with CH. YE was found to be an effective adjuvant in both the varieties with all the three media to induce calli favourably.

Induced calli in both the varieties displayed different morphotypes. Majority of the calli were dry, compact, friable, fast growing, bright white in colour while a few were slimy, wet surfaced and slow growing. The former was found to be containing more embryogenic structures with an abundance of embryos and proembryos, while the latter contained less as evident from stereomicroscopic observations. Many calli had rhizogenetic capacity only, and produced profuse roots while placed on RM. Within 10–15 days of culture, most of the embryogenic calli routed through ElM differentiated into somatic embryos. They eventually developed shoot and root meristems. Coleoptile and roots were simultaneously developed there-from. Histological observations displayed the presence of a communicating vascular strand in between them that reaffirms their somatic embryogenetic origin (photograph not presented). No calli could regenerate on the medium containing 0.5 mg L^{-1} IAA, 1 mg L^{-1} BAP, and 250 mg L^{-1} CH probably due to the incompatibility of the osmoticum as well as unfavourable nutrient profile. The route B-b (in all the three media) showed rhizogenesis. Sucrose 2% and sorbitol 3% were found to be the best combination for optimum plant regeneration in both the varieties. The regeneration percentage was found to be maximum of 68.9% for PTB 10 on N_6 in A-a route followed by 66.5% on MS and 55.8% on LS. Use of 6% sucrose in N_6 medium resulted in maximum plantlet regeneration of 55.6% (B-b) for Pokkali and 63.5% (B-b) for PTB 10. The route C-c and D-d in all the media were not suitable for plantlet regeneration since they failed to regenerate any plants. Calli from individual seeds of PTB 10 produced an average of 6.65 plants while Pokkali produced 3.83 plantlets across successful treatments in MS, LS and N_6 (A-a and B-b routes) media. A few albino and yellow mutants with transient chlorophyll deficiency were also regenerated in Pokkali. Somatic embryo derived plants produced fertile healthy seeds though in vitro developed plantlets showed more spikelet sterility than control and were visually found to be true to type.

DISCUSSION

The results indicate that genotype and hormones played a vital role in callus induction and plantlet regeneration under in vitro culture system. Specificity and dose of hormones for each genotype were evident earlier also (Paulas and Sree Rangasamy 1995). The levels of embryo induction and plant regeneration in in vitro tissue cultures are also influenced by the physiological status of the donor plant, culture medium used and the interactions thereof between them (Lazar et al., 1984; Mathias and Simpson 1986; Bregitzer 1992). The level of 2,4-D required for optimum callus induction varied with the genotypes (Maheswaran and Sree Rangaswamy 1989; Paulas and Sree Rangaswamy 1995). Results also indicate that 2,4-D alone is sufficient for callus induction. CH and YE showed good embryo formation. However, they produced less regenerants probably due to the inapporpriate constituents of the RM. A clear trend was discernible that callus induction was medium dependent. This observation is in congruence to Rueb et al., (1994). This study concludes that N_6-N_6 route was the most appropriate in facilitating somatic embryogenesis in rice, which may be used confidently in genetic transformation at least with the present set of varieties. The routes C-c and D-d for all media were not suitable for plantlet regeneration. This may due to the incompatibility of the osmoticum as well as inappropriate nutrient and hormonal profile. The present observations clearly suggest that genotype also play an important role in callus induction and plant regeneration. A comparative study of three media revealed that N_6 was found to be the most appropriate than other in inducing maximum calli. Regarding adjuvants, YE was found to be more effective followed by CW in facilitating callus induction. It was reported earlier by Suprasanna et al., (1995) that addition of 0.1% CH to medium was found to play a crucial role in the development of prominent somatic embryos. Callusing response with CW was moderate. A clear trend emerged that callus induction was medium dependent (Table 14.1). This observation is corollary to Rueb et al., (1994). Supplements of CH and YE showed good embryo formation, however, produced less regenerants probably due to inappropriate constituents of the RM. The combination of callus induction medium (A) and regeneration medium (a) yielded maximum plantlets per regenerative calli. Calli from individual seed of PTB 10 produced average 6.65 plants, whereas it was only 3.83 in Pokkali. This advocates less potential of Pokkali in plantlet regeneration or inappropriateness of the medium for the latter used under the present study. The results further elucidates that genotype and hormonal level played an important role in callus induction and plantlet regeneration. Specificity of hormonal level and kind for each genotype was evinced (Paulas and Sree Rangasamy 1995). Besides green plants, a few albino and yellow mutants with transient chlorophyll deficiency were regenerated in Pokkali. Somatic embryo derived plants produced fertile healthy seeds though in in vitro developed plantlets, spikelets sterility was higher than control plants.

The study conclude that N_6-N_6 route was found to be most appropriate in facilitating somatic embryogenesis in rice which may be used confidently in genetic transformation work to recover more transformants at least with these two *indica* varieties.

Table 14.1 Somatic embryogenesis in salt-tolerant and salt-susceptible *indica* varieties

Variety	*Medium used**	*Callus induction medium (CIM)*		*Regeneration*		
Designation		*Treatment*	*Callus induction (%)*	*Medium and treatment*	*Regeneration (%)*	*No. of plantlets /individual seed calli*
		A = Control	98.0	a = 2% sucrose + 3% sorbitol	50.6	2.8
	MS	B = 10% CW	80.0	b = 6% sucrose	41.8	3.2
		C = 250 mg L^{-1} CH	86.5	c = 0.5 mg L^{-1} IAA + 2 mg L^{-1} kinetin	-	-
		D = 250 mg L^{-1} YE	92.0	d = 0.5 mg L^{-1} IAA + 1 mg L^{-1} BAP + 250 mg L^{-1} CH	-	-
		A = Control	62.0	a = 2% sucrose + 3% sorbitol	52.5	3.6
Pokkali	LS	B = 10% CW	66.0	b = 6% sucrose	44.5	3.9
		C = 250 mg L^{-1} CH	90.0	c = 0.5 mg L^{-1} IAA + 2 mg L^{-1} kinetin	-	-
		D = 250 mg L^{-1} YE	74.0	d = 0.5 mg L^{-1} IAA + 1 mg L^{-1} BAP + 250 mg L^{-1} CH	-	-
		A = Control	90.0	a = 2% sucrose + 3% sorbitol	62.2	4.6
	N_6	B = 10% CW	88.0	b = 6% sucrose	55.6	4.9
		C = 250 mg L^{-1} CH	86.0	c = 0.5 mg L^{-1} IAA + 2 mg L^{-1} kinetin	-	-
		D = 250 mg L^{-1} YE	84.0	d = 0.5 mg L^{-1} IAA + 1 mg L^{-1} BAP + 250 mg L^{-1} CH	-	-
		A = Control	88.6	a = 2% sucrose + 3% sorbitol	66.5	5.8
	MS	B = 10% CW	76.5	b = 6% sucrose	61.0	6.7
		C = 250 mg L^{-1} CH	78.0	c = 0.5 mg L^{-1} IAA + 2 mg L^{-1} kinetin	-	-
		D = 250 mg L^{-1} YE	94.0	d = 0.5 mg L^{-1} IAA + 1 mg L^{-1} BAP + 250 mg L^{-1} CH	-	-
		A = Control	88.0	a = 2% sucrose + 3% sorbitol	55.8	6.3
PTB-10	LS	B = 10% CW	81.4	b = 6% sucrose	49.6	6.8
		C = 250 mg L^{-1} CH	65.0	c = 0.5 mg L^{-1} IAA + 2 mg L^{-1} kinetin	-	-
		D = 250 mg L^{-1} YE	59.5	d = 0.5 mg L^{-1} IAA + 1 mg L^{-1} BAP + 250 mg L^{-1} CH	-	-
		A = Control	95.6	a = 2% sucrose + 3% sorbitol	68.9	6.4
	N_6	B = 10% CW	89.9	b = 6% sucrose	63.5	7.9
		C = 250 mg L^{-1} CH	87.3	c = 0.5 mg L^{-1} IAA + 2 mg L^{-1} kinetin	-	-
		D = 250 mg L^{-1} YE	95.0	d = 0.5 mg L^{-1} IAA + 1 mg L^{-1} BAP + 250 mg L^{-1} CH	-	-

*MS (Murashige and Skoog 1962); N_6 (Chu 1975); LS (Linsmaier and Skoog 1965).
-denotes no regeneration of plantlets

ACKNOWLEDGEMENT

We thank Director, Central Agricultural Research Institute, Port Blair for his keen interest, constant encouragement and facilities provided to carry out this work.

ABBREVIATIONS

2,4-D—2,4-dichlorophenoxy acetic acid; MS medium—Murashige and Skoog's medium (1962). BAP—6 benzylamino purine; kin—kinetin; SE—somatic embryogenesis. CH—casein hydrolysate; YE—yeast extract; CW—coconut water.

REFERENCES

Abe, T., Futsuhara, Y., 1986, Genotypic variability for callus formation and plant regeneration in rice (*Oryza sativa* L.). Theor. Appl. Genet. **72:** 3-10.

Bajaj, S., Rajam, M.V., 1995, High frequency plant regeneration from long term callus cultures of rice by spermidine. Plant Cell Rep. **14:** 717-720.

Bregitzer, P., 1992, Plant regeneration and callus type in barley: Effects of genotype and culture medium. Crop Sci. **32:** 1108-1112.

Chopra, V.L., Narasimhulu, S.B., 1990, Biotechnology for crop improvements. ln: Chopra, V.L. and Nasim Anwar (Eds). Genetic Engineering and Biotechnology Concepts, Methods and Applications : Pages 159-180. Oxford IBH Publishing Co. Pvt. Ltd, New Delhi, Mumbai, Kolkata.

Chowdhury, C.N., Tyagi, A.K., Maheswari, N., Maheswari, S.C., 1993, Effect of L-proline and L-Tryptophan on somatic embryogenesis and plantlet regeneration of rice (*Oryza sativa* L. cv Pusa 169). Plant Cell Tissue and Organ Cult. **32:** 357-361.

Chu, C.C., Wang, C.C., Sun, C.S., Hsu, C., Yin, K.C., Chu, C.Y., 1975, Establishment of an efficient medium for anther culture of rice through comparative experiments on the nitrogen sources. Sci. Sin. **18:** 659-668.

Dey, S.K., Sharma, D.K., Shrivastava. M.V., 1995, In vitro regenreration and somaclonal variation in rice. *Oryza.* **32:** 146-152.

Henke, R.R., Mansur, M.A., Constantin, M.J., 1978., Organogenesis and plantlet formation from organ and seedling derived calli of rice (*Oryza sativa* L). Physiol. Plant. **44:** 11-14.

Kavikishore, P.B., Reddy, G.M., 1986, Regeneration of plants from long term root and embryo derivative callus cultures. Curr. Sci. **55:** 664-665.

Khanna, H.K., Raina, S.K., 1997, Enhanced plant regeneration in basmati rice (*Oryza sativa* L. cv Karnal Local) embryo-calli through modifications of NO_3 + NH_4 concentrations. Biochem. Biotech. **6:** 75-80.

Lazar, M.D., Schaeffer, G.W., Baenziger, P.S., 1984, Cultivar and cultivar × environment effects on the development of callus and polyhaploid plants from anther cultures of wheat. Theor. Appl. Genet. **67:** 273-277.

Linsmaier, E.M., Skoog, F., 1965, Organic growth factor requirement of tobacco tissue cultures. Physiologia. Pl. 18P: 100-127.

Maheswaran, M., Sree Rangaswamy, S.R., 1989, Effects of 2,4-D and kinetin on callus induction and plant regeneration form somatic cell culture of rice. *Oryza.* **26:** 320.

Mathias, R.J., Simpson, E.S., 1986, The interaction of genotype and culture medium on the tissue culture response of wheat (*Triticum aestivum* L. em. Thel) callus. Plant Cell Tissue and Organ Cult. **7:** 31-37.

Murashige, T., Skoog, F., 1962, A revised medium of rapid growth and bioassays with tobacco tissue cultures. Physiol. Plant. **15:** 473-497.

Potrykus, I. 1993, Gene transfer to cereals: as assessment. Bio/Technology. **8:** 535-542.

Paulas, S.D., Sree Rangasamy, S.R., 1993., Hormonal and genetic influence on callus induction in rice. *Oryza.* **42:** 190-192.

Rueb, S., Leneman, M., Schilperoort, R.A., Hensgens, L.A.M., 1994, Efficient plant regeneration through somatic embryogenesis from callus induced on mature rice embryos (*Oryza sativa* L). Plant Cell Tissue and Organ Cult. **36:** 259-264.

Singh, D., Ravi, Minocha, J.L., 1993, Efficiency of different explants and media for callus induction and plant regeneration in Basmati rice. Crop Improv. **20(2):** 139-142.

Suprasanna, P., Ganapathi, T.R., Rao, P.S., 1995, Establishment of embryogenic callus, somatic embryos and plant regeneration in *indica* rice. J. Genet. and Breed. **49:** 9-14.

Suprasanna, P., Ganapathi T.R., Rao, P.S., 1996, Artificial seeds in rice (*Oryza sativa* L.). Encapsulation of somatic embryos from mature embryo callus cultures. AsPac J. Mol. Biol. and Biotechnol. **4(2):** 90-93.

Tsukahara, M., Hirosawa, T. 1992., Simple dehydration treatment promotes plantlet regeneration of rice (*Oryza sativa* L.) callus. Plant Cell. Rep. **11:** 550-553.

15

In vitro Variablility in Tissue Culture: A Fresh Look

A. Mujib
Department of Botany, Hamdard University, New Delhi 110 062, India

ABSTRACT

Tissue culture based propagation has been used successfully, sometimes commercially, for a wide range of plants. Although the propagation by in vitro method assures desired characters in its clone, tissue culture-induced off-type variants are also common in culture which is categorized into 3 different types: (i) chimeral; (ii) somaclonal; and (iii) temporary or physiological variation. The variant types and possible mechanistic basis of variation has been reviewed.

INTRODUCTION

Genetic variability has been the raw material and is important because of its unlimited scope. This phenotypic and DNA variations are determined by both genetic and epigenetic factors. Under in vitro system, where Petri dish is considered to be the breeder's field, variation has been noticed spontaneously in cultures in a large number of plant groups and several improved cell lines/crops were isolated by a selection of somatic variants (Hammerschlag et al., 1995; Karp 1995; Duncan 1997; Brar and Jain 1998; Veilleux and Johnson 1998). The utility has been well acknowledged because it often increases the possibility and efficiency of variation (Gavazzi et al., 1987). The in vitro use of mutagens (physical and chemical) has also been used simultaneously (Novak 1991) and some successful variants (Plate 15.2) with new traits are available (Brunner and Keppl 1991; Schmidt 1994). In this chapter, in vitro tissue culture variations and its various aspects have been reviewed.

In vitro variation in culture: The variation in tissue culture-derived plants and their bases of occurrence have been broadly divided into 3 sections. Chimeral or pre-existing variations are those where the plant contains more than 1 genotypic constitution. The second group constitutes permanent heritable change of material, which includes both the genetic and epigenetic materials. While the last off-type variant group mimick true variations, where the changes are temporary and/or physiological.

E-mail: mujibabdul@hotmail.com

A. CHIMERAL VARIATION

A chimera is an individual with 2 or more distinct genotypes. The meristem consists of 2 histogenic (outer and inner) layers. Several types of chimeras exist in plants, depending upon the change of layers. When the inner or outer layer is changed, it is called as periclinal chimera, whereas in a sectorial chimera, some part of the inner or the outer layer has been replaced. Both the forms of chimeras are very common in nature and can be found in clonal crops like *Dianthus, Hemerocallis, Malus, Acalypha, Caladium, Philodendron, Pelargonium, Rhododendron* and many others.

Periclinal chimeras remain stable most of the time in in vivo propagation, while tissue culture has led to the separation of chimeras in grapevine (Skene and Barlass 1983), tobacco (Marcotrigiano and Gouin 1984), blackberry (McPheeters and Skirvin 1983), and a number of other crop species. Sometimes, this separation has been easily detected in plants by the loss or gain of morphological characters, as in *Philodendron erubescens* cv. Pink Prince (Mujib and Jana 1995) where the nodal stem callus regularly induced a large number of variants (Table 15.2) that were easily distinguishable morphologically from the normal seedlings. In many others it is hardly detected even though changes at different levels were observed (Cassells et al., 1986; Preil 1986). Yet, in another group, where pre-existing chromosomal variation was noted in non-chimeric plants that may or may not be separated during culture. For example, no increased ploidy variation was reported in *Poa* although they were reported to contain diploid meristematic and endopolyploid cells (Wu and Jampates 1986). The mechanism of this variation varies in different genotypes; however, the source of explant is always important, among which younger tissues are generally more stable.

B. SOMACLONAL VARIATION

Somaclonal variation is another type of variation occurs during culture and is defined as phenotypic and genetic changes among the clonally-propagated somaclones. It was noted earlier as quantitative and qualitative change in plant phenotypes (Larkin and Scowcroft 1981; Larkin et al., 1984), and later reported in other putative plant clones (Ahloowalia 1986; Sun and Zheng 1990; Kaeppler and Phillips 1993). In a study by Thomas et al., (1982) variations were found amongst the progenies originated from the same initial, which further support the change in genetic constitution in culture. This variation is described as either somatically or meiotically stable event, and may or may not be transmitted to subsequent generations (Kaeppler et al., 2000).

Table 15.1 Normal shoots with some variants in *Caladium bicolor*

Hormone (mg/L)		*No of shoots/ culture*	*No of variants/ culture*	*Frequency*
BAP	NAA			
0.5	-	13.25 ± 1.75	0.80 ± 0.83	6.03
1.0	-	11.75 ± 2.50	0.60 ± 0.89	5.10
0.5	0.5	3.75 ± 1.75	0.20 ± 0.44	5.33
1.0	0.5	12.25 ± 1.70	1.20 ± 0.83	9.79
2.0	0.5	6.0 ± 2.90	0.40 ± 0.89	6.66
4.0	0.5	1.25 ± 0.75	0	0

Some Mechanisms

I. Karyotypic alteration: A large number of possible genetic mechanisms have been suggested for somaclonal variation which includes karyotypic alterations, expression of cryptic transposable elements, rearrangement of organelle's genome, mitotic recombination, gene amplification, and single base pair change and virus elimination (Larkin and Scowcroft 1981; Lee and Phillips 1988). Structural abnormalities, including rearrangements and ploidy number changes have been the most prevalent change among the tissue culture regenerants. In a recent report, Mujib et al., (2000) reported about 5–10% regenerants were off-type which exhibited noticeable chromosomal anomalies (Table 15.1, Plate 15.1) among the regenerated progenies in *Caladium*. Among rearrangement, translocations were the most frequent changes observed with deletion and inversion. Other studies (Benzion et al., 1986; Hang and Bregitzer 1993; Kaeppler et al., 2000) suggested that frequent chromosome breakage across the centromere and heterochromatin regions led to different types of cytogenetic

Table15.2 Shoot regeneration along with variants in *Philodendron erubescens* cv. Pink Prince

Hormone (mg/L)			*No. of shoots/ culture*	*No. of variants/ culture*	*Frequency*
BAP	NAA	CH			
1.0	–	–	10.2 ± 1.83	2.6 ± 0.48	25.49
0.5	–	–	16.6 ± 2.65	3.8 ± 0.74	22.89
0.5	–	100	16.8 ± 2.40	4.4 ± 0.80	26.19
			Medium without agar or liquid medium		
0.5	0.1	–	13.6 ± 2.65	3.4 ± 1.01	25.0
0.5	0.2	–	13.2 ± 2.13	3.6 ± 1.20	27.27

All the values are expressed as Mean (M) ± Standard deviation (SD)

abnormalities. Besides, several minor genetic changes have been noted in the regenerated somaclones. The tissues such as callus, single cell or protoplast-derived callus that are considered to be useful sources for dedifferentiation, show more off-type variations than organized tissues. The basic mechanism of such variations, especially for ploidy number changes, have been due to the failure of mitotic apparatus (mitotic asynchrony), marked by the appearance of bridges, laggards, bi-, multi-, or micro-nucleate conditions, is perhaps caused by several plant growth regulators commonly used for in vitro growth. Phytohormones, 2,4-D and 2,4,5-T in particular, often induce polyploidy by affecting normal DNA and post-replication mechanisms (Deambrogio and Dale 1980; Pederson and Minocha 1988). Other biologically-active compounds such as antibiotics, alkaloids, ethylene alcohol, DMSO, EDTA, either added or evolved during culture has been shown to alter DNA synthesis in producing mutations and even sometimes off-type variants (Griesbach 1983; Gecheff 1989). Except for a few (Groose and Bingham 1984; Vasil et al., 1984), several other lines of research indicate observed and non-expressed variations that may also arise if the cultural sources are old and are maintained for a long time with or without regular subculturing (Sutter and Langhans 1981; Westerhops et al., 1984; Stimart 1986; Binarova and Dolezel 1988). These noted changes generally occur due to different recessive mutations being accumulated over the period. As in usual mutation process, in vitro changes in regenerant's genetic constitution is not always reflected in phenotypes which show no effect to immense changes to the regenerants. In general, plants with higher ploidy number hardly bother about the loss or gain of individual chromosomes since polyploids are complemented by extra set of chromosomes. For example, in polyploid *Chrysanthemum*, normal morphology and behavior was noticed, although it possessed 1–3 extra chromosomes to its basic set (De Jong and Custers 1986). More surprisingly, individual base pair changes in nucleic acid, where the basic number and structural configurations of all chromosomes remained intact, produced plants of altered morphology and nature in many occasions (Groose and Bingham 1984; Larkin et al., 1984).

II. Sequence variation: Sequence change has been detected in other tissue culture induced change using different DNA markers like RFLP, RAPD, etc. Although the success report of sequence change and altered protein has been very limited, analyses of specific mutants and others indicate that beside gene deletion, sequence change—particularly A-T transversion—is more common in tissue culture-derived mutants (Brettell et al., 1986; Dennis et al., 1987). Different electrophoresis profiles of variant proteins further support this sequence change incidence.

III. Variation in DNA methylation: DNA methylation variation or instability has been postulated for several tissue culture-induced variations

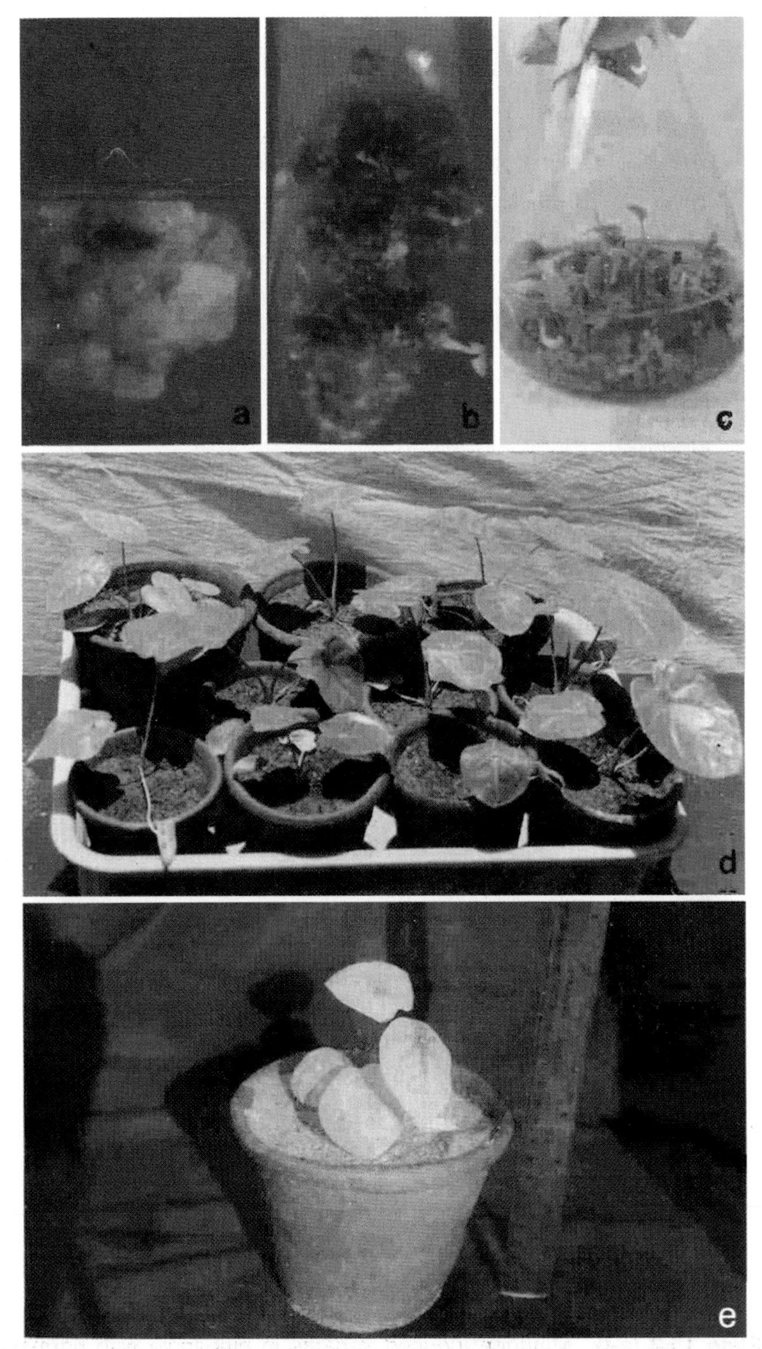

Plate 15.1 (a-e) Callus, normal plantlet regeneration (d) and in vitro raised variant (e) in *Caladium bicolor*

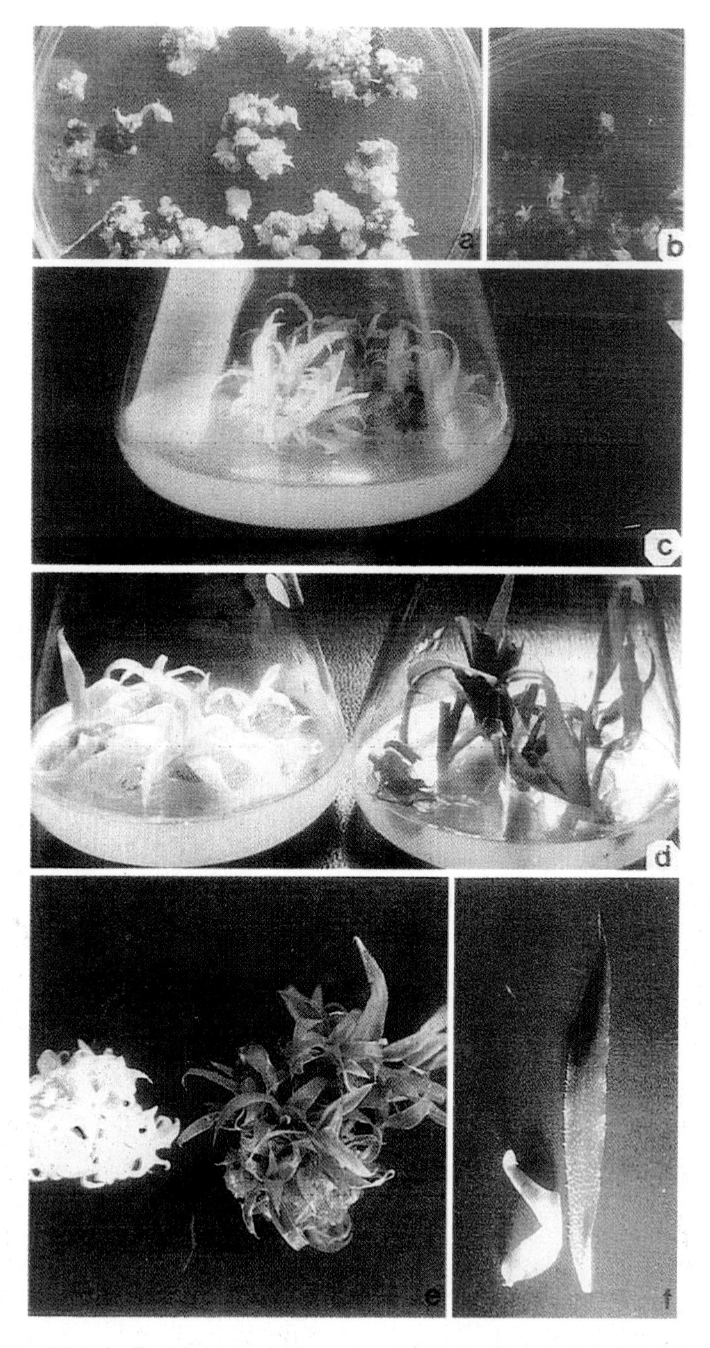

Plate 15.2 (a-f) Mutagen induced variants in pineapple with normal

such as quantitative change, chromosome breakage and other types of alteration in the culture process. Alteration of methylation pattern of specific sites was noted and reported in different plant genera like rice and maize (Brown et al., 1991). Simultaneously, global methylation level was also reported to vary in response to different cultural and chimeral conditions (Arnholt- Schmitt 1995). Both kinds of analysis suggest that DNA methylation variation is almost a general rule during the process of in vitro culture.

IV. Somaclonal variation involving **epigenetic change** has also been suggested when Kaeppler et al., (2000) reported the activation of quiescent transposable elements and transposons as epigenetic changes during culture process. As a result and after heterochromatin modification a large-scale cytogenetic instabilities, which are unique in somaclonal variation, have been noticed that eventually show phenotypic variation through the modulation of gene function. Their observation suggests that DNA methylation among the regenerated plants and their progeny are highly variable and modifications of DNA are less stable compared to seed grown plants. Similar observation was earlier noticed in which genomic shock produced from chromosome breakage events activate transposable elements in other system (McClintock 1984). Extensive research (Peschke et al., 1987; Peschke and Phillips 1991) corroborated the same result with enhanced Ac and Spm/en activity during maize culture process. Hirochika et al., (1996) added to the existing information and showed that retrotransposon—another class of DNA element that are generally remain quiescent in plant genome also have activated transcription rates in tissue culture. Transposons and quiescent retrotransposon in plant genome and their activation indicated that the earlier silenced epigenetic segment is released by the culture process. However, none of the enhanced transcription or insertion of transposon/retrotransposon into genetic region is not always correlated with phenotypic change.

C. TEMPORARY OR PHYSIOLOGICAL VARIATION

Many of the in vitro raised plant's performance, such as strawberry, potato, lily, and apple, are temporary. They show morphological (leaf, phyllotaxy, spine, and branching number change, etc.), physiological (altered flowering, seedset, sex determination, etc.), or other biochemical and cytogenetical alterations. This changed morphology and behavior is due to rejuvenation, promoted by the use of various phytohormones commonly used in vitro. Although the mechanism of rejuvenation is yet to be clear, the phytohormones such as GA, cytokinin, and ABA (not very certain) are frequently used for promoting juvenile phase. Ex vitro temporary performance of some other tissue cultured plants are influenced by the in vitro use of phytohormones, in which plants receive treatments early on but with their long-term memory exerts phenotypic effect when grown outdoor only (Evans et al., 1986; Detrez et al., 1989). These temporary changes revert back to normal both in morphology and behavior with time with not much influence on progenies (Swartz et al., 1981; Evans et al., 1986; Smith and Bhaskaran 1988; Verga et al., 1988).

CONCLUSIONS

Several kinds of off-type variations have been noticed in tissue culture propagated plants. Those variations, which are genetic in nature, may be induced by biochemicals and stresses or by the expression of chimeral (i.e. pre-existing) variations. Generally, this occurs from dedifferentiated cells and in organogenesis and embryogenesis, which utilizes single or multiple cells leads to enhanced expression of pre-existing variations. Further, many of the changes are temporary and sometimes mimick morphological and physiological variations, while some of them are silent. Irrespective of sources, most of these variations are deleterious and a major problem to nurserymen; plant breeders, biotechnologists, and other scientific communities are, however, very optimistic about their potential uses in plant improvement.

REFERENCES

Ahloowalia B.S., 1986, Limitations to the use of somaclonal variation in crop improvement. In: Semal J (ed.). Somaclonal Variation And Crop Improvement, Martinus Nijhoff, Boston, pp. 14-27.

Arnholdt-Schmitt B., 1995, Physiological aspects of genome variability in tissue culture. II. Growth phase dependent quantitative variability of repetitive BstN1 fragments of primary cultures of *Daucus carota* L. Theor. Appl. Genet. **91:** 816-823

Benzion G., Phillips RL and Rines HW, 1986, Case histories of genetic variability in vitro; oats and maize, In: Vasil IK (ed). Cell Culture And Somatic Cell Genetics of plants, Vol. 3, Academic Press, New York, pp. 435-448.

Binarva P. and Dolezel J., 1988, Alfalfa embryogenic cell suspension culture: growth and ploidy level stability. Plant Physiol., **133:** 561-566.

Brar DS and Jain SM, 1998, Somaclonal variation: mechanism and applications in crop improvement. In: Jain SM, Brar DS and Ahlowalia BS (eds). Somaclonal Variation And Induced Mutations in Crop Improvement, Kluwer Academic Publishers, Dordrecht.

Brettell RIS, Dennis ES, Scowcroft WR and Peacock WJ., 1986, Molecular analysis of a somaclonal variant of alcohol dehydrogenase. Mol. Gen. Genet. **202:** 335-344.

Brown PTH, Gobel E and Lorz H., 1991, RFLP analysis of *Zea mays* callus cultures and their regenerated plants. Theor. Appl. Genet. **81**:227-232.

Brunner H and Keppl H., 1991, Radiation induced apple mutants of improved commercial value. In: Plant Mutation Breeding for Crop Improvement. Vol. IAEA, Vienna, pp. 547-552.

Cassells AC, Farrell G and Goetz EM., 1986, Variation in the tissue culture progeny of the chimeral potato (*Solanum tuberosum*) variety Golden wonder. In: Semal J (ed.) Genetic Manipulation In Plant Breeding, Martinus Nijhoff, Dordrecht, pp. 202-213.

Deambrogio E and Dale PJ., 1980, Effect of 2,4-D acid on the frequency of regenerated plants in berley and on genetic variability between them. Cereal Res. Commun., **8:** 417-422.

DeJong J and Custer JBM., 1986, The effect of explant source *in vitro* regeneration and irradiation on variation in yield induced in *Chrysanthemum morifolium*. In: Semal J (eds), Genetic Manipulation In Plant Breeding. Martinus Nijhoff, Dordrecht, pp. 607-609.

Dennis ES, Brettell RIS and Peacock WJ., 1987, A tissue culture induced Adh1 null mutant of maize results from a single base change. Mol. Gen. Genet., **210:** 181-183.

Detrez C, Sangwan RS and Norrel HJ., 1989, Phenotypic and karyotypic status of *Beta vulgaris* plants regenerated from direct organogenesis in petiole culture. Theo. Appl. Genet., **77:** 462-468.

Duncan RR., 1997, Tissue culture induced variation and crop improvement. Adv. Agron., **58:** 201-240.

Evans DA, Chu IYE, Hartman RD and Swartz HJ., 1986, Summary of panel discussion on phenotypic and genotypic stability of tissue cultured plants. In: Zimmerman RH, Griesbach RJ, Hammerchlag FA and Lawson RH (eds). Tissue Culture As A Plant Production System For Horticultural Crops, Martinus Nijhoff, Dordrecht, pp. 95-96.

Gavazzi G, Tonnelli C, Todesco G, Arrehini E, Raffaldi F, Veechio F, Barbuzzi G, Biasini MG and Sala F., 1987, Somaclonal variation versus chemically induced mutagenesis in tomato (*Lycopersicon esculentum* L). Theor. Appl. Genet., **74:** 733-738.

Gecheff KI., 1989, Position specific effects in the mutagenic action of mitomycin C on the chromosomes of *Hordeum vulgare* L. Theor. Appl. Genet., **77:** 705-710.

Griesbach RJ., 1983, Somatic meiosis. Comb. Proc. Intl. Plant Prop. Soc., **33:** 398-403.

Groose RW and Bingham ET., 1984, Variation in plants regenerated from tissue cultures of tetraploid alfalfa heterozygous for several traits. Crop. Sci., **24:** 655-658.

Hammerschlag FA, Ritchie D, Werner D, Hashmi G, Krusberg L, Meyer R and Huettel R., 1995, In vitro selection of disease resistance in fruits trees. Acta Hort., **392:** 19-26.

Hang A. and Bregitzer P., 1993, Chromosomal variation in immature embryo derived calli from six barley cultivars. J. Hered., **84:** 105-108.

Hirochika H., Sugimoto K, Otsuki Y, Tsugawa H and Kanda M., 1996, Retrotransposons of rice involved in mutations induced by tissue culture. Proc. Natl. Acad. Sci., USA, **93:** 7783-7788

Kaeppler SM, Kaeppler HF and Rhee Y., 2000, Epigenetic aspects of somaclonal variation in plants. Plant Mol Biol., **43(2-3):** 179-188.

Kaeppler SM and Phillips RL., 1993a, DNA methylation and tissue culture induced variation in plants. In Vitro Cell Dev. Biol., **29:** 125-130.

Karp A and Bright WJS., 1985, On the causes and origin of somaclonal variation. Oxford Surveys. Plant Mol. Cell Biol., **2:** 199-234.

Larkin PJ, Ryan SA, Brettell RIS, Scowcroft WR., 1984, Heritable somaclonal variation in wheat. Theor. Appl. Genet., **67:** 443-445.

Larkin PJ and Scowcroft WR., 1981, Somaclonal variation- a novel source of variability from cell cultures for plant improvement. Theor. Appl. Genet., **60:** 197-214.

Lee M and Phillips RL., 1988,. The chromosomal basis for somaclonal variation. Ann. Rev. Plant Physiol. Plant Mol. Biol., **39:** 413-437.

Marcotrigiano M and Gouin FR., 1984,. Experimentally synthesized plant chimeras I. In vitro recovery of *Nicotiana tabaccum* L. from mixed callus cells. Ann. Bot., **54:** 503-511.

McClintock B., 1984,. The significance of responses of the genome to challenge. Science **226:** 792-801.

McPheeters K and Skirvin R., 1983,. Histogenic layer manipulation in chimeral 'Thornless Evergreen' tailing blackberry. Euphytica, **32:** 351-360.

Mujib A, Bandyopadhyay S and Ghosh PD., 2000,. Tissue culture derived plantlet variation in *Caladium bicolor* L an important ornamental. Plant Tissue Cult., **10(2):** 149-155.

Mujib A and Jana BK., 1995, Variation in tissue culture derived seedlings of *Philodendron erubescens* cv. Pink Prince. Plant Tiss. Cult. **5(2):** 113-118

Novak FJ., 1991, In vitro mutation system for crop improvement. In: Plant Mutation Breeding for Crop Improvement, Vol. 2, IAEA, Vienna, pp. 327-342.

Pedersen TJ and Minocha SE., 1988, Effects of n-sodium butyrate on cell division in Jerusalem artichoke (*Helianthus tuberosum* L) tuber explants cultured in vitro. J. Plant Physiol., **130:** 623-630.

Peschke VM, Phillips RL., 1991, Activation of the maize transposable element suppressor- mutator(spm) in tissue culture. Thoer. Appl. Genet., **81:** 90-97.

Peschke VM, Phillips RL and Gengenbech BG., 1987, Discovery of trasnposable element activity among progeny of tissue culture derived maize plants. Science, **238:** 804-807.

Preil W., 1986, In vitro propagation and breeding of ornamental plants: advantages and disadvantages of variability. In: Semal J(ed.) Genetic Manipulation In Plant Breeding. Martinus Nijhoff, Dordrecht, 377-403.

Skene KGM and Barlass M., 1983, Studies on the fragmented shoot apex of grapevine iv separation of phenotypes in a periclinal chimera in vitro. J. Exp. Bot., **34:** 1271-1280.

Smith RH and Bhaskaran S., 1988, Sorghum cell culture: somaclonal variation/ screening, Iowa State J. Res., **62:** 571-585.

Stimart DP., 1986, Commercial micropropagation of florist flower crops. In: Zimmerman RH, Griesbach RJ, Hammerschlag FA and Lawson RH (eds). Tissue Culture As A Plant Production System For Horticultural Crops, Martinus Nijhoff, Dordrecht, pp. 301-316.

Sun ZX and Zheng KL., 1990, Somaclonal variation in rice. In: Bajaj YPS (ed.). Biotechnology in Agriculture and Forestry, Vol. 3, Springer-Verlag, Berlin, pp. 288-325.

Sutter E and Langhans RW., 1981, Abnormalities in *Chrysanthemum* regenerated from long term cultures. Ann. Bot., **48:** 559-568.

Swartz HJ, Galletta GJ and Zimmerman RH., 1981, Field performance and phenotypic stability of tissue culture propagated strawberries. J. Amer. Soc. Hort. Sci., **106:** 667-673.

Thomas E, Bright SWJ, Franklin J, Lancaster VA, Miflin BJ and Gibson R., 1982, Variation amongst protoplast derived potato plants (*Solanum tuberosum* cv Maris Bard). Theor. Appl. Genet., **62:** 65-68.

Varga A, Thomas LH and Bruinsma J., 1988, Effects of auxins and cytokinins on epigenetic stability of callus propagated *Kalanchoe blossfeldiana* Poelln. Plant Cell Tiss. Org. Cult., **15:** 223-231.

Vasil V, Vasil IK and Lu CY., 1984, Somatic embryogenesis in long term callus cultures of *Zea mays* L. Amer. J. Bot., **71:** 158-161.

Veilleux RE and Johnson AT., 1998, Somaclonal variation: molecular analysis, transformation interaction and utilization. Plant Breed. Rev. **16:** 229-268.

Westerhop J, Hakkart FA and Versluijs JMA., 1984, Variation in two *Begonia X hiemalis* clones after *in vitro* propagation. Scientia Hort., **24:** 67-74.

Wu L and Jampates R., 1986, Chromosome number and isoenzyme variation in Kentucky bluegrass cultivars and plants regenerated from tissue culture. Cytologia, **51:** 125-132.

16

Tissue Culture Studies in Lentil (*Lens culinaris* Medik)

M. Imdadul Hoque[1], Fathi Hassan[2], Heiko Kiesecker[3] and Hans Joerg Jacobsen[3]
[1]*Department of Botany, University of Dhaka, Dhaka - 1000, Bangladesh*
[2]*Directorate of Agricultural Scientific Research, DASR-Douma-P.O. Box 113, Syria*
[3]*Department of Molecular Genetics, University of Hannover, Herrenhauser Str. 2, 30419 Hannover, Germany*

ABSTRACT

Lentil (*Lens culinaris* Medik) has been considered as one of the main sources of dietery protein for the common people of most of the developing countries. The main objective of the present investigation was to develop in vitro regeneration system from six different explants, namely, shoot tip, leaf segment, cotyledonary node, cotyledon segment, epicotyl and decapitated embryo. MS, B5 and MSB media with different concentrations and combinations of auxins (NAA, IBA) and cytokinins (BAP, Kn) were used for this purpose. It has been possible to obtain a large number of shoot buds from epicotyl, cotyledonary node and decapitated embryo explants on MS/MSB medium supplemented with BAP either alone or in combination with Kn. Best multiple shoot regeneration was obtained from decapitated embryo explants on MSB medium supplemented with 2.22 µM BAP + 2.32 µM Kn + 1.07 µM NAA followed by a TDZ shock (5–10 µM) for three days. The shoot buds were developed mostly from the pre-existing meristems.

INTRODUCTION

In most of the developing countries, lentil is the main source of dietery proteins for the common people who cannot afford to buy red meat, fish and eggs. Lentil contains as much as proteins (about 24.2%) present in red meat, fish and egg, 60% carbohydrate and 2.4 – 4.2% mineral matter (Hulse 1994). Despite the fact that lentil is one of mankind's oldest cultivated crop plants, its agronomic performances are similar to a wild rather than a cultivated plant. As with other food legume crops, lentil production and productivity are both affected by many biotic and abiotic stress factors. The main constraints for lentil breeding and production are the poor yield stability and high susceptibility to fungal diseases.

The major fungal diseases on the Indian subcontinent are *Stemphylium sarciniformae, Uromyces fabae* and root rot caused by *Fusarium oxysporum/ Sclerotium rolfsii*. These diseases cause variable degrees of yield loss, which ranges from zero to 100%, depending on the severity of the infection (Rahman and Bakr 1998). Due to the rather narrow genetic base and lack of resistant genes against the major fungal diseases in the lentil germplasm, the conventional breeding methods could not fulfill the expectation of the end users.

In fact, while for several of the constraints conventional breeding has offered suitable solutions, for some of the major pests and diseases conventional breeding technology has to be complemented by biotechnology-based approaches, whenever the required resistance genes have not been detected in the gene pools of the crop plants and closely-related genera. Therefore, the improvement of the agronomic performance of lentil through genetic engineering may be one of the methods of choice. However, before embarking upon such a programme on genetic engineering in any crop, the in vitro plant regeneration protocol of that particular crop must be developed.

Lentil tissue culture and transformation are still in their infancy if compared to grain legumes from Europe and North America. Although there are some reports on lentil regeneration experiments (Williams and McHughen 1986; Saxena and King 1987; Polanco et al., 1988; Singh and Raghuvanshi. 1989; Malik and Saxena. 1992; Warkentin and McHughen.1993; Khanam et al., 1995; Ahmad et al., 1997; Polanco and Ruiz. 1997, 2001; Gulati et al., 2001; Fratini and Ruiz. 2002), there is no convincing report on the development of an efficient and stable routine regeneration system for this crop available. In this chapter the results of our work on regeneration will be highlighted.

PLANT MATERIALS

Three varieties of lentil, namely, 'BARIMASUR-1' (BM-1), 'BARIMASUR-2' (BM-2), and 'BARIMASUR-3' (BM-3) were obtained from Bangladesh Agricultural Research Institute (BARI), Joydebpur, Gazipur, Bangladesh have been used in the present investigation.

METHODS

(a) ***Surface Sterilization of Seeds***: Seeds of the working material were first soaked in 70% ethanol for one minute, surface sterilized with 6.0% sodium hypochlorite for five minutes and then washed with sterilized distilled water for 3–4 times. The surface sterilized seeds were cultured on water-agar medium and kept in dark upto their germination.

(b) ***Preparation of Explants***: Shoot tip (ST), leaf segment (LS), epicotyl (EC), cotyledonary node (CN), cotyledon segment (CS), decapitated embryo explants were used in this investigation. Shoot tip, epicotyl and cotyledonary nodes were excized from aseptically grown three days old seedlings. Whereas for culture of cotyledon segments and decapitated embryo, surface sterilized seeds were grown on distilled water-agar medium overnight and explants were excized in the following day. In case of embryo explants, the overnight grown sprouted seeds were split open and both root and shoot parts of the mature zygotic embryos were decapitated and subsequently cultured on different media, containing various hormonal supplements for callus induction and shoot regeneration.

(c) ***Media Composition***: MS (Murashige and Skoog 1962), B_5 (Gamborg et al., 1968) and MSB (macro- and micro-salts of MS and vitamins of B_5 (Gamborg et al., 1968) and B_5 media were used routinely. The pH of the medium was adjusted to 5.8 with 1N KOH or 1N HCl before autoclaving. Various concentrations and combinations of growth regulators were tested: BAP (6-benzyl amino purine); kinetin (Kn) (6-furfuryl amino purine); TDZ (N-phenyl-N′-1,2,3, thidiazol-5-yl urea); zeatin (6-[4-hydroxy-3-methyl-but-2-enylamino] purine); NAA (naphthalene acetic acid); IBA (indole-3-butyric acid), IAA (indole-3-acetic acid), 2,4-D (2,4 dichlorophenoxyacetic acid), GA_3 (gibberellic acid). Filter sterilized hormonal supplements were added to the medium before dispensing into plastic petri dishes (9 cm ϕ) or into plastic boxes (10 cm ϕ).

RESULTS

In vitro Plant Regeneration

Six different explants, namely, shoot tip (ST), leaf segment (LS), epicotyl (EC), cotyledonary node (CN), cotyledon segment (CS), decapitated embryo were used for in vitro callus induction and multiple shoot regeneration on MS, MSB and B_5 media containing various concentrations and combinations of BAP (6-benzyl amino purine), kinetin (Kn) (6-furfuryl amino purine), TDZ (N-phenyl-N′-1,2,3, thidiazol-5-yl urea), zeatin, 6-(4-hydroxy-3-methyl-but-2-enylamino) purine, NAA (naphthalene acetic acid), IBA (indole-3-butyric acid), and GA_3 (gibberellic acid).

Effect of various concentrations of BAP and Kn on callus induction and shoot regeneration from different explants: For this experiment, 10 different concentrations of benzyl amino purine (BAP) ranging from 2.22 to 8.87 μM either alone or in combination with kinetin (1.0 – 4.65 μM) were added to MS medium to observe their effect on callus induction and shoot

Table 16.1 Response of different explants on in vitro shoot regeneration

Explants	*Response to shoot regenerations*	*Origin of shoot*
Shoot tip	Mostly single shoot	PEM*
Leaf segment	No shoot development	-
Cotyledonary node	Multiple shoot	PEM
Cotyledonary segment	No shoot development	-
Epicotyl	Multiple shoot	PEM
Decapitated embryo	Multiple shoot	PEM

*Pre-existing meristem

regeneration from the above explants of three varieties of lentil. Results of this study have been presented in Table 16.1.

It has been observed that shoot tip (ST) explant of all the varieties normally produced a single shoot. Both epicotyl (EC) and cotyledonary nodal (CN) explants produced multiple shoot buds only when there were pre-existing meristematic tissues (Fig. 16.1). On the other hand the explants devoid of any meristematic tissue induced development of callus. The callus mass induced on BAP and kinetin supplemented medium and after subculture on the same medium of the same composition, produced large number of embryoid like multiple shoot buds. Shoot buds regenerated on MS/MSB medium containing only BAP were very compact and stunted in growth (Fig. 16.2). On the other hand, shoot buds regenerated on MS/MSB medium supplemented with BAP and kinetin were not compact and growth was faster than the shoots regenerated on BAP supplemented medium (Fig. 16.3). It may be mentioned that the callus induced on 2,4-D containing media did not differentiate.

Effect of different concentrations of TDZ in MS/MSB medium on shoot regeneration from decapitated embryo explant: Since TDZ is being effectively used for shoot regeneration in different leguminous as well as in some tree species, we decided to use this growth regulator in the present investigation for enhancing multiple shoot regeneration. Four different concentrations of TDZ (5, 10, 15 and 20 μM) were used in MS and MSB media for multiple shoot regeneration from epicotyl and decapitated embryo explants of two varieties of lentil ('BM-1' and 'BM-2'). Results of this study have been presented in Table 16.2. Among the two varieties used, 'BM-2' showed comparatively better responses on shoot initiation on MSB medium containing 5–15 μM TDZ (Fig. 16.4).

Regenerated shoot buds were subcultured on MSB medium containing two concentrations of BAP (2.22 and 4.44 μM), as well as on the same media where initial shoot buds were developed in order to observe their effect on shoot proliferation and their subsequent development. On the above three tested media, the initial shoot buds produced new callus and formed numerous new shoot buds. The shoot buds developed on either 2.22 μM BAP containing medium were green in color, compact and stunted in growth. Shoots developed on 4.44 μM BAP were light green in color. Although shoots regenerated on 2.22. μM BAP were green, they subsequently became vitrified and died at different stages of development.

Effect of BAP, Kn and NAA on shoot developement followed by TDZ shock/pulse

Since it was difficult to recover fully-developed shoots from MS/MSB medium supplemented with BAP alone or in combination of kinetin, it was decided to culture the decapitated embryo explants on MSB medium supplemented with 5–10 μM TDZ only for three days, to have a TDZ

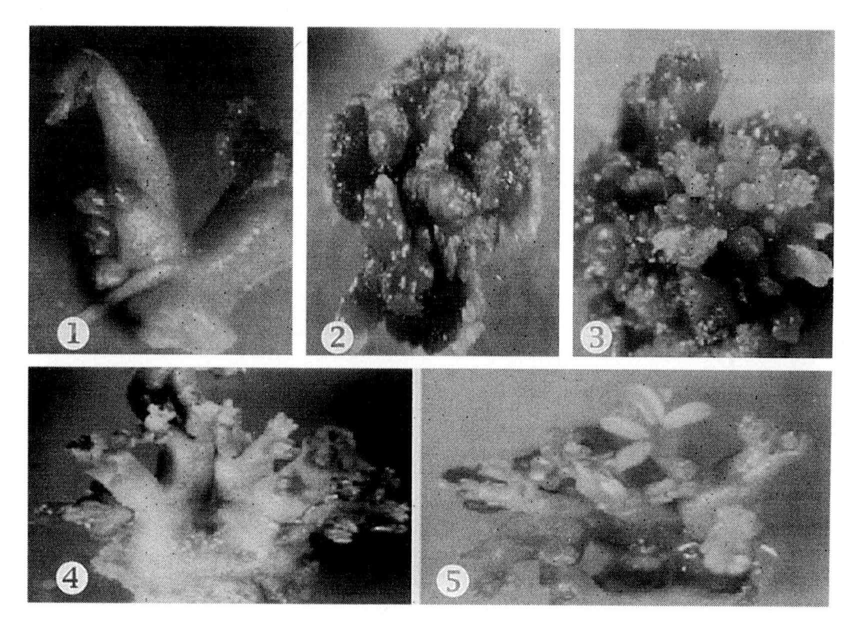

Fig. 16.1-5 In vitro shoot development in lentil. 1. Multiple shoot regeneration from epicotyl explants having pre existing meristem (note: no shoot development at the other side of the explant where there was no PEM; 2. Multiple shoot bud induction on MSB medium supplemented with 2.22 μM BAP (shoot buds are very compact); 3. Multiple shoot bud induction on MSB medium supplemented with 2.22 μM BAP + 2.32 μM Kn; 4. Profuse shoot bud induction on MSB medium containing 10 μM TDZ; 5. Elongation of TDZ induced shoots on MSB medium supplemented with 2.22 μM BAP, 2.32 μM Kn and 1.07 μM NAA.

Table 16.2 Effect of different concentrations of TDZ in MS and MSB media on shoot development from decapitated embryo explants of two varieties of lentil (BM-1 and BM-2).

Variety	*Media composition*	*No. of explants inoculated*	*No. and % of shoot initiation*	*Mean No. of shoot buds/explant*	*Nature of shoot initiation*	*Remarks*
BM-1	MS + 5.0 µM TDZ	60	39.0 (65.0)	4.37	Direct as well as via small amount of callus formation	Produce new callus and new multiple shoot buds upon subculture
BM-2	MS + 5.0 µM TDZ	60	18.0 (30.0)	3.36	"	"
BM-1	MSB + 5.0 µM TDZ	60	34.0 (56.66)	4.81	"	"
BM-2	MSB + 5.0 µM TDZ	60	35.0 (58.33)	4.61	"	"
BM-1	MS + 10.0 µM TDZ	60	15.0 (25.0)	3.0	"	"
BM-2	MS + 10.0 µM TDZ	60	37.0 (61.66)	3.05	"	"
BM-1	MSB + 10.0 µM TDZ	60	30.0 (50.0)	4.04	"	"
BM-2	MSB + 10.0 µM TDZ	60	37.0 (61.66)	3.80	"	"
BM-1	MS + 15.0 µM TDZ	60	29.0 (48.33)	3.39	"	"
BM-2	MS + 15.0 µM TDZ	60	35.0 (58.33)	2.67	"	"
BM-1	MSB + 15.0 µM TDZ	60	29.0 (48.33)	4.14	"	"
BM-2	MSB + 15.0 µM TDZ	60	39.0 (65.0)	4.19	"	"
BM-1	MS + 20.0 µM TDZ	60	25.0 (41.66)	2.89	"	"
BM-2	MS + 20.0 µM TDZ	60	21.0 (35.0)	2.75	"	"
BM-1	MSB + 20.0 µM TDZ	60	27.0 (45.0)	3.63	"	"
BM-2	MSB + 20.0 µM TDZ	60	33.0 (55.0)	3.96	"	"

shock, and then subcultured on MSB medium containing different concentrations of BAP, Kn and NAA. It has been observed that optimum shoot development was possible from the above explant when MSB medium was supplemented with 2.22 μM BAP, 2.32 μM Kn and 1.07 μM NAA (Fig. 16.5).

ROOT INDUCTION IN THE REGENERATED SHOOTS

In vitro regenerated shoots were cultured on MS and half strength of MS medium, containing various concentrations of IBA alone, or in combination with other auxins, namely, NAA or, IAA. Following the report of Khanam et al., (1995) the base of the regenerated shoots was soaked for a few minutes at high concentrations of IBA (0.49 μM) and afterwards subcultured on half the strength of MS containing 1–49 μM IBA. Although there were root induction at the base of the shoot of control (shoots isolated from germinated seed), there was no sign of root initiation at the base of the in vitro regenerated shoots. In some cases, root induction was observed at the stem region, and occasionally produced in vitro flower buds. To overcome this serious problem of rooting, in vitro micro-grafting onto the stock of in vitro raised seed derived seedlings may be one of the choices.

DISCUSSION

In lentil, shoot regeneration through in vitro techniques has been reported previously by many workers (Williams and McHughen 1986; Saxena and King 1987; Singh and Raghuvanshi 1989; Malik and Saxena 1992; Warkentin and Hughen 1993; Khanam et al., 1995; Ahmad et al., 1997; Polanco and Ruiz 1997, 2001). Recently, Fratini and Riuz (2002) reported in vitro plant regeneration in lentil from cotyledonary nodal explants following micro-grafting. In our experiment, we repeated the media composition to regenerate shoots from different explants. However, some of the results reported in the above publications could not be reproduced in the present study. It may be mentioned that the above workers did not use same media composition for in vitro shoot regeneration studies. This might be explained by the differences frequently reported concerning the effect of genotypes and culture conditions used by different researchers. Hulse (1994) reported that seed composition is influenced by genetic background, conditions of crop cultivation, state of maturity at harvest, method of harvest, and conditions. Composition can vary among seeds from the same crop harvest and among seeds from the same plant.

The above workers mainly used shoot tip, cotyledonary node, stem node, epicotyl explants for shoot regeneration. However, our experience on transformation experiments shows that the cotyledonary nodal explants are not readily amenable to *Agrobacterium*-mediated transformation. It is in this background in addition to the above explants we have used decapitated zygotic embryo explants for shoot regeneration. The results of this investigation present a reproducible shoot regeneration system through direct organogenesis using cotyledonary node, epicotyl and decapitated embryo explants. Persistent stunting over several subculture periods had been reported to be one of the major problems when BAP was the only cytokinin in the medium (Fasolo et al., 1989). We have also observed the above problem in this study, even when the explants subcultured to different media supplemented with GA_3 for stimulating elongation of shoots. In the present study, a combination of BAP, kinetin and NAA were used in order to overcome the problem of stunt-growth on young regenerants, or TDZ alone were used for stimulating shoot regeneration and elongation in lentil line tested. On the other hand, Nordstorm and Eliasson (1986) reported that the continuous presence of cytokinins in the medium is of utmost importance for the formation of new shoots.

Thidiazuron (TDZ), which has been reported to be very efficient for multiple shoot regeneration in different legumes as well as in some tree species (Murthy et al., 1998; Sarwar and Skirivin 1997; Malik and Saxena 1992; Kaneda et al., 1997) was also used in the present study. Addition of TDZ accelerates in vitro shoot regeneration. In the presence of TDZ, the explants of lentil line tested produced numerous adventitious shoots. The optimum level of TDZ used in this study was found to be 5 μM, which is lower than the concentration recommended by Malik and Saxena (1992), who reported that 10 μM TDZ induced greater and healthier shoots as compared to 50 μM TDZ. They also reported that at higher concentration of TDZ, shoots were deformed and growth was stunted. The result of the present investigation is in agreement with the report of Malik and Saxena (1992).

In the present study, multiple shoot buds were also induced in a medium containing a combination of BAP and kinetin, which is in agreement with the findings by Polanco and Ruiz (1997). Therefore, it can be concluded that the function of the continuous cytokinin supply is to stimulate growth of new shoots. Another obvious effect exercized by the presence of cytokinin in the medium was the complete or partial inhibition of root formation, which might explain the difficultness found in rooting regenerated shoots.

Response of different explants used (cotyledonary node, epicotyl and leaf segments) to hormones was significantly different in terms of number of shoots regenerated, where cotyledonary nodes showed better response than epicotyl segments with higher shoot numbers recorded for cotyledonary node explants, while no regeneration could be observed in case of leaf segments. This might be ascribed to the fact that regeneration occurred only from pre-existing meristems (PEM). This finding is in agreement with conclusion of Williams and McHughen (1986) that grain legumes in general have proven to be recalcitrant in terms of *de novo* regeneration in vitro.

Cotyledonary nodes have many PEM, which produce more shoots compared to epicotyl segments. This can be explained, since no regeneration was obtained when epicotyl segments without PEM were used, where they produced only callus. This result is in agreement with the findings by Ignacimuthu and Franklin (1999).

It has been observed that BAP is less effective than TDZ in terms of induction of shoot regeneration. This result is in agreement with the report of Sarwar and Skirvin (1997). When explants have been maintained for a prolong time on TDZ-containing medium, numerous shoot buds were observed, the old shoot buds started to die and new shoot buds were formed, but without elongation and their growth was stunted. Kiesecker (1999) also observed similar results in case of chickpea, where the regenerated shoots did not elongate on TDZ-containing medium, but he could obtain multiple shoots in subsequent cultures. This result is also similar to the findings of Ignacimuthu and Franklin (1999). On light of these results, TDZ was used in the further experiments only for three days just to give TDZ pulse to induce multiple shoot buds, after which the explants were subcultured onto media supplemented with 1.22 μM BAP+ 2.32 μM kin+ 1.07 μM NAA. This result is however, in contrast with that of Malik and Saxena (1992), who reported that the optimal concentration of TDZ (10 μM) was required for maximal shoot induction of lentil cv. Eston after 4 days' culture.

Callus could be induced in all explants cultured onto media supplemented with 2,4-D. Explants response was maximum (100%) in terms of callus induction where all explants showed swelling in the first week and developed callus within three to four weeks. The calli derived from 2,4-D either alone or in combination with BAP + Kn did not induce differentiation. Callus derived from medium supplemented with these hormones was green or brownish green. Upon subculturing, these calli onto different media supplemented with BAP + Kn + NAA or BAP + Kn+ NAA+ GA_3, the calli mass increased in volume, but no shoot regeneration was observed and the color of the calli turned brown and finally died.

Williams and McHughen (1986) obtained a rooting frequency of 11%, while Khanam et al., (1995) obtained a frequency of 70%. However, in the present study, rooting proved to be difficult. According to Polanco and Ruiz, (2001) BAP and Kn added to shoot regeneration media could be the reason for inhibition of subsequent root formation when transferring shoots into rooting medium which is true in the present case. This also can be referred to the known after-effect of cytokinins in inhibiting root formation in vitro. Furthermore, such an effect of cytokinins on root inhibition is more obvious when the regenerated shoots are transferred to the rooting medium after being maintained for a long period of time in the regeneration medium in the presence of cytokinins (Polanco and Ruiz 1997).

In the present study, it has been possible to induce a large number of shoots from epicotyl, cotyledonary node and decapitated embryo explants on MSB medium supplemented with 5–10 μM TDZ for only three days and subsequently subcultured on MSB medium containing various concentrations of BAP, Kin and NAA. Addition of BAP, Kin and NAA in the regeneration and maintenance media greatly reduced the elongation problem. Although root induction protocol could not be fully developed, this shoot regeneration protocol could be used for lentil transformation experiments. It is suggested to give more emphasis for developing an efficient in vitro root induction protocol for different lentil cultivars.

ACKNOWLEDGEMENTS

M.I. Hoque is grateful to Alexander von Humboldt Foundation for offering him the Georgge Forster Fellowship to conduct this research at the University of Hannover, Germany. Fathi Hassan also received a scholarship from DAAD to obtain M.Sc degree at Hannover University, which is gratefully acknowledged. The authors would like to thank the Director General of BARI for providing the plant material used in this study.

REFERENCES

Ahmad, M., Fautrier, A.G., McNeil, D.L., Hill, D.G., and Burritt, D.G., 1997, In vitro propagation of Lens species and their F_1 interspecific hybrids. Plant Cell Tissue and Organ Culture. **47:** 169-176.

Fasolo, G.H., Zimmerman, R.H. & Fordham, I. (1989): Adventitious shoot formation on excised leaves of in vitro grown shoots of apple cultivars. *Plant Cell Tissue and Organ Culture,* **16:** 75-87.

Fratini, R., and Ruiz. M.L., 2002, Comparative study of different cytokinins in the induction morphogenesis in lentil (*Lens culinaris* Medik). In Vitro Cellular & Development Biology-Plant. **38(6):** 46-51.

Gamborg, O.L., Miller, R.A. & Ojima, K. (1968): Nutrient requirement of suspension cultures of soybean root cells. *Exp. Cell. Res.,* **50:** 151-158.

Gulati. A., Schryer, P. and McHughen, A., 2001, Regeneration and micrografting of lentil shoots. In Vitro Cellular & Development Biology-Plant. **37(6)** : 798-802.

Hulse, J.H., 1994. Nature, composition and utilization of food legumes. In: Muehlbauer, F.J., Kaiser, W.J., (eds.) Expanding the Production and Use of Cool Season Food Legumes. pp. 77-97. Kluwer Academic Publishers, The Netherlands.,

Ignacimuthu, S., and Franklin, G., 1999. Regeneration of plantlets from cotyledon and embryonal axis explants of *Vigna mungo* L. Hepper. Plant Cell, Tissue and Organ Culture **55:** 75-78.

Kaneda, Y., Tabei, Y., Nishimura, S., Harada, K., Akihama, T., and Kitamura, K., 1997. Combination of thidiazuron and basal media with low salt concentrations increases the frequency of shoot organogenesis in soybeans (*Glycine max* L. Merr.). Plant Cell Reports **17:** 8-12.

Khanam, R., Sarker, R.H., Hoque, M.I., and Hoque, M.M., 1995, In vitro root morphogenesis in lentil (*Lens culinaris* Medik). Plant Tissue Culture **5:** 35-41.

Kiesecker, H., 1999, Development of transgenic chickpea (*Cicer arietinum* L.). Ph.D. Dissertation, University of Hannover.

Malik, K.A., and Saxena, P.K., 1992, Thidiazuron induces high-frequency shoot regeneration in intact seedlings of pea (*Pisum sativum*), chickpea (*Cicer arietinum*) and lentil (*Lens culinaris*). Aust. J. Plant Physiology **19:** 731-740.

Murashige, T. & Skoog, F. (1962): A revised medium for rapid growth and bioassays with tobacco tissue cultures. *Physiol. Plant.*, **15:** 473-497.

Murthy, B.N.S., Murch, S.J., and Saxena, P.K., 1998, Thidiazuron: a potential regulator of in vitro plant morphogenesis. In Vitro Cell. Dev. Biol.-Plant **34:** 267-275.

Nordstorm, A.C. and Eliasson, L., 1986, Uptake and translocation of C14 -labeled benzylaminopurine in apple shoots grown in vitro in relation to shoot development. Physiol. Plantarium **68(3):** 431-435.

Polanco, M.C., and Ruiz, M.L., 1997, Effect of benzylaminopurine on in vitro and in vivo root development in lentil (*Lens culinaris* Medil). Plant Cell Reports. **17:** 22-26.

Polanco, M.C., and Ruiz, M.L., 2001, Factors that affect plant regeneration from in vitro culture of immature seeds in four lentil cultivars. Plant Cell, Tissue and Organ Culture **66:** 133-139.

Polanco, M.C., Pelaez, M.I. and Ruiz, M.L., 1988., Factors affecting callus and shoot formation from in vitro cultures of *Lens culinaris* Medik. Plant Cell, Tissue and Organ Cult. **15(2)** : 175-182.

Rahman, M.L., and Bakr, M.A., 1998, Current status of Lentil Diseases and Future Needs. (Eds). K.B. Singh and M.M. Rahman In: Proc. Workshop on Diseases Research Breeding in Pulses, 24-25 March, 1998, BARI, Joydebpur, Gazipur, Bangladesh, pp. 23-31.

Sarwar, M. and Skirvin, R.M. 1997, Effect of thidiazuron and 6-benzylaminopurine on adventitious shoot regeneration from leaves of three strains of 'McIntosh' apple in vitro. Scientia Horticulture **68:** 95-100.

Saxena, P.K. and King, J. 1987, Morphogenesis in lentil: plant regeneration from callus culture of *Lens culinaris* Medik, via somatic embryogenesis. Journal of Plant Scince. **52:** 223-227.

Singh, R.K., and Raghuvanshi, S.S., 1989, Plant regeneration from nodal segments and shoot tip derived explants of lentil. Lens Newsletter **16:** 33-35.

Warkentin, T.D., and McHugen, A., 1993, Regeneration from lentil cotyledonary nodes and potential of this explant for transformation via *Agrobacterium tumefaciens*. Lens Newsletter **20:** 26-28.

Williams, D.J., and McHughen, A., 1986, Plant regeneration of legume *Lens culinaris* Medik. (lentil) in vitro. Plant Cell, Tissue and Organ Culture **7:** 149-43.

17

Development of Transgenic Chicory (*Cichorium intybus* L.)

M.Z. Abdin[1], R.U. Rehman[1], M. Israr[1], P.S. Srivastava[1] and K.C. Bansal[2]

[1]*Centre for Biotechnology, Faculty of Science, Hamdard University, New Delhi 110 062, India.*

[2]*National Research Centre for Plant Biotechnology, Indian Agricultural Research Institute, New Delhi 110 012, India.*

ABSTRACT

An efficient protocol has been developed for regeneration of plantlets from leaf explants of chicory (*Cichorium intybus* L.). Regeneration via callus was obtained on the modified Murashige and Skoog's semi-solid medium (MS) containing 2.0 µM Indole-3-acetic acid (IAA) and 5.0 µM 6-Furfurylaminopurine (KN). Each callus regenerated at least five or more shoots. Regeneration occurred by both somatic embryogenesis and organogenesis. The regenerated shoots were excized, and root formation was induced with 0.2 µM Indole-3-butyric acid (IBA). After rooting, the plantlets were successfully established in soil after hardening. *Agrobacterium tumefaciens* was used to genetically transform *Cichorium intybus* L. leaf explants from the in-vitro grown seedlings. The *Agrobacterium* strain GV2260, carrying the binary vector Bin AR under the control of constitutive 35SCaMV promoter, was used. The plasmid osm/BinAR contains kanamycin resistant gene nptII to act as an antibiotic selection marker. A number of transformation conditions were tested and a total of 44 transformed events were produced from 1175 leaf explants. The transformed calli were regenerated on the medium (MS + 2 µM IAA and 5 µM KN) containing Kanamycin (150 mg/l). Kanamycin resistant shoots were rooted on Indole Butyric acid (0.2 µM). The integration of osmotin gene in the genome of transformants was confirmed by Polymerase Chain Reaction (PCR).

INTRODUCTION

Cichorium intybus L. (Asteraceae) is a popular crop that provides a variety of edible products from both leaves and storage roots. Nutritional aspects such as hypocaloric value, high fibre, vitamin and mineral salt content increase the importance of chicory for the diet of human beings in developed countries. In addition, chicory is interesting from an agronomic point of view because of the low terminal input required. Chicory is commonly used as salad and some varieties are cultivated for their sweet roots which, when roasted, become a coffee substitute. The chicory powder increases the volume of coffee and improves the flavour. Added to coffee, it counteracts caffeine and helps in digestion. A tea made from chicory is beneficial in curing an upset stomach. In addition, chicory is a potential industrial crop for the production of fructose syrup from its roots. In traditional system of medicine, the plant is used for the treatment of diseases like jaundice, gout and rheumatism. Chicory (*Cichorium intybus* L.) has been regenerated from primary explants such as storage roots (Margara and Rancillac 1966), green leaves (Toponi 1963), etiolated leaves (Vasseur 1979b), floral stems (Bouriquet and Vasseur 1973), suspension cultures (Yassseen and Splittstoesser 1995), leaf and petiole (Eung et al., 1999), leaf midrib (Mix-Wagner and Eneva T, 1998), leaf (Mix-Wagner and Eneva T, 1996), pollen (Castano and De Proft, 2000) and protoplasts (Varotto et al., 1997; Binding et al., 1981: Crepy et al., 1982; Saski et al., 1986). Plants have also been regenerated from tissue cultures of mature tap roots (Heirwiegh et al., 1985). A regeneration protocol from the leaf explants has earlier been described by Vermeulen et al., (1992) and Genga et al., (1994). Inspite of these regeneration protocols published, there is an urgent need to develop an efficient regeneration protocol of chicory for its genetic transformation since with current protocols, the number and dimension of regenerated shoots generated and their height achieved with earlier protocols were not appropriate for genetic transformation experiments. Hence, this study was conducted to develop an efficient regeneration protocol for in vitro culture of *Cichorium intybus* L. and an *Agrobacterium tumefaciens* mediated genetic transformation technique, in order to integrate osmotin gene in the genome of chicory plant, inducing osmotin protein so that it became resistant/tolerant to drought and salt stress. The gene product, osmotin, is a 24 KD protein, and originally isolated from tobacco cells adapted to NaCl stress and desiccation. The stress-induced synthesis and accumulation of the osmotin protein is correlated with osmotic adjustment in tobacco cells. The overexpression of osmotin gene in transgenic tobacco has been reported to impart tolerance against both salinity and drought stress (Barthakur et al., 2001). It is, therefore, expected that chicory plants carrying osmotin gene will also tolerate both salinity and water stress. Hence, these plants can be profitably cultivated in saline and rainfed areas of the country.

MATERIALS AND METHODS

Seeds of chicory (*Cichorium intybus* L.) were obtained from the Herbal Garden of Hamdard University, New Delhi, India. The seeds were washed with the detergent (0.5% cetrimide) for 7 min. and surface sterilized by immersion in 0.1% (w/v) mercuric chloride ($HgCl_2$) for 7–10 min. The seeds were then germinated on MS (Murashige and Skoog 1962) basal medium with 2% sucrose. Leaf discs were excized from four-week-old seedlings and used for in vitro culture.

In vitro Culture Conditions and Shoot Regeneration

Leaf discs from the four-week-old seedlings were taken and inoculated on full strength MS medium supplemented with vitamin-free Casein hydrolysate (1000 mg/l), sucrose (3%) and agar (6 g/l). The different growth regulators: Indole-3-acetic acid (IAA), Kinetin (KN) and 6-Benzylaminopurine (BAP) were used in different combinations. Vitamin-free Casein hydrolysate (CH) was only tested between the combinations of IAA and KN media. The pH was adjusted to 5.8 for all media before autoclaving, which was performed for 20 min. at 120°C. A total of 40–50 tubes per treatment, with one explant each, were inoculated/treatment. The cultures were incubated at 27°C under a 16/8 hours (light/dark) photoperiod with a white fluorescent light flux of 150–200 $\mu E\ m^{-2}\ s^{-1}$ with the relative humidity of 50–60%. The effects of different combinations of the growth regulators used in this experiment were studied. Basal medium supplemented with 2.0 µM IAA and 5.0 µM KN, induced the best shoot-forming response and was, therefore, used in further experiments. Shoot cuttings, excized from the regenerated calluses were transferred to the BM supplemented with 0.2 µM Indole-3-butyric Acid (IBA) to induce root formation. Rooted plantlets were transferred to sterile sand or soilrite supplemented with half strength MS salts and incubated under saturated humidity conditions. After two weeks of hardening, plants were transferred to pots and grown at 30± 2°C under natural light conditions in a growth chamber.

Scanning Electron Microscopy (SEM)

Samples were fixed in 2.5% glyceraldehyde overnight; then washed with 0.1 M phosphate buffer (pH-7.2). The samples were then dehydrated in gradual series of acetone; critical point dried and sputter coated with gold. Thereafter, samples were observed under SEM Leo 435 VP.

Light Microscopy

One-week-old calli were fixed in FAA, embedded in paraffin, sectioned and stained with safranine. The slides were observed and photographed in Olympus Vanox S AHB 2 (Japan) light microscope.

Transformation of Chicory

The *Agrobacterium tumefaciens* strain GV2260 harbouring the plasmid pBinAR was inoculated into YEM (yeast extract 1 g/l;, mannitol 10 g/l;

NaCl 1 g/l; $MgSO_4 \cdot 7H_2O$ 2 g/l and K_2HPO_4 5 g/l; pH 7.0) medium containing 50 mg/l kanamycin, 75 mg/l rifampicin and grown at 28°C and 150 rpm for 24 h to obtain a A600 of 0.3–0.4 (A600 1.0 corresponds to 1 × 108 cells/ml). The bacterial cells were pelleted at 1000 × g at room temperature and resuspended in liquid MS basal medium without sugar. The primary leaf explants subcultured for 4–6 days were infected with the bacteria for 7 minutes and then placed on co-cultivation medium (MS+2 μM IAA+5 μM KN+1000 mg/l CH) and cultured for three days. The co-cultivated leaf discs were then transferred to the selection medium (MS+2 μM IAA+5 μM KN+1000 mg/l CH+150 mg/l kanamycin+500 mg/l cefotaxime). Portions of proliferating callus were then transferred to regeneration medium (MS+2 μM IAA+5 μM KN+1000 mg/l CH+ cefotaxime 250 mg/l) and cultured under the same conditions. Plants regenerated from the same resistant callus were regarded as clones of the same line. The regenerated primary transformants were transferred to a mixture of soil and soilrite and grown in growth chambers under 16 h photoperiod at 25 ± 2°C.

PCR Analysis of Transgenic Plants

The genomic DNA was isolated from the leaves of the transformants and untransformed control plants using CTAB protocol (Murray and Thompson 1980). PCR was performed using a Techne PCR machine (UK).The PCR amplification was carried out in 40 μL reaction mixture containing 2 mM $MgCl_2$, 10 mM dNTP mixture and buffer (pH 8.3). The samples were denatured initially at 94°C for 3 min followed by 30 cycles of 1 min denaturation at 92°C, 1 min. primer annealing at 52°C, 1 min. synthesis at 72°C with a final extension step for 3 min. The PCR products were analysed on 1% agarose gel.

Data Analysis

The treatments were replicated, and the standard error was calculated. Experiments were repeated three times with the treatment parameters remaining unchanged.

RESULTS

Callusing

Callus formation could be observed on all the combinations. Leaf discs from the four-week-old seedlings exhibited better responses on BM supplemented with IAA, BAP and KN. It took 1–2 weeks for the calluses to develop from the leaf discs. The callus formation was compact and pale yellow on combinations of MS+IAA (0.5–2.0 μM) and KN (5.0 μM). However, the calli were friable on combinations of MS+IAA (0.5–2.0 μM) and BAP (5.0 μM). The frequency of callus percentage with the combination of IAA (0.5–2.0 μM) and KN (5.0 μM) ranged between 75.52 and–84.4 percent. The maximum callus percentage of 84.4 was observed with the combination MS+IAA (2.0 μM) and KN (5.0 μM). However the callus percentage was quite low, with the combinations of IAA (0.5–2.0 μM) and BAP (5.0 μM), which ranged from 30.0 to–70.0%. (Fig 17.2; Plate 17.1).

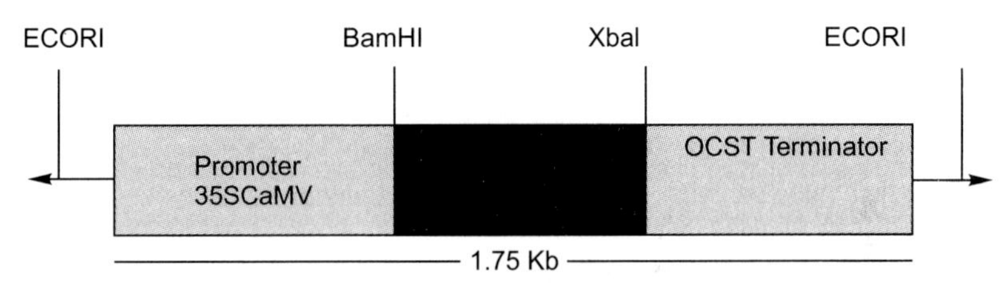

Fig. 17.1 The osmotin gene construct. The gene is cloned under the control of CaMV 35S promoter in binary vector Bin AR containing nptII gene as selection marker

Regeneration

Regeneration occurred by both somatic embryogenesis and organogenesis (Plates: 17.1 and 17.2) . Shoot buds started arising on two-week-old callus. The buds transformed into green and healthy shoots in the presence of wide range of growth regulators: viz., IAA (0.5-2.0 μM) and KN (5.0 μM) supplemented with CH (1000 mg). The responses on all combinations were monitored in terms of number of shoots per callus and the length of regenerated shoots after four weeks. The number of regenerated shoots ranged from 4.76 to 5.92 with IAA and KN combinations (IAA 0.5–2.0 μM and KN 5.0 μM, respectively). The number of shoots was maximum (5.9) with the combination of IAA (0.5 μM) and KN (5.0 μM), whereas, the shoot length ranged from 4.25 to 6.44 cm. The maximum value of 6.44 cm was observed with the combination of IAA (2.0 μM) and KN (5.0 μM). However, with the combinations of IAA (0.5–2.0 μM) and BAP (5.0 μM), the number of shoots ranged from 2.2 to 3.9 and the shoot height from 0.76 to 2.75 cm. As the maximum height, along with green, healthy shoots with compact calli were obtained with the combination of IAA (2.0 μM) and KN (5.0 μM), it was subsequently and repeatedly used for subculturing (Fig 17.3; Plate 17.1B).

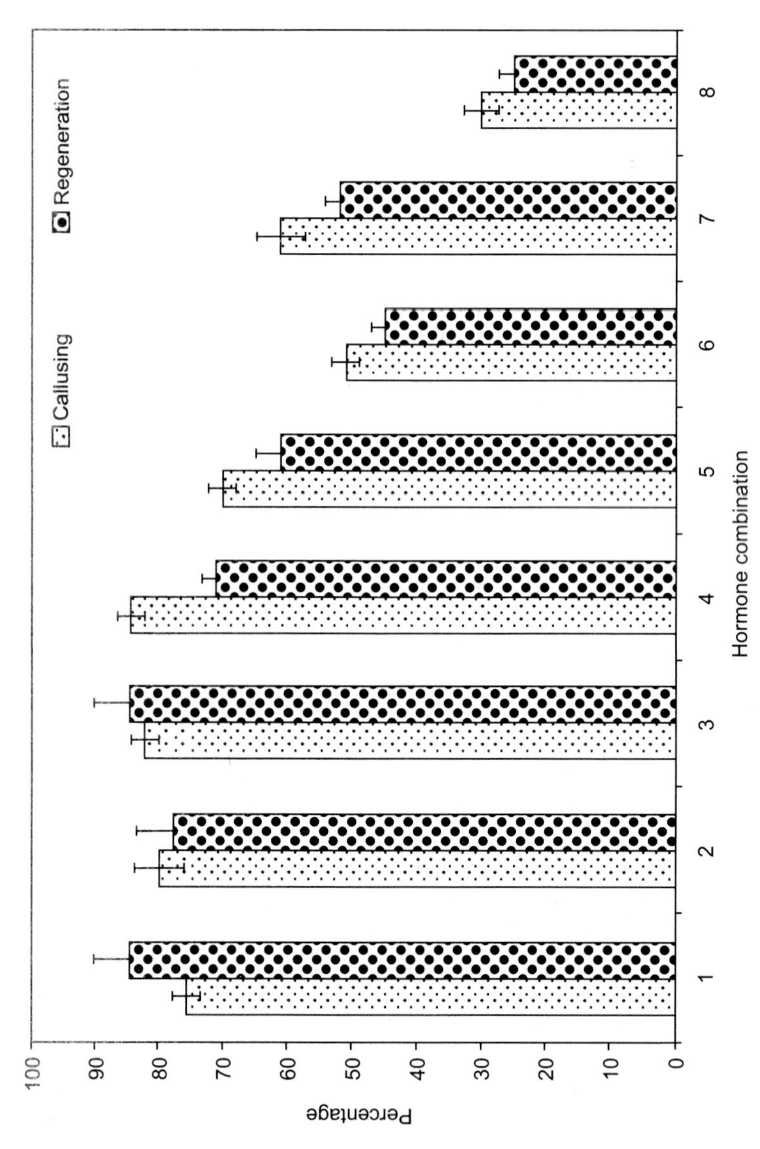

Fig. 17.2 Effect of different hormone combinations on callusing and regeneration percentage in *Cichorium intybus* L.: 1. IAA (0.5) + KN (5.0): 2. IAA (1.0) + KN (5.0):3. IAA (1.5) + KN (5.0): 4. IAA (2.0) + KN (5.0):5. IAA (0.5) + BAP (5.0):6. IAA (1.0) + BAP (5.0):7. IAA (1.5) + BAP (5.0): 8. IAA (2.0) + BAP (5.0).

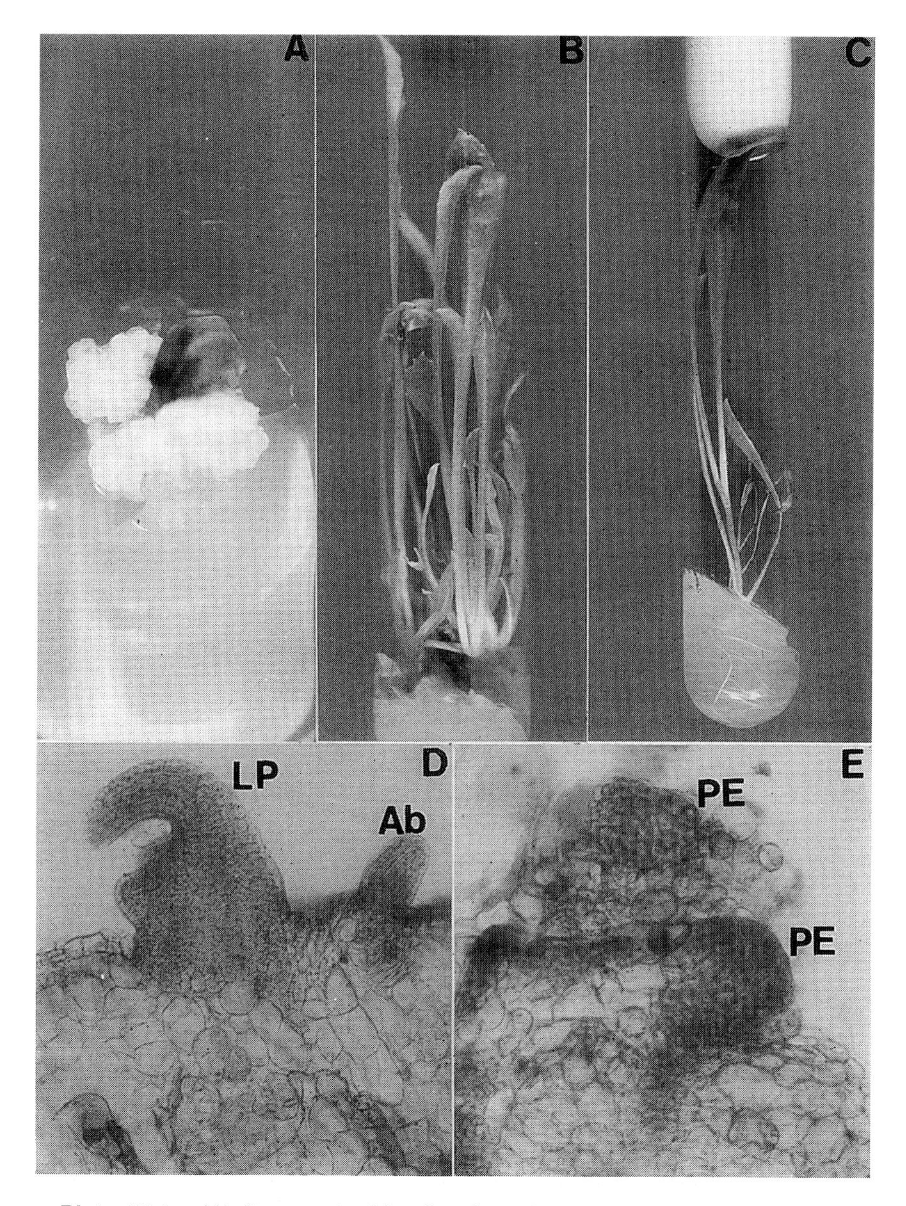

Plate 17.1: (A) One-week-old callus from the leaf explant. (B) Regenerated shoots from four-week-old callus. (C) Rooting of the shoot after one week. (D, E) Light microscopy of *Cichorium* leaves cultured for 7 days. (D) Apical bud (Ab) arising from surface; Apical bud with leaf primordia (LP). (E) Direct somatic proembryo (PE); Bar = 10 µm; D, E.

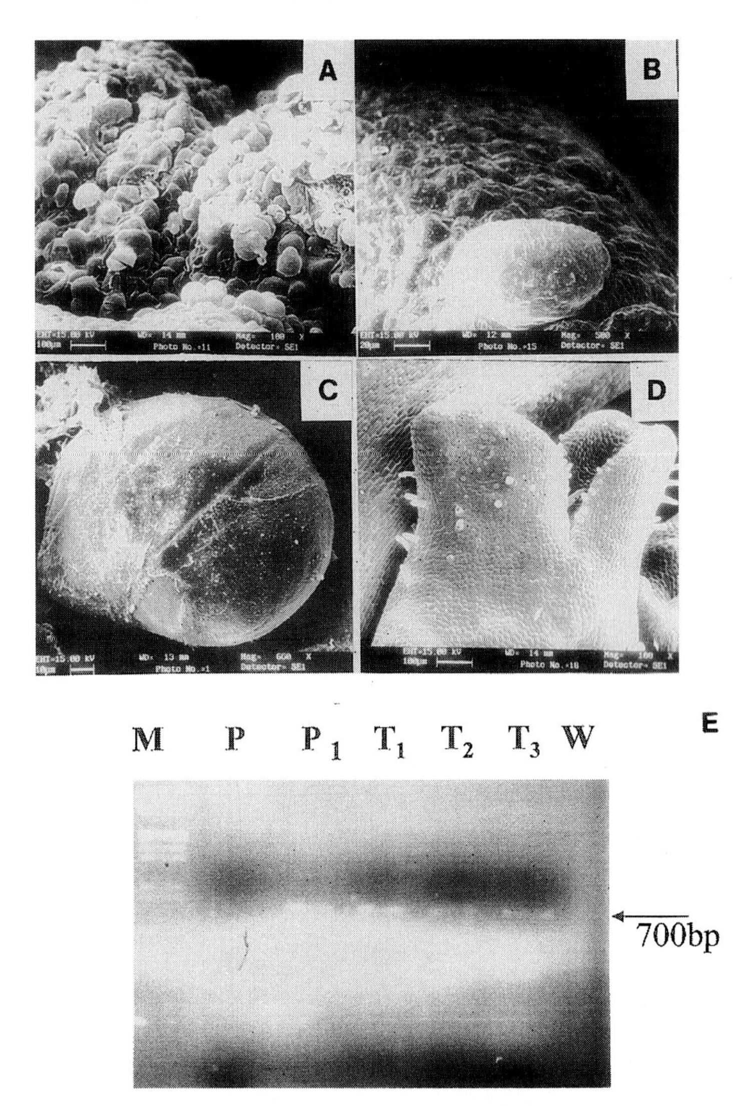

Plate 17.2 SEM of *Cichorium* leaves cultured for 10 days (A, B, C, D). A: Globular proembryos emerging out from the surface of callus; B: Globular proembryo linked with a fibrillar network; C: Proembryo with a slit in the centre; D: Embryonic bud developed on the callus. Bar = 100 µm: A,D; Bar =20 µm : B; Bar =10 µm: C. (E) PCR analysis of plant using Osmotin gene-specific primer. A fragment of about 700 bp corresponding the coding region of about 700 bp of the osmotin gene was amplified in transgenics (T1-T3) and Plasmid (P1). Wild Type: W, Plasmid: P, Marker: M.

Rooting

The shoots from 4-5-week-old regenerating callus were taken and transferred to the MS medium supplemented with IBA (0.2 μM). The roots were normal and lengthy. These rooted plantlets were successfully placed on the sterile soilrite for 2.0 weeks for hardening, and then transferred to pots and grown at 30 ± 2°C under natural light conditions in a plant growth chamber (Plate 17.1C)

Transformation of Chicory

Transformation of chicory was carried out with the osmotin gene using the construct as shown in the Fig. 17.1. About 44 transformed events were produced from about 1175 explants (%). The transformed calli were resistant to kanamysin at 150mg/l. These calli were placed on the regeneration medium (MS+2 μM IAA+5 μM KN+1000 mg/l CH+ cefotaxime 250 mg/l).The regenerating calli were further subcultured on the same medium after one month. The regenerated shoots were placed on the rooting medium eight weeks after subculturing. The presence of the osmotin gene in the transgenic plants was detected by PCR amplification of the osmotin gene using gene-specific primers. An amplified product of about 700 bp was detected.

Discussion

Micropropagation of *Cichorium intybus* L. provides an opportunity to produce clonal populations, which can be multiplied and rooted readily for the production of mature plants under Indian conditions. Micropropagation from leaf explants has been achieved previously by many authors (Genga et al., 1994, used BA 4.4 μM; Vermeulen et al., 1992, used IAA, NAA, IBA in the concentrations ranging from 0.0 to 2.0 μM with or without BAP 1.0 μM; Yasseen et al., 1995, used 2,4-D 1.3 μM and KN 1.3 μM). In the present work, however, a wide range of growth regulators were used in different combinations and among them IAA (2.0 μM) and KN (5.0 μM) was found to be the best. The better response could be due to the enhancement in cell division when auxin and cytokinin were applied together at relatively higher concentrations. This positive effect on growth could be explained by an enhanced RNA synthesis as reported by Vasseur (1979a). The regeneration on this medium as observed by both organogenesis and somatic embryogenesis (Plates 17.1 and 17.2) is a new finding in the clonal propagation of this plant.

Chicory was transformed with *Agrobacterium tumefaciens* (GV2260) containing the binary vector pBinAR. In the earlier experiments by Vermeulin et al., (1992) a self-compatible French population of witloof

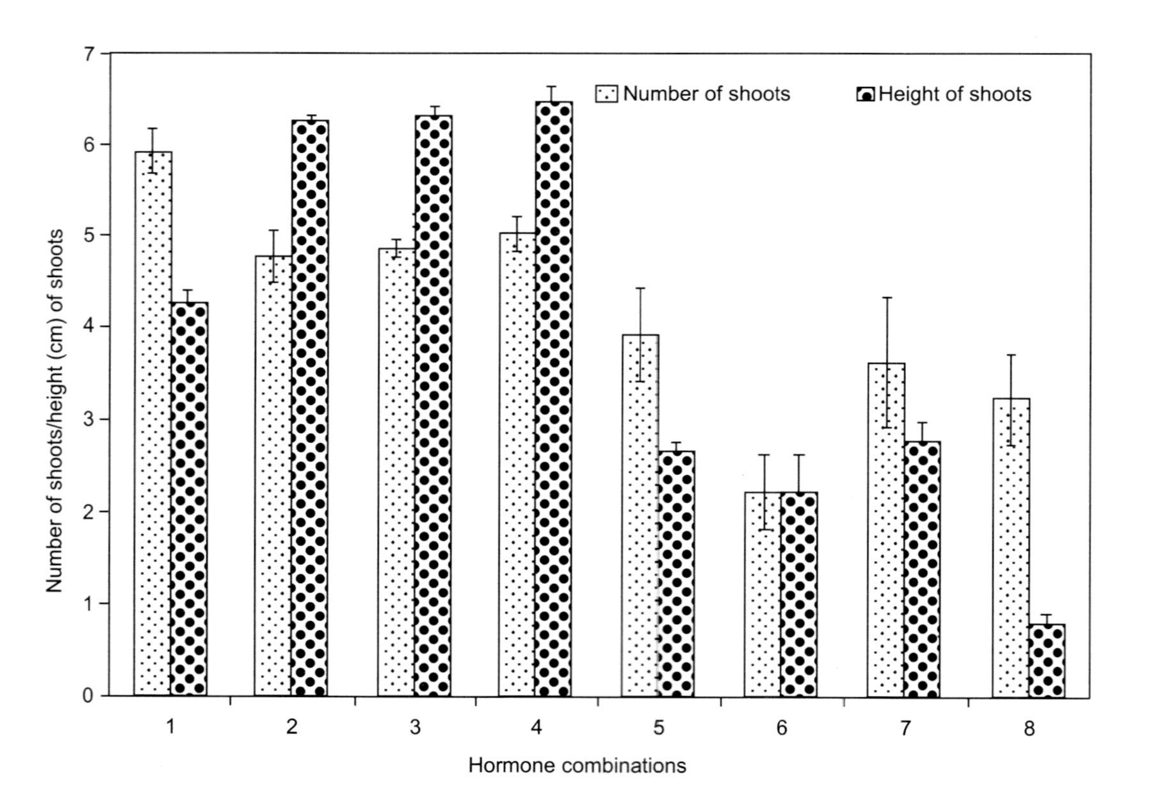

Fig 17.3 Effect of different hormone combinations on number of regenerating shoots/culture and height of regenerating shoots in *Cichorium intybus* L.; 1. IAA (0.5) + KN (5.0); 2. IAA (1.0) + KN (5.0); 3. IAA (1.5) + KN (5.0); 4. IAA (2.0) + KN (5.0); 5. IAA (0.5) + BAP (5.0); 6. IAA (1.0) + BAP (5.0); 7. IAA (1.5) + BAP (5.0); 8. IAA (2.0) + BAP (5.0).

ecotype of chicory was found to be susceptible to the infection with *Agrobacterium tumefaciens*. In addition, the Belgian endive hybrid flash of *Cichorium intybus* was transformed with *Agrobacterium rhizogenes* and *Cichorium intybus* var. *foliolus* was transformed with *Agrobacterium tumefaciens* (Genga et al., 1994). The results of these experiments, together with those herein reported, suggest that *Cichorium intybus* is an amenable species for *Agrobacterium*-mediated genetic transformation.

The genetically-transformed chicory plants with osmotin gene can be cultivated on marginal lands exposed to various kinds of abiotic stresses, viz., moisture and salt stress. As chicory is used in herbal formulations to treat acne, ophthalmia, inflamed throat, fevers, vomiting, diarrhoea, enlargement of spleen and liver ailments (Wealth of India), the cultivated chicory plant can provide authentic raw material to pharmaceutical companies for making herbal formulations.

ACKNOWLEDGEMENTS

R.U. Rehman is thankful to Hamdard National Foundation for providing the fellowship. The authors are thankful to Dr Deepshika Pande, Research Associate, Centre For Biotechnology, Hamdard University and Dr Sharmishta Balathakur, Scientist, National Research Centre For Plant Biotechnology, IARI, New Delhi, for their valuable suggestions and technical help.

REFERENCES

Binding H., Nehlis R., Kock R., Finger J., 1981, Comparative studies on protoplast regeneration in herbaceous species of the dicotyledoneae class. Zpflanzenphysiol. **101**: 119-130.

Bouriquet R., Vasseur J., 1973, Croissance et bourgeonnement des tissus de feuilles d'endive en fonction de l'age et du lieu de prelevement des explantats. Bull. Soc. Bot. Fr. **120**: 27-32.

Castano C.I., De Proft M.P., 2000, In vitro pollination of isolated ovules of *Cichorium intybus* L. Plant Cell Reports **19**: 616-621.

Crepy L., Chupeau M.C., Chupeau Y., 1982, The isolation and culture of leaf protoplasts of *Cichorium intybus* and their regeneration into plants. Zpflanzenphysiol **107**: 123-131.

Eung J.P., Tae L.H., Park E.J., Lim H.T., Myung L.J., 1999, Establishment of an efficient in vitro plant regeneration system in chicory (*Cichorium intybus* L. var. *sativus*). Acta Horticulturae **483**: 367-370.

Genga A., Giansante L., Bernacchia G., Allavena A., 1994, Plant regeneration from *Cichorium intybus* L. Leaf explants transformed by *Agrobacterium tumefaciens*. J. Genet and Breed. **48**: 25-32.

Heirwegh K.M.G., Banerjee N., Nerum K., Van, De Langhe E., 1985, Somatic embryogenesis and plant regeneration in *Cichorium intybus* L. (witloof Compositae). Plant Cell Rep. **4**: 108-111.

Margara J., Rancillac M., 1966, Recherches experimentales sur la neoformation de bourgeonsg inflorescentiels ou vegetatifs in vitro a partir d explantats d endive *Cichorium intybus* L. II. Observations sur la vernalization prealable de la racine. Ann. Physiol. Veg. **8**: 39-47.

Murashige T., Skoog F., 1962, A revised medium for rapid growth and bioassay with tobacco tissue cultures. Physiol. Plant **15**: 473-497.

Murray M.G., Thompson W.F., 1980, Rapid isolation of high molecular weight plant DNA. Nucleic Acid Research **8**: 4321-4325.

Saksi N., Dubois J., Millicamps J.L., Vasseur J., 1986, Regeneration de plantes de chicoree witloof cv. Zoom a partir de protoplastes: influence de la nutrition glucidique et azotee. C. R. Acad. Sci. Paris **302**: 165-170.

Toponi M., 1963, Action combinee de la kinetine et de l'acide indolylacetique sur la neoformation d'organes par des fragments de feuilles d'endive (*Cichorium intybus* L.) cultives in vitro. C. R. Acad. Sci. Paris **257**: 3030-3033.

Varotto S., Lucchin M., Parrini P., 1997, Plant regeneration from protoplasts of Italian red chicory (*Cichorium intybus* L.). J. Genet. and Breed. **51**: 17-22.

Vasseur J., 1979a, Action de l'acide indolyl-acetique, de la kinetine et de l'hydrazide maleique sur la neoformation des bourgeons et la synthese d'ARN obsevees au cours de la culture in vitro de fragments de feuilles etiolees d'endive. C. R. Acad. Sci. Paris **189**: 93-96.

Vasseur J., 1979b, Etude du bourgeonnement de fragments de feuilles etiolees d'Endive. II. Formation des bourgeons en fonction de caracteristiques biochimiques. Rev. Gen. Bot. **86**: 113-190.

Vermeulen A., Vaucheret H., Pautot V., Chupeau Y., 1992, Agrobacterium mediated transfer of a mutant *Arabidopsis* acetolactate synthase gene confers resistance to chlorsulfuron in chicory (*Cichorium intybus* L.). Plant Cell. Rep. **11**: 243-247.

Wagner G.M., Eneva T., 1996, Positive effect of cefotaxime on plant regeneration from *Cichorium intybus* L. leaf material. Landbauforschung-Volkenrode (FAL), Germany, **46**: 4, 166-168.

Wagner G.M., Eneva T., 1998, Plant regeneration from *Cichorium intybus* L. var. *sativum* leaf midrib explants induced by ancimidol supplemented culture medium. Landbauforschung-Volkenrode (FAL), Germany, **48**: 2, 53-55.

Wealth of India: Phondke GP (Director) Dictionary of Indian Raw Materials and Industrial Products. Raw Materials. Vol. 3. Publication and Information Directorate. Council of Scientific and Industrial Research. Dr K.S. Krishnan Marg, New Delhi 110 012.

Yasseen M.Y., Splittstoesser W.E., 1995, Somatic embryogenesis from leaf of witloof chicory through leaf of suspension culture. Plant Cell. Reports **14:** 804-806.

18

Engineering Defense Responses in Crops for Improvement and Yield: Recent Advancements in in vitro Gene Transfer Technology

Shakir Ali[1] and A. Mujib[2]

[1]Department of Biochemistry & [2]Department of Botany, Faculty of Science, Hamdard University, New Delhi 110 062, India

ABSTRACT

The presence of a wide variety of pathogens—including the microbes and insects which are abundant in nearly all ecosystems—results in low crop productivity. One of the practical means of increasing crop production is, therefore, to minimize the pathogen-associated losses. Plants have evolved defense strategies to attempted infections by microbial pathogens, which include not only the surface protectants (cuticle and periderm) or the products of plant's secondary metabolism (terpenes, phenolics, or nitrogen-containing compounds like alkaloids and glycosides) but also the strategies such as gene silencing, and HR or SAR responses. The isolation of the genes involved in disease resistance in plants (R genes) and the technological breakthroughs in the in vitro gene transfer technology during the last decade have allowed for cloning and stitching the disease resistant genes into economically-important crop plants for better yield. The following chapter is an attempt to review these advancements in gene transfer technology and metabolic engineering in field crops aimed to confer resistance to pathogens and improve crop yield.

INTRODUCTION

Low crop productivity due to losses caused by the pathogens is a serious problem and has far-reaching implications on human population. Irish famine due to the late blight of potato (1845-1847) is a notorious example. According to a report, yield losses due to pathogens are estimated at 14% of the total agricultural production (Oerke et al., 1994). There are additional costs in the form of pesticides applied for pest control, currently valued at US$ 10 billion per annum. The losses made by insects, pests, diseases and weeds are estimated upto US$ 234 billion in eight major field crops (42%), out of total attainable production of US$ 568.7 billion worldwide (Sharma et al., 2001).

Plant diseases with the potential to wipe out crops are exploding. The reasons are numerous. Efficient global travel is helping the spread of pathogens, while global warming is allowing insect vectors to expand their ranges. At the same time, the use of some effective chemicals like fungicide methyl bromide is being banned because of environmental concerns. Furthermore, many pathogens (and the insects that spread them) adapt to crop resistance, introduced by traditional breeding practices. Plants do try to supercede the pathogens, resulting in a mad race of co-evolution. The stakes in this battle are high, because when the plants lose, a huge proportion of agricultural produce is lost, resulting in food insecurity and poor nutrition of human population. Genetic enrichment of crops via introducing alien disease-resistant genes, by using the in vitro technique of gene transfer technology, is an attempt to give an edge to plants over their pathogens in the process of "artificial" co-evolution.

Plant engineers like Roger Beachy, at Donald Danforth Plant Science Centre in St. Louis, have been successful in transforming various plants with genes from many of the 40 or more families of plant viruses, producing (viral disease) resistant melon, tomato, tobacco and papaya. Indeed, a modified strain of papaya, produced by the group of Dennis Gonsalves and Carol Gonsalves, has saved the Hawaiian papaya industry. Recently, their team, for the first time, has been able to introduce a multiple virus resistance by putting a chimeric transgene (containing sequences from the turnip mosaic and tomato spotted with viruses) into tobacco. Plant engineers have also been successful in engineering potatoes with a chimeric gene encoding segments of two insects' protein: a cecropin (an anti-microbial peptide made by moths and other organisms) and melittin (a component of bee venom). The plants proved resistant to the potato blight fungus and against a bacterium that causes soft rot in stored potato. Development of various DNA delivery methods and optimization of gene expression cassettes have produced transformation protocols for many crops with success, resulting in resistance to abiotic as well as biotic stress.

R GENES

Biotic stresses result from a battery of potential pathogens like fungi, bacteria, nematodes and insects that intercept the photosynthate produced by plants, and viruses use the replication machinery at the expense of host plant. To perceive such attacks and to translate that perceptions into an adaptive response, plants have evolved highly polymorphic, sophisticated recognition-dependent diseases resistance mechanisms. Plant resistance (R) genes constitute gene families involved in pathogen recognition pathways, which are important for "gene for gene" interactions with pathogens.

Plant pathogen interaction is generally governed by specific interactions between pathogen's avr (avirulence) gene loci and alleles of the corresponding R (disease resistance) locus in plant. When corresponding avr and R genes are present in both host and pathogen, the result is disease resistance. If either is inactive or absent, the result is a disease. In simple words, R products recognize avr-dependent signals and trigger the chain of signal transduction events that culminate in activation of defense mechanisms and an arrest of pathogen growth.

Classes of R Genes

Many R genes have been identified in the past decade (Table 18.1) (Bent, 1996). Functional R genes isolated so far encode sequences inducing resistance to bacterial, fungal, viral, oomycete and even nematode and insects. It is interesting to note that despite such a wide range of pathogen taxa, R genes encode only five classes of proteins namely NB-LRR, Xa21 and Cf-X proteins, Pto gene product and recently-cloned RPW8 gene product (Dangl and Jones 2001).

Of the five classes of R genes two Xa21 and Cf-x proteins, carry transmembrane domains and extracellular LRRs (leucine rich repeats) regions, which play important roles in plant defenses (Jones and Jones 1996). The LRR domains are the major determinants of recognition specificity for avr factors and may be involved in direct protein-protein interactions with avr gene products of pathogen.

The classes of R genes are quite diverse. The complete *Arabidopsis* sequence permits a comprehensive analysis of the diversity of NB-LRR gene sequences in one plant (TAGI Initiative 2000). Annotation revealed approximately 150 sequences with homology to the NB-LRR class of R genes, distributed between different chromosomes with 49R homologues on chromosome 1, two on chromosome 2, sixteen on chromosome 3, twenty-eight on chromosome 4 and fifty five on chromosome 5. A continuously-updated annotation of *Arabidopsis* R genes by Mayers and colleagues can be found at http://mips.gsf.de/proj/thal/db/index.html.

Functions of R Genes: 'Guard Hypothesis'

R gene products act as a guard of cellular machinery, 'guard hypothesis' (van der Biezen and Jones 1998a). R genes are predicted to encode receptors for pathogen-derived molecules (Feys and Parker 2000; Baker et al., 1997). Interestingly, homologies between plant resistance gene products and regulators of cell death in animals have been reported

Table 18.1 Different classes of plant disease resistance genes

Class	*R-gene*	*Plant*	*Pathogen*	*Structure*	*Cellular location*	*Size (a.a.)*
Class 1	*RPS2*	*Arabidopsis*	*Pseudomonas syringae*	LZ-NBS-LRR	Cytoplasmic	909
	RPS5			LZ-NBS-LRR	Cytoplasmic	889
	RPS4			TIR-NBS-LRR	Cytoplasmic	1217
	N	Tobacco	*Tabacco mosaic virus*	TIR-NBS-LRR	Cytoplasmic	1144
	L^{6}	Flax	*Melampsora lini*	TIR-NBS-LRR	Cytoplasmic	1294
	Mi	Tomato	Nematode Aphid and	NBS-LRR	Cytoplasmic	1257
	Prf	Tomato	*Pseudomonas syringae*	NBS-LRR	Cytoplasmic	1824
Class 2	*Pto*	Tomato	*Pseudomonas syringae*	Protein Kinase	Cytoplasmic	321
Class 3	*Cf-9*	Tomato	*Cladosporium fulvum*	eLRR-TM	Transmembrane	863
	Cf-4			eLRR-TM		806
	Cf-2			eLRR-TM		1112
	Cf-5			eLRR-TM		968
Class 4	*Xa21*	Rice	*Xanthomonas oryzae*	*LRR-P. Kinase*	Transmembrane	1025
Class 5	*Hm1*	Maize	*Cochliobolus carbonum*	Toxin reductase	Cytoplasmic	357
Class 6	*mlo*	Barley	*Erisiphe graminis*	*G-protein* coupled receptor	Transmembrane	553
Class 7	*HS1*$^{Pro\text{-}1}$	Sugar beet	Nematode	LRR—unknown domains	Transmembrane (?)	282
	pC131	Chickpea	*Ascochyta (?)*	*LRR—unknown* domains	Transmembrane (?)	458

(van der Beizen and Jones 1998b). The striking parallels between plant R genes and mammalian MHC complex suggest that understanding the evolutionary dynamics of disease-resistance in plants will have applicability in other organisms, including human.

The recently-cloned RPW8 gene products carries a putative signal anchor at the N terminus, and the Pto gene encodes a cytoplasmic Ser/Thr kinase, which may be associated with membrane through its N terminal myristoilated site. The NB (nucleotide binding) LRR, the largest class of R proteins, is presumably cytoplasmic (although they could be membrane associated) and carry distinct terminal domains (Boyes et al., 1998). Some members of different classes of R gene families except NB-LRR, have demonstrated functions in cellular and developmental processes unrelated to defense. Pto from tomato encodes a Ser/Thr kinase that confers resistance to *Pseudomonas syringae* strains carrying avr Pto. Pto might function through a phosphorylation cascade, triggered by Avr Pto-Pto interaction (Scofield et al., 1996; Tang et al., 1996). Pto function requires the NB-LRR protein Prf (Salmeron et al., 1996). The rice Xa21 gene encodes a transmembrane receptor carrying a large extracellular LRR domain and an intracellular protein kinase domain (Song et al., 1995). Chimeras of Xa21 and related LRR receptor-like kinase that recognizes the brassinosteroid hormone show that specificity for this class of R protein also resides in LRRs (He 2000). The tomato Cf-X genes encode single pass membrane proteins with extracellular LRPs (Jones et al., 1994). Whether these other structural classes of R proteins use signal transduction cascades similar to those used by the NB-LRR family is not yet known, although a few findings suggest they do. An important example is that NB-LRR protein Prf requires Pto protein kinase to activate defense upon recognition of Avr Pto, indicating the involvement of signal transduction cascade similar to those used by the NB-LRR. The function of Avr Pto for the pathogen *P. syringae* is to target Pto and suppress this non-specific defense pathway of the host. Prf is thus an NB-LRR protein that 'guards' Pto, detects its interdiction by the Avr Pto (or any other bacterial effector) and then activates defense.

However, despite the 7 years that have elapsed since the isolation of the first R genes much remains elusive about their functioning. Even now, there is a great deal to learn about how R proteins function to confer Avr recognition. There is a great need to conduct more field experiments to study how R genes work in the natural populations and to test approaches using genetic polymorphism to provide more durable disease resistance in crops (Dangl and Jones 2001).

IN VITRO GENE TRANSFER TECHNOLOGY AND CROP YIELD

The Need for Developing Transgenic Crops

Natural resistance, which can be transferred between species by cross-breeding, is probably the most valuable defense against plant diseases. But cross-breeding is time consuming and microbial pathogens evolve more quickly against such type of resistance. Plant engineers have offered a new tool in the form of genetic engineering by introducing disease resistant genes directly into plants to keep the plants say a step ahead of rapidly-evolving plant pathogens. Advances in genetic transformation and gene expression during the last decade (Sharma et al., 2001) have resulted in a rapid progress in using genetic engineering for crop improvement, of which protection of crops against insects is a major goal. Genetically modified (GM) crops have been successfully produced and are found to be more disease resistant. Once efficient protocols for tissue culture and transformations are developed, the production of transgenic plants with different genes becomes fairly routine (Sharma et al., 2001).

Resistance to Insect Pesticides: Bt-modified Crops

The most prominent example of genetically-modified crop is the Bt-modified crops. Here, the transfer of a gene encoding an insect killing protein from the bacterium *Bacillus thuringiensis* (Bt) into crop like corn, cotton and potato etc. has been successfully achieved. Bt-modified plant was the first transgenic plant derived from *B. thuringiensis*. Plantation of Bt-modified crops has resulted in reduced application of chemical insecticides and a high yield is obtained. Successful expression of Bt genes against the lepidopterous pests has also been obtained in tomato, potato, brinjal, groundnut and chickpea (Sharma et al., 2001). Efforts are being made to transform crop plants with several Bt. toxin genes, as different strains of Bt produce different toxins, at once. However, resistance to such multiple toxins is unlikely to develop rapidly, according to entomologist Leigh English of Moansanto Crop in St. Louis.

Non-Bt Transgenic Crops

Many studies are underway to use non-Bt genes such as genes for protease inhibitors, chitinase, secondary plant metabolites and lectin. The products of these genes interfere with the nutritional requirements of insects. Genes conferring resistance to insects have been inserted into crop plants such as maize, cotton, potato, tobacco, rice, broccoli, lettuce, walnut, apples, alfalfa and soybean. Transgenic tobacco, maize and rice expressing lectin genes have exhibited adverse effects against several insect pests feeding on these crops (Maddock et al., 1991). Transgenic tobacco plants expressing chitinase gene have shown an increased resistance to lepidopteran insects. Activity of Bt can also be increased in combination with tannic acid and proteinase inhibitors (Cornu 1996).

The Benefits of Transgenic Crops

Deployment of transgenic plants in pest management has tremendous benefits, both to the environment as well as to the economics. Adoption of insect-resistant crops has, for example, led to a reproduction of one million kg of pesticides in the USA in 1999 as compared to 1998 (NRC report 2000). Papaya ringspot virus resistant papaya is being grown in Hawaii since 1996. Rice yellow mottle virus (RYMVY), which is difficult to control with conventional approaches, can now be controlled through transgenic rice (Pinto et al., 1999). Other examples of successful pest control through GM crops include papaya resistant to ringspot virus, blight-resistant potatoes and leaf blight-resistant rice (Sharma et al., 2001). Several candidate genes are evaluated for their biological efficacy against some major crop pests in the semi-arid tropics at ICRISAT, India.

Other Uses of Transgenic Technology

Transgenic technology has great potential to increase the yield of medicines derived from plants (example, salicylic acid). Anti-cancer antibodies expressed in rice and wheat can be useful in diagnosis and treatment of cancer in future. It has been possible, using trasgene technology, to produce "edible vaccines" against infectious disease of gastrointestinal tract in potatoes and bananas. Transgenic rice, with a capacity to produce beta carotene and rice with elevated levels of iron, has been produced using genes involved in the production of an iron binding protein that facilitates iron availability in the human diet (Lucca 1999).

Transfer technology is also useful in abiotic stresses. Development of crops to have an inbuilt capacity to withstand abiotic stresses would help stabilize the crop production. Plants with the ability to produce more citric acid in roots are tolerant to aluminium in acid soils. A salt-tolerant gene isolated from mangrove (*Avicennia marina*) has been cloned and can be transferred to other crop plants. The gut D gene from *E. coli* can also be used to provide salt tolerance. These genes would have a great potential in cultivation of marginal lands.

METHODS TO DELIVER DNA INTO PLANT TISSUES AND CALLUS

Particle Bombardment (Biolistics)

Biolistics is the most commonly-used methods to deliver DNA into plant tissues and callus. Initially, DNA coated tungsten particles were accelerated in vacuum to velocities that allowed penetration into biological tissues. After tungsten particles were shown to have phytotoxic effects, they were replaced with gold particles. Among the various original particle guns, the Helios Gene Gun can be used as a handheld device as it does not require a vacuum chamber to hold the target tissue.

A number of physical and biological parameters such as the size and the number of particles, the amount of DNA coated onto them, the type of explants, the osmotic conditions, the incubation time of culture medium before bombardment, affect the transformation efficiency of biolistics. Embryonic somatic tissues are the most common target tissues used for particle bombardment. Expression and inheritance of the multiple transgene in rice has been made possible by this method.

Agrobacterium-mediated Transformation

The soil microorganism *Agrobacterium tumefaciens* has the unique ability to transfer and integrate transferred DNA (T-DNA) into the genome of wounded plant cells. Studies on *Agrobacterium*-mediated transformation of maize cells suggest that integration of the T-DNA into the host genome is the bottleneck of the whole transformation process. Incorporation of foreign DNA into the host genome is likely to occur via illegitimate recombination but little is known about the factors involved in the process. *Agrobacterium tumefaciens*-mediated DNA delivery has been successfully used to obtain a number of transgenic plants such as mint (Veronese et al., 2001).

Other Transformation Methods

Electroporation of DNA and silicon carbide whiskers-mediated techniques have been successfully used to transform cell suspension cultures. However, none of these techniques appears to have any technical advantage over biolistics or *Agrobacterium*-based systems (Repellin et al., 2001). Imbibition, microinjection, electrophoresis, ultrasound treatment, laser treatment or transfer of DNA into plant cells via pollen tubes are other simple and potential methods for transformation, which could have broad applications if developed. The identification of transformed cells and tissues is possible using various selection and scorable marker systems (Repellin et al., 2001).

SOME PROBLEMS OF IN VITRO GENE TRANSFER TECHNOLOGY AND THEIR SOLUTIONS

Theoretically speaking, it seems possible to beef up disease defenses in crop plants by giving them additional copies of the resistant genes, or by altering the genes regulatory sequences to make them more active. However, it is not so and a number of practical problems do arise and remain to be solved.

An important practical problem of transgenic plants is the unpredictable placement of gene. Furthermore, unknown plant mutations (somaclonal variation) are generated during tissue culture, which is needed to grow transformed cells into whole plants, can induce unexpected traits or loss of desirable traits. Other important issues that need to be addressed while considering the deployment of transgenic crops include (i) development of resistance by the pathogen; (ii) gene escape (introgression of transgenes into the wild relatives) into the environment; (iii) effects on non-target organisms; and (iv) biosafety of transgenic foods. An answer to these problems may be in the use of an alternative approach to crop protection, such as the use of externally applied chemical stimulators or elicitors that elicit plant's natural defense mechanisms, which include HR (hypersensitive response) and SAR (systemic-acquired resistance).

HR and SAR: An Approach to Overcome the Limitations of Gene Transfer Technology

HR causes cells to die in the immediate vicinity of the infectious sites, thereby preventing pathogen spread, whereas SAR first results in necrotic lesions and then activates a signal system that results in a marked reduction of disease symptoms following subsequent infection (Morel and Dangl 1997; Stuiver and Custers 2001).

Researchers are trying to identify the compounds that elicit defense responses in plants. Harpin, a glycine-rich protein from the bacterium *Erwinia amylovora*, which causes fire blight in apple and pear trees, is one such compound which was identified to elicit HR and SAR to pathogens and insects. As an added advantage, it also enhances growth. The harpin technology was licensed to Eden BioScience (Bothell, Washington), last summer its harpin product (trade name "messenger") went on sale for use on a broad array of plants like strawberry, cotton and tomato.

In a similar effort, Alison Tally and his colleagues identified an isonicotinic acid derivatives ("Actigard") for its ability to induce SAR. Once activated by "Actigard", a plant's defense mechanism may remain active for many weeks, even when the elicitor is degraded. This compound was registered by US Environmental Protection Agency in August 2000 and is now being introduced into the market place to prevent bacterial spot and speck infections of tomatoes, downy mildew on spinach, and blue mold on tobacco.

METABOLIC ENGINEERING FOR IMPROVED DISEASE RESISTANCE AND CROP YIELD

Plants elaborate a vast and diverse array of over 100,000 low molecular mass natural products (or secondary metabolites), most of which are derived from the isoprenoid, phenyl propanoid, alkaloid or fatty acid/ polypeptide pathways (Dixon, 2001; Grayer and Harbone 1994; Harbone 1999). All the antimicrobial plant natural products (plant antibiotics) can be put into classes, viz., phytoalexins (compound synthesized de novo) and phytoanticipins, which are the pre formed infectional inhibitors (Van etten et al., 1994). The distinction between phytoalexin and phytoanticipin is not always obvious. Some compounds may be phytoalexin in one species and phytoanticipins in other. Methylated flavonone sakuranetin is one such example (Kodama et al., 1998), which accumulates constitutively in leaf glands of blackcurrant, but is a major inducible antimicrobial metabolite in rice leaves. It is interesting to note that various pathways leading to phytoalexin or phytoanticipins are interrelated through intermediary (primary) metabolism. Some of the genes of these pathways have been cloned (Dixon 2001), opening avenues to improve resistance by pathway engineering or metabolic engineering.

In recent years, genetic and genomic approaches have successfully been applied to identify genes of plant natural products biosynthesis. The 2-HIS cytochrome P450 that catalyzes the entry point reaction into isoflavonoid phytoalexin biosynthesis, has been identified by using comparative EST (expressed sequence tag) database mining. Mass sequencing of complementary DNA libraries corresponding to metabolically-specialized cells has been used to identify several of the genes or monoterpene biosynthesis and the associated Rhomer pathway for formation of the isoprenoid precursor isopentyl diphosphate in mint glandular trichomes (Lange et al., 1998; Crock et al., 1997). Due to these and other similar approaches, metabolic engineering for plant disease resistance has become possible.

Introduction of resveratrol (a novel phytoalexin) into alfalfa by the constitutive expression of a grapevine stilbene synthase gene has resulted in reduced symptoms, following infection by the leaf spot pathogen *Phoma decicagines* (Hipskind and Paiva 2000). Constitutive overexpression of isoflavone O methyltrasferase in transgenic alfalfa has resulted in more rapid and increased production of the pterocarpan phytoalexin medicarpin after infection by *P. medicaginis*, resulting in amelioration of symptoms (He and Dixon 2000).

The concept of improving disease resistance by engineering natural product pathways has, however, met with several objections (Dixon 2001). An important objection concerns the large number of genes that may have to be transferred—and coordinately regulated—to introduce effective anti-microbial activity, although several single-step conversions can generate anti-microbial compounds from ubiquitous common metabolic intermediates. Increased production of an endogenous anti-microbial compound through over expression of a rate limiting enzyme is conceptually a simple strategy, but in most cases, the flux control points in the pathways are not fully understood. All these technical problems make metabolic engineering a difficult task.

ENGINEERING DISEASE RESISTANCE THROUGH 'MASTER SWITCHES'

Plants use several different defense pathways against different pathogens (Thomma et al., 2001). In general, these pathways are characterized by the signaling molecules that are crucial in the regulation of expression of defense proteins (Delaney, 1994; Pieterse and van Loon, 1999). Treatment of plants with, for example, salicylic acid or its analogues, induces expression of a subset of plant defense responses against certain (but not all) pathogens (Thomma 1998). Small signaling molecules like ethylene and jasmonic acid also induce resistance to a different group of pathogens (Feys and Parker 2000). Recently, evidence suggesting the role of reactive oxygen species (ROS) as signaling molecules have been obtained (Grant et al., 2000; Pieterse, 1998). Treatment of plants with one or more of these signaling molecules causes the coordinated induction of antifungal proteins, phytoalexins and enzymes involved in plant cell wall reinforcement or breakdown of pathogen infection structures. Involvement of signaling molecules in the expression of defense related genes have prompted research to identify and use signal transduction "master switches"' to engineer disease resistance.

The use of these key regulators of defense pathways (master switches) to "tweak" resistance provide an excellent opportunity to achieve "absolute yield" or at least "yield stability" in agricultural produce. This has been successful in some cases like barley (Jarosch et al., 1999). The *Arabidopsis* NIMI/NPRI gene seems to be crucial in salicylic acid mediated resistance, overexpression leads to resistance against several pathogens (Cao and Dong 1998). NIMI/NPRI is an isolog of Ik-B, and interacts with a bZIP family of transcription factors (Delaney 2000). Other enhanced disease resistance (edr) mutants have also been identified (Bowling et al., 1994: Frye et al., 2001).

IMPROVING CROP YIELD BY TINKERING WITH A PLANT'S OWN DISEASE RESISTANCE GENES

The introduction of genes directly into plants from entirely different species and the subsequent production of GM crops has sparked a huge international backlash. Much hue and cry raised against GM foods has compelled researchers to look for alternative strategies. Tinkering with a plant's own disease resistance genes is an effort in this direction.

The team of Brian Williamson, at the Scottish Crop Research Institute in Dundee, found that the gray mold *Botrytis cinerea*, the most serious cause of disease in ripe raspberries, can be controlled by PGIP (polygalactouronase inhibiting protein), which is normally expressed only in immature green raspberries. Altering the PGIP gene's regulatory sequences in raspberry in such a way that the protein PGIP is produced all the times (i.e even in ripe fruits raspberry) would be an effective way to control the *Botrytis cinerea* in ripe raspberries.

Such attempts have not been successful in raspberry plants so far. However, scientists have been able to introduce the raspberry PGIP gene (with regulatory element from cauliflower mosaic virus, which keep the PGIP gene continuously active) in chickpea, a valuable crop in India. The germplasm is now being tested at ICRISAT, India.

GENE SILENCING FOR CROP IMPROVEMENT

Gene silencing, which was perceived initially as an unpredictable and inconvenient side effect of introducing transgenes into plants, is now perceived as a defense strategy of plants against viruses and transferable elements (Waterhouse et al., 2001). This inbuilt adaptive defense mechanism of plants help them to fight against an attack by the microbial pathogens. If the plants own defense is revived by exposing it to the pathogen's proteins (a process somewhat akin to vaccination), plants resistance to specific disease-causing pathogens may be increased. In this process, researchers have successfully stitched viral genes (or their fragments) into the plant genome. It is suggested that in gene silencing, the foreigner (viral) gene (stitched into plant genome) triggers a plant defense mechanism that degrade the viral RNAs, thus disabling the infectious agent. This process, also referred to as RNA silencing, seems to have tremendous potential in providing resistance to plants against pathogen attack.

Mechanisms of Gene Silencing

Gene silencing/RNA silencing is a remarkable type of gene regulation based on sequence specific targeting and degradation of RNA (Kooter et al., 1999; Convey et al., 1997). First discovered in transgenic plants, RNA silencing has been termed as co-suppression on posttranscriptional gene silencing (PTGS). In the process of RNA silencing, a double stranded (ds) RNA acts as trigger or intermediate. The dsRNA is cleaved into small

(21-25 nucleotides long) interfering RNAs (siRNAs), which act as guides to direct the RNA degradation machinery of the target RNA molecules (Zamore et al., 2000). In plants, RNA silencing is considered as a general antiviral defense mechanism, which can be triggered locally and then spread throughout the organism via an unknown mobile silencing signal. The signaling molecule is, however, expected to contain a nucleic acid component to account for the sequence specificity. Finally, in plants, RNA silencing is correlated with methylation of homologous transgene DNA in the nucleus. Systemic spread of silencing also occurs in other organisms, though the mechanism may not be the same as in plants. In fact, development of an in vitro RNA silencing system from *Drosophila* has allowed a biochemical analysis of some steps in the silencing pathway. Further, mutant analysis has identified a number of genes that are required for RNA silencing in multiple organisms.

The first line of research, indicating towards RNA silencing, come from studies of pathogen-derived resistance (PDR) in plants. In PDR, resistance to a particular virus is organized by stably transforming plants with a transgene derived from the virus. Once RNA silencing of the (viral) transgene had been established, all RNAs with homology to the transgene were degraded including those derived from an infecting virus. It is now clear that the plants, which recover from certain plant viral infections become resistant to reinfection by the initial virus (and to closely-related viruses) because of the silencing mechanism. It is interesting to note that many plant viruses encode proteins that suppress RNA silencing. Suggesting a co-evolution of defense and counter defense between the host and the invading pathogen (virus). These viral suppressors of silencing have provided a new tool to understand the mechanism of RNA silencing in plants. HC-pro (the helper component-proteinase) of potyviruses and the p25 (a protein encoded by potato virus X or PVX) are the two plant viral suppressors' which represent viral strategies to suppress silencing (Anandalakshmi et al., 2000).

Several cellular proteins controlling RNA silencing in plants have been identified through genetic screens of *Arabidopsis* mutants impaired in transgene-induced RNA silencing, RDRP (RNA dependent RNA polymerase) is one such protein. There are currently 72 different defined genera of plant viruses, containing over 500 species. The genome of some plant viruses are encoded using single-stranded (ss) or double-stranded (ds) DNA or RNA. However, over 90% of plant viruses have ssRNA genomes that require a (virus encoded) RDRP (Dalmay et al., 2000). Plants defend themselves by exploiting this requirement of most plant viruses to replicate using a ds replicative machinery (Waterhouse et al., 2001). Morris and Dodds (1979) had isolated and analyzed dsRNA virus-infected plant and fungal tissue.

It is important to know that the RNA silencing pathway is branched and that the branches converge in the production of ds RNA. In a branched model of RNA silencing, dsRNA is proposed to be the common intermediate linking the various ways of initiating RNA silencing. Once the longer dsRNAs is formed, it is acted upon by a dsRNase thus generating the dsRNA (siRNAs) of both polarities (Zamore et al., 2000). The siRNAs incorporate into a multicomponent silencing complex, where they act as guides to target complementary RNAs (Yang and Erickson 2000). Thus, dsRNA acts as a key inducer to RNA silencing, dsRNA also induces transcriptional gene silencing (TGC), a process that acts as a defense against certain transposable elements.

CONCLUSION

Recognition of microbe invasion by plants triggers various responses, that result in host plant defense, which may include activation of certain genes, or programed cell death. The emerging revolution in the in vitro gene transfer technology and the identification/isolation of numerous genes responsible for providing resistance against plant pathogens, serve as a powerful tool to improve livelihood. Pioneering efforts of plant engineers to strengthen the defense responses of crop plants by gene manipulation can play an important role in integrated gene management, integrated pest management, and efficient post-harvest management, to ensure food security and quality nutrition to the exploding human population, which, according to UN projections, will increase by 25% to 7.5 billion in 2020. Development and deployment of the knowledge of in vitro gene transfer technology in an effective manner and is important prerequisite for sustainable use of biotechnology for crop improvement. Plant with insecticidal genes stitched in them, or the use of gene silencing, or metabolic engineering are set to feature prominently in pest management in both the developed and the developing world in future.

REFERENCES

Anandalakshmi R., 2000, A calmodulin related protein that suppresses posttranscriptional gene silencing in plant. Science, **290:** 142-144

Baker B., Zambryski P., Staskawiez B. and Dinesh Kumar S.P., 1997, Signalling in plant microbe interactions. Science, **276:** 726-733.

Bent A., 1996, Function meets structure in the study of plant resistance genes. Plant Cell, **8:** 1757-1771.

Bowling S.A., 1994, A mutation in *Arabidopsis* that leads to constitutive expression of systemic acquired resistance. Plant Cell, **6:** 1845-1857.

Boyes D.C., Nam J. and Dangl J.L., 1998, The *Arabidopsis thaliana* RPM1 disease resistance gene product is a peripheral plasma membrane protein that is degraded coincident with the hypersensitive response. Proc. Natl. Acad. Sci. USA. **95:** 15849-15854.

Cao H., Li X., and Dong X., 1998, Generation of broad spectrum disease resistance by over-expression of an essential regulatory gene in systemic acquired resistance. Proc. Natl. Acad. Sci. USA, **95:** 6531-6536.

Cornu D., 1996, Expression of a proteinase inhibitor and a *Bacillus thurigiensis* δ endotoxin in transgenic poplars. In: Somatic Cell Genetics and Molecular Genetics of Trees (Eds. Ahuja M.R., Boerjan W. and Neale D.B.), pp. 131-136. Kluwer Academic Publishers, Dordrecht, The Netherlands.

Crock J., Wildung M., and Croteau R., 1997, Isolation and bacterial expression of sesquiterpene synthase cDNA clone from pepperment (*Mentha piperita.* L.) that produces the aphid alarm pheromone (E) B-farnesene. Proc. Natl. Acad. Sci. USA, **94:** 12833-12838.

Dangyl J.F., and Jones J.D.G., 2001, Plant pathogens and integrated defense responses to infection. Nature, **411:** 826-833.

Delaney T.P., 2000, New mutants provide clues into regulation of systemic acquired resistance. Trends Plant Sci., **5:** 49-51.

Delaney T.P. 1994, A central role of salicylic acid in plant disease resistance. Science, **266:** 1247-1250.

Dixon R.A., 2001, Natural products and plant disease resistance. Nature, **411:** 843-847.

Feys B., and Parker J.E., 2000, Interplay of signalling pathways in plant disease resistance. Trends Genet., **16:** 449-455.

Frye C.A., Tang D., and Innes R.W., 2001, Negative regulation of defense responses in plants by a conserved MAPKK kinase. Proc. Natl. Acad. Sci. USA, **98:** 373-378.

Grant J.J., Yun B.W., and Loake G.J., 2000, Oxidative burst and cognate redox signalling reported by luciferase imaging: identification of a signal network that functions independently of ethylene, SA and Me-JA but is dependent on MAPKK activity. Plant J., **24:** 569-582.

Grayer R.J., and Harborne J.B., 1994, A survey of antifungal compounds from higher plants. Phytochemistry, **37:** 19-42.

Harborne J.B., 1999, The comparative biochemistry of phytoalexin induction in plants. Biochem. System. Ecol., **27:** 335-367.

He X.Z., and Dixon R.A., 2000, Genetic manipulation of isoflavone 7 Omethyltrasferase enhances the biosynthesis of 4′-methylated isoflavonoid phytoalexins and disease resistance in alfalfas. Plant Cell, **12:** 1689-1702.

He Z., 2000, Perception of Brassinosteroids by the extracellular domain of the receptor kinase BRII. Science, pp. 2360-2363.

Hipskind J.D., and Paiva N.L., 2000, Constitutive accumulation of a resveratol glucoside in transgenic alfalfa increases resistance to *Phoma medicaginis*. Mol. Plant-Microbe Interact., **13:** 551-562.

Jarosch B., Kogel K.H., and Schaffrath U., 1999, The ambivalence of the barley Mlo locus; mutations conferring resistance against powdery mildew (*Blumeria graminis* f. sp. *hordei*) enhance susceptibility to the rice blast fungus *Magnaporthe grisea*. Mol. Plant Microbe Interact., **12:** 508-514.

Jones D.A., and Jones J.D.G., 1996, The roles of leucine rich repeats in plant defenses. Adv. Bot. Res. Adv. Plant Pathol., **24:** 90-167.

Jones D.A., Thomas C.M., Hammond-Kosack K.E., Balint -Kurti P.H., and Jones J.D.J., 1994, Isolation of the tomato Cf-9 gene for resistance to *Cladosporium fulvum* by transposon tagging. Science, **266:** 789-793.

Kodama O., Suzuki T., Miyakawa J., and Akatsuka T., 1988. Ultraviolet induced accumulation of phytoalexins in rice leaves. Agric. Biol. Chem., **52:** 2469-2473.

Kooter J.M., Matzke M., and Meyer P., 1999, Listening to the silent genes: trasgene silencing, gene regulation and pathogen control. Trends. Plant Sci., **4:** 340-347.

Lange B.M., Wildung M.R., McCaskill D., and Croteau R., 1998, A novel family of transketolases that directs isoprenoid biosynthesis via a mevalonate independent pathway. Proc. Natl. Acad. Sci. USA, **95:** 2100-2104.

Lucca P., 1999, In: General Meeting of the International Programme of Rice Biotechnology, Thailand, September, 1999, pp. 20-24.

Maddock S.E., Hufman G., Isenhour D.J., Roth B.A., Raikhel N.V., Howard J.A., and Czapla T.H., 1991, In: 3rd International Congress of Plant Molecular Biology, Tucson, Arizona, USA.

Morel J.B., and Dangl J.L., 1997, The hypersensitive response and the induction of cell death in plants. Cell Death Differ., **4:** 671-683.

Morris T.J., and Dodds J.A., 1979, Isolation and analysis of double stranded RNA from virus infected plant and fungal tissue. Phytopathology, **69:** 854-858.

Oerke E.C., Dehne H.W., Schonbeck F., and Weber A., 1994, Crop Production and Crop Protection: Estimated Losses in Major Food and Cash Crops. Elsevier Publishing Co, Amesterdam, The Netherlands.

Pieterse C.M., 1998, A novel signaling pathway controlling induced systemic resistance in Arabidopsis. Plant Cell, **10:** 1571-1580.

Pieterse C.M., and Van Loon L.C. 1999, Salicylic acid independent plant defense pathways. Trends Plant Sci., **4:** 52-58.

Pinto Y.M., Kok R.A., and Baulcombe D.C., 1999, In: World Food Prospects: Critical issues for the early twenty-first century. International Food Policy Research Institute, Washington, DC, USA.

Repellin A., Baga M., Jauhar P.P., and Chibbar R.N., 2001, Genetic enrichment of cereal crops via alien gene transfer: new challenges. Plant Cell Tissue and Organ Culture, **64:** 159-183.

Salmeron J.M., 1996, Tomato Prf is a member of the leucine rich repeat class of plant disease resistance genes and lies embedded within the Pto kinase gene cluster. Cell, **86:** 123-133.

Scofield S.R., Tobias M.C., Rathjen J.P., Chang J.H., Lavelle D.T., Michelmore R.W., and Staskawicz B.J. 1996, Molecular basis of gene for gene specificity in bacterial speck disease of tomato. Science, **274:** 2063-2065.

Sharma H.C., Sharma K.K., Seetharama N., and Ortiz R., 2001, Genetic transformation of crop plants: Risks and opportunities for the rural poor. Current Science, **80:** 1495-1508.

Song W.Y., Wang G.L., Chen L.L., Kim H.S., Pi L.Y., Holsten T., Gardner J., Wang B., Zhai W.X., Zhu L.H., Fauquet C., Ronald P., 1995, A receptor kinase-like protein encoded by the rice disease resistance gene Xa21. Science, **270:** 1804-1806.

Stuiver M.H., and Custer J.H.V. 2001, Engineering disease resistance in plants. Nature, **411:** 865-868.

TAGI (The Arabidopsis Genome Initiative) 2000, Analysis of the genome of the flowering plant *Arabidopsis thaliana.* Nature, **408:** 796-815.

Tang X., Frederick R.D., Zhou J., Halterman D.A., Jia Y., and Martin G.B. 1996, Physical interaction of avr Pto and the Pto kinase defines a recognition event involved in plant disease resistance. Science, **274:** 2060-2063.

Thomma B.P., 1998, Separate jasmonate dependant and salicylate dependent pathways in *Arabidopsis* are essential for resistance to distinct microbial pathogens. Proc. Natl. Acad. Sci., USA, **95:** 15107-15111.

Van der Biezen E.A., and Jones J.D.J., 1998a, Plant disease resistance proteins and the gene for gene concept. Trends Biochem. Sci., **23:** 454-456.

Van der Beizen E.A., and Jones J.D.J., 1998b. Homologies between plant resistance gene products and regulators of cell deaths in animals. Current Biol., **8:** R226-R227.

Van Etten H., Mansfield J.W., Baily J.A., and Farmer E.E., 1994, Two classes of plant antibiotics: phytoalexins versus "phytoanticipins". Plant Cell, **6:** 1191-1192.

Veronese P., Li X., Niu X., Weller S.C., Bressan R.A., and Hasegawa P.M. 2001, Bioengineering mint crop improvement. Plant Cell Tissue and Organ Culture, **64:** 133-144.

Waterhouse P.M., Wang M., and Lough T., 2001, Gene silencing as an adaptive defense against viruses. Nature, **411:** 834-842.

Yang D., Lu H., and Erickson J.W., 2000, Evidence that processed small dsRNAs may mediate sequence specific mRNA degradation during RNAi in *Drosophila* embryos. Curr. Biol., **10:** 1191-1200.

Zamore P.D., Tush I.T., Sharp P.A., and Bartel D.P., 2000, RNSi: double stranded RNA directs the ATP dependent cleavage of mRNA at 21-23 nucleotide intervals. Cell, **101:** 25-33.

Index

About the Editors

Dr. A. Mujib

Dr. A. Mujib, Ph D is a Lecturer in the Department of Botany at the Hamdard University, New Delhi. Dr. Mujib currently teaches molecular genetics, cytogenetics and plant biotechnology. His current research interests include micropropagation, somatic embryogenesis, somaclonal variation, in vitro mutagenesis, etc. using plant tissue culture techniques. Dr Mujib has over 50 publications as book chapters, review articles and original research papers. Dr Mujib is presently supervising four PhD students one of whom has already been awarded the degree.

Dr Mujib received his PhD degree in Botany from the University of Kalyani, West Bengal, India, working on in vitro studies on bulbous ornamentals. Subsequently he held scientific positions in the Dept of Biotechnology, Indian Institute of Technology, Kharagpur before joining his current post.

Dr. M.-J. Cho

Dr. Myeong-Je Cho has been leading crop transformation and gene expression at the University of California-Berkeley. He developed efficient tissue culture and transformation methods for commercial varieties of major monocot crop species. Using a new tissue culture system, major cereal crops and forage grasses have been successfully transformed with high frequencies. He also developed a stable gene expression system using seed-specific promoters to produce pharmaceuticals and to improve nutritional quality in cereals, and isolated and characterized trx and ntr genes from barley, which are being used for generation of hypoallergenic and hyperdigestible wheat and other cereals. Dr. Cho also leads the Plant Biotechnology and Genomics Programs at Byotix, Richmond, California. He is responsible for developing novel transformation technologies and gene expression systems for major crop species. He obtained his PhD in Agronomy (Soybean Physiology and Biochemistry) from the University of Illinois at Urbana-Champaign. He did postdoctoral study in the field of Soybean Molecular Biology at the University of Illinois. He has been a consultant/scientific advisor to 4 biotech companies. Dr. Cho is an author of 50 scientific publications and 3 book chapters and has 4 issued patents and 1 pending regarding plant transformation, gene expression, and gene isolation/characterization.

Dr. Predieri Stefano

Dr. Predieri Stefano is a researcher at the Italian National Council of Research (CNR) with research interests in the areas of ecophysiology of woody plants exposed to stress factors, in vitro study of environmental stress, induction of genetic variability and selection of genotypes with increased tolerance to stress. Dr Stefano is currently working in Bologna, Italy, and is in charge of the IBIMET_BO Tissue Culture Lab. He is responsible for the research projects funded by Regione Emilia-Romagna on "In vitro mutagenesis and field selection of pear" and on improvement of quality of fruit crops. He has conducted postdoctoral research in Belgium (CRA, Gembloux) and in UK (IHR, East Malling) on plant tissue culture, physiology, and biotechnology applied to fruit trees, and in USA (USDA, Beltsville) on developing in vitro models for studying UV-B effects on plants. Has been a member of "EU Tematic group AGROFOOD" as an expert on sensory evaluation of fruits, and of the Scientific Committees of the "8th International Pear Symposium ISHS" (2000) and "6th International Peach Symposium ISHS" (2005).

Dr. S. Banerjee

Dr. Saumitra Banerjee was awarded a research scholarship of the Indian National Science Academy and obtained his PhD degree in Botany (cell, tissue and protoplast culture of wheat and triticale, somatic embryogenesis, and biochemical changes and isozyme patterns during *in vitro* morphogenesis) from the University of Kalyani in India. Dr Banerjee did postdoctoral studies on tissue culture and micropropagation techniques of bulbous ornamental plants at the Agri-Horticultural Society of India, Calcutta. He joined the All India Institute of Hygiene and Public Health, Calcutta as a Research Officer and was responsible for research and teaching at post-graduate levels. After migrating to Australia, he continued his research at the Department of Biological Sciences at Monash University, Melbourne and worked on developing a reliable and efficient *in vitro* plant regeneration system and agrobacterium-mediated gene transfer methods in eucalyptus (*E. nitens* and *E. globulus*). In 1999, Dr Banerjee made a career move and joined CSIRO Publishing. He is currently the Associate Editor of two highly reputed international journals (viz. *Australian Journal of Agricultural Research* and *Australian Journal of Soil Research*). He has been a consultant/ scientific advisor to 2 companies. Dr Banerjee is author of over 15 scientific publications, reviews, and popular articles.